The New Weibull Handbook

Reliability & Statistical Analysis for Predicting Life, Safety, Risk, Support Costs, Failures, and Forecasting Warranty Claims, Substantiation and Accelerated Testing, Using Weibull, Lognormal, Crow-AMSAA, Probit, and Kaplan-Meier Models

Fifth Edition by Dr. Robert B. Abernethy
Author and Publisher

The New Weibull Handbook, Fifth Edition

Library of Congress Cataloging in Publication Data
96-202898 TA169.A35 1996 620/.0045 21
Abernethy, Robert B. The New Weibull Handbook, 2nd ed.
1v. (various pagings) :ill.; 28 cm
Includes bibliographical references and index.
ISBN 0-9653062-3-2
1. Reliability (Engineering). 2. Statistical Methods
3. Weibull distribution
Book source: ad

ISBN-10 0-9653062-3-2 ISBN-13 978-0-9653062-3-2

Published and distributed by Nancy S. Embury
Available at Amazon https://www.amazon.com/dp/0965306232/

Also distributed by:
1. SAE International Ph: (1) 877-606-7323, or
Phone: (1) 724-776-4970, Web Page: www.sae.org, E-mail: customerservice@sae.org
2. Wes Fulton, Phone: (1) 310-548-6358, Web Page: WeibullNews.com, FultonFindings.com
3. Carl Tarum, Phone: (1) 989-780-5088, Web Page: BathtubSoftware.com
4. Ronald Schop, Phone: +31 6-3835-8663, Web Page: Weibull.nl, E-mail: Weibull@weibull.nl
5. Nancy Embury, Phone: (1) 865-604-1964, E-mail: nsembury@gmail.com

Printing History (Each printing is updated to the latest technology)
First Edition, December 1993, All Printings to June 1996, All edits made by Dr. Abernethy.
Second Edition, July 1996, 9 Printings to January 1998, All edits made by Dr. Abernethy.
Third Edition, September 1998, 13 Printings to May 2000, All edits made by Dr. Abernethy.
Fourth Edition, September 2000, 12 Printings to November 2003, All edits made by Dr. Abernethy.
Fifth Edition, December 2006 through April 2010, All edits made by Dr. Abernethy.
Beginning October 2013 through December 2016, Minor edits made by Paul Barringer (no content changes).
Beginning October 2017 through December 2024, Minor clarifications made by Wes Fulton, Carl Tarum, and Rich Unkle (no significant content changes).

Dedication:

This book is dedicated to all my friends and colleagues that helped so much. Finally, it is dedicated to the reader with the hope it will have "some utility," to use Professor Weibull's words.

And lastly we have a special note for the Weibull Workshop students for whom we designed this Handbook…

Weibull Workshop Students:

If you are fortunate enough to have obtained or borrowed the Handbook before your Workshop, studying the Handbook before the Workshop is very helpful. **Please** read at least Chapters 1-4. We will cover the basic standard method the first day, Chapters 1-4. For the second day, study Chapters 5-7 (7 is least important). For the third day, Chapters 8, 9, & 10. (Most important material for the third day is the Crow-AMSAA Model.) Chapter 11 is case studies which you should browse. Please bring a clear plastic ruler, a calculator, and a problem from your work. The data should be in a single column, positive values are failures, negative are suspensions. You may bring the data in ASCII format, or a spreadsheet file (Excel) or SuperSMITH Weibull files. The file may be on a diskette or in your laptop.

Editor's Note:

Dr. Robert B. Abernethy (Dr. Bob) passed away in 2021 at the age of 91 in North Palm Beach, Florida. His daughter, Nancy Embury, maintains copyright and publication of this handbook. References to Dr. Bob's deleted web site and to his contact information have been removed from the text in 2022. The course that he developed and corresponding software are available from the following (see contact information listed previously above):

Wes Fulton
Carl Tarum
Ronald Schop

Paul Barringer, mentioned several times in the text, passed away in 2016. Any queries about his methods or applications should be directed to Wes Fulton or Carl Tarum or the current editor of this handbook.

"When you can measure what you are speaking about, and express it in numbers, you know something about it; but when you cannot measure it, when you cannot express it in numbers, your knowledge is of a meager and unsatisfactory kind; it may be the beginning of knowledge, but you have scarcely, in your thoughts, advanced to the stage of science, whatever the matter may be."

Lord Kelvin.... 1883

A quotation suggested by Walter Thomas, NASA Goddard.

"Now in building of chaises, I tell you what,
There is always *somewhere* a weakest spot,-
In hub, tire, felloe, in spring or thill,
In panel, or crossbar, or floor, or sill,
In screw, bolt, thoroughbrace,-lurking still,
Find it somewhere you must and will,-
Above or below, or within or without,-
And that's the reason, beyond a doubt,
A chaise *breaks down*, but doesn't *wear out*.

Oliver Wendell Holmes..."The One-Hoss Shay"

PREFACE

Background A quarter of a century has passed since the *USAF Weibull Analysis Handbook* was published. In the years that followed, the author frequently taught Weibull analysis using the USAF Handbook as the text, encouraged by both the United States Air Force and Pratt & Whitney. In 1987, the author retired early to devote full time to research and lecturing in Weibull Analysis. It has been most enjoyable and interesting. The best part has been meeting outstanding engineers from many industries with new and different applications for Weibull methods. The students inspire additional research and methods development. This in turn required rewriting and updating the USAF Handbook producing *The New Weibull Handbook.* This document is the fifth edition. It contains the latest methods, techniques, and applications that have been developed by the author, Wes Fulton, Paul Barringer, Joe Wauben, Bob Rock, Wayne Nelson, Randy Pirtle, Todd Marquart, Carl Tarum, Geoff Cole and David Nevell and many others.

The New Weibull Handbook has become the national and international standard for Weibull analysis. It is in daily use throughout the world. Thousands of copies have been distributed. Users include military, automotive, medical, electronics, materials & composites research, aerospace, electrical power, nuclear power, dental research, advertising, bearings, compressors, v-belts, chain drives and on and on. Many organizations have adopted the Handbook and software implementing its methods as standard including the Federal Aviation Administration, Harley Davidson, Motorola, TVA, Honeywell, BICC General, GKN Automotive, Dupont, Meritor Automotive, Teleflex, Guidant Cardiac Pacemaker (CPI), Daimler Chrysler, General Electric, Siemens, Behr, Borg-Warner, NASA, Rolls Royce, Gates Rubber, the US Naval Air Arm (NAVAIR), the US Air Force at SAALC and WPAFB, and Dow Chemical. The author has provided many free copies (shipping charges only) of the Handbook to universities in Europe, North America, and Australia. The author hereby extends this offer to university libraries worldwide. Wes Fulton provides free demonstration copies of the SuperSMITH software on his web site.

2009 Update: A new world standard, IEC 61649 Edition 2, has been published. As the author served on the IEC Committee, much of the IEC standard is taken verbatim from this Handbook. Further, the two documents are absolutely consistent with each other.

The **SuperSMITH** software is the most significant technical breakthrough. Wes Fulton has created friendly, easy-to-use software that does everything in the Handbook. *The New Weibull Handbook* provides an understanding and reference for SuperSMITH. The computer software has made application so much easier. It has also made teaching Weibull easier, more efficient, and more fun. Within the Weibull workshops, many more applications and case studies can be treated without the drudgery of hand calculations and plotting. As this document is the workbook for the Weibull Workshops, many references to the software capabilities are included. This referencing may appear to be inappropriate, but the reliance on software in the workshops requires it. If the student using *The New Weibull Handbook* desires, he may avoid almost all hand calculations and hand plotting by using SuperSMITH. The new "*PlayTIME with SuperSMITH*" computer tutorial enhances the Handbook and provides many step-by-step case studies illustrating all the methods. Alternatively, the reader may use this Handbook to write "home-grown" software. Outputs are easily validated with benchmark case studies found in both demonstration and full versions of the SuperSMITH software.

The **Fifth Edition** features new technology:

- Goodness of fit, illustrating for the first time the "power" of the three best methods for goodness of fit: Critical Correlation Coefficient, likelihood ratio and Anderson-Darling. "Power" is the probability that the method makes the correct choice identifying the parent distribution for a given data set.
- Crow-AMSAA modeling has become so popular that it has been moved to a new chapter. The new write-up includes the latest methods from the new IEC standards with best practices recommended.
- A new section on warranty analysis has been added by Joe Wauben of Honda
- Life Cycle Costs are presented in a new section by Paul Barringer.
- Wayne Nelson has updated his graphical repair analysis with a case study in Appendix M.
- Bob Rock of PACCAR updated his extremely useful life data analysis logic diagram which takes you step-by-step to the best methods and analysis for your particular data. The logic diagram is also available in Spanish translated by Gerardo Vargas of Carbones de Cerrejon.

- Todd Marquart of Micron Technology has completed an extensive comparison of the three methods for interval and grouped data. Which is most accurate?
- The New Weibull Handbook is unique in recommending "best practices" wherever there are alternative methods available. Some of the "best practices" have changed based on new research.
- Cost analysis is described for production process reliability, failure forecasts, Crow-AMSAA, Life Cycle Costs and optimal parts replacement.

To make the Handbook more readable, all the heavy mathematics will be found in the appendixes. Asterisks on section subtitles indicate advanced topics. The novice reader may skip these advanced sections. As the workbook is designed for the Weibull Analysis Workshops, the format includes much white space for notes and calculations. Dozens of case studies are employed to present the methodology. Paul Barringer took the most important reference texts and papers from my library and put them on his website for download as indicated herein by a preceding "$". His website may be unavailable until new adequate website hosting can be established.

June 2008: The data sets for the Weibull plots herein are now available as SuperSMITH software files distributed with the software if you would like to follow the analysis step-by-step.

Thank You: Many friends, students, experts, and clients have helped and contributed much to the Handbook. For me, all this support has been overwhelming. Everyone helped! Thank you all. Some are mentioned here:

- First, my associate, Wes Fulton, not only for his friendly software, but also for his research and development of new Weibull methods.
- Second, our associate, Paul Barringer, for his help, comments, reviews, research, and contributions.
- Third, our associate, Carl Tarum, for new solution methods and software improvements, especially on mixtures and pve% goodness-of-fit.
- A board of industry experts has reviewed the fifth edition and provided many comments and contributions: Walter Thomas, Wes Fulton, Ronald Schop, Bob Rock, Charles DeSica, Randy Pirtle, Todd Marquart, Carl Tarum, Geoff Cole, Paul Barringer, Jim Breneman, and Joe Wauben. I am indebted to them. April 2010: Thanks to Richard Unkle for his many comments and suggestions.
- Many contributed sections of the Handbook: Bob Rock, PACCAR; David B. Smith, Detroit Edison; Marie Stanton, FPL; David Weber, Consultant; Carl Tarum, Delphi; Drs. Memis and Lautenschlager, Dentists at Northwestern University; Richard Rudy, Chrysler; Jim Keener and Mel Thomas, FPL; Bill Pritchard, TVA; Shashank Kolhatkar, Borg Warner; Charlie Williams, Florida Power; Geoffrey Cole and David Nevell, Rolls Royce; David Williamson, Dow Chemical; M.A. Vasan, Meritor Automotive; David B. Smith, Detroit Edison; Todd Marquart, Micron; David Langanke and Randy Pirtle, Honeywell Engines.
- This edition is improved by incorporating the suggestions from the excellent constructive review of the Handbook by Dr. Alan Winterbottom for the Royal Statistical Society Journal A (1997), 160, Part 2. The author is indebted.
- None of the above are in any way responsible for errors in the Handbook. The author takes all the blame.

Please call or write with questions, comments, and constructive criticism about *The New Weibull Handbook*.

Dr. Robert B. Abernethy

Table of Contents

* Asterisks on section subtitles indicate advanced topics. The novice reader may skip these advanced sections.

* Asterisks on section subtitles indicate advanced topics. The novice reader may skip these advanced sections.

* Asterisks on section subtitles indicate advanced topics. The novice reader may skip these advanced sections.

CHAPTER 1. AN OVERVIEW OF WEIBULL ANALYSIS

1.1 Objective

This handbook will provide an understanding of life data analysis. Weibull and lognormal analysis will be emphasized particularly for failure analysis. There are new applications of this technology in medical and dental implants, warranty analysis, life cycle cost, materials properties and production process control. Related quantitative models such as the binomial, Poisson, Kaplan-Meier, Gumbel extreme value and the Crow-AMSAA are included. The author intends that a novice engineer can perform Weibull analysis after studying this document. A secondary objective is to show the application of personal computers to replace the laborious hand calculations and manual plotting required in the past.

1.2 Background

Waloddi Weibull invented the Weibull distribution in 1937 and delivered his hallmark American paper on this subject in 1951. He claimed that his distribution applied to a wide range of problems. He illustrated this point with seven examples ranging from the strength of steel to the height of adult males in the British Isles. He claimed that the function "...may sometimes render good service." He did not claim that it always worked. Time has shown that Waloddi Weibull was correct in both of these statements. His biography is in Appendix N.

The reaction to his paper in the 1950s was negative, varying from skepticism to outright rejection. The author was one of the skeptics. Weibull's claim that the data could select the distribution and fit the parameters seemed too good to be true. However, pioneers in the field like Dorian Shainin and Leonard Johnson applied and improved the technique. The U.S. Air Force recognized the merit of Weibull's method and funded his research until 1975. Today, Weibull analysis is the leading method in the world for fitting and analyzing life data.

Waloddi Weibull 1887-1979
Photo by Sam C. Saunders

Dorian Shainin introduced the author to statistical engineering at the Hartford Graduate Center (RPI) in the mid-fifties. He strongly encouraged the author and Pratt & Whitney Aircraft to use Weibull analysis. He wrote the first booklet on Weibull analysis and produced a movie on the subject for Pratt & Whitney Aircraft.

Leonard Johnson at General Motors improved on Weibull's plotting methods. Weibull used mean ranks for plotting positions. Johnson suggested the use of median ranks which are slightly more accurate than mean ranks. Johnson also pioneered the use of the Beta-Binomial confidence bounds described in Chapter 7.

E.J. Gumbel showed that the Weibull distribution and the Type III Smallest Extreme Values distributions are the same. He also proved that if a part has multiple failure modes, the time to first failure is best modeled by the Weibull distribution. This is the "weakest-link-in-the-chain" concept. See page 3-26 for more on Dorian Shainin and E.J. Gumbel.

The author found that the Weibull method works with extremely small samples, even two or three failures for engineering analysis. This characteristic is important with aerospace safety problems and in development testing with small samples. (For statistical relevance, larger samples are needed.) Advanced techniques such as failure forecasting, substantiation of test designs, and methods like Weibayes and the Dauser Shift were developed by the author and others at Pratt & Whitney. (In retrospect, others also independently invented some of these techniques like Weibayes in the same time period.) Such methods overcome many deficiencies in the data. These advanced methods and others are presented in this Handbook.

1.3 Examples:

The following are examples of engineering problems solved with Weibull analysis:

- A project engineer reports three failures of a component in service operations during a three-month period. The Program Manager asks, "How many failures will we have in the next quarter, six months, and year?" What will it cost? What is the best corrective action to reduce the risk and losses?
- To order spare parts and schedule maintenance labor, how many units will be returned to depot for overhaul for each failure mode month-by-month next year? The program manager wants to be 95% confident that he will have enough spare parts and labor available to support the overall program.
- A state Air Resources Board requires a fleet recall when any part in the emissions system exceeds a 4% failure rate during the warranty period. Based on the warranty data, which parts will exceed the 4% rate and on what date?
- After an engineering change, how many units must be tested for how long, without any failures, to verify that the old failure mode is eliminated, or significantly improved with 90% confidence?
- An electric utility is plagued with outages from superheater tube failures. Based on inspection data, forecast the life of the boiler based on plugging failed tubes. The boiler is replaced when 10% of the tubes have been plugged due to failure.
- The cost of an unplanned failure for a component, subject to a wear out failure mode, is twenty times the cost of a planned replacement. What is the optimal replacement interval?

1.4 Scope

Weibull analysis includes:

- Plotting the data and interpreting the plot
- Failure forecasting and prediction
- Evaluating corrective action plans
- Test substantiation for new designs with minimum cost
- Maintenance planning and cost effective replacement strategies
- Spare parts forecasting
- Warranty analysis and support cost predictions
- Controlling production processes
- Calibration of complex design systems, i.e., CAD\CAM, finite element analysis, etc.
- Recommendations to management in response to service problems

Data problems and deficiencies include:

- Censored or suspended data
- Mixtures of failure modes
- Nonzero time origin
- Unknown ages for successful units
- Extremely small samples (as small as one failure)
- No failure data
- Early data missing
- Inspection data, both interval and probit

Failure types include:

- Development, production and service
- Mechanical, electronic, materials, and human failures
- Nature: lightning strikes, foreign object damage, human error, woodpecker holes in power poles
- Quality control, design deficiencies, defective material
- Warranty claims

Math modeling for system analysis includes:

- Explicit models for independent failure modes
- Monte Carlo simulation for dependent failure modes
- Reliability growth, repairability, and management tracking using Crow-AMSAA models
- Exponential, binomial and Poisson models
- Kaplan-Meier Survivor Models
- Warranty Claims models
- Production Process Control Models

Statistical derivations are in the appendixes to keep the main body of the handbook more readable. The author leans toward simple methods as being most useful and easily understood. Complex methods that require sophisticated mathematics are academically interesting, but they are difficult to communicate and explain. Engineers are reluctant to use methods they do not understand. However, many of these complex methods such as confidence intervals are included, as the student may be required to employ them. Qualitative reliability methods are not included such as failure mode and effects analysis, failure analysis, and fault trees. These are important and recommended, but they are not described herein as the emphasis is on quantitative methods. See [O'Connor] for treatment of qualitative methods.

1.5 Advantages of Weibull Analysis

The primary advantage of Weibull analysis is the ability to provide reasonably accurate failure analysis and failure forecasts with extremely small samples. Solutions are possible at the earliest indications of a problem without having to "crash a few more." Small samples also allow cost effective component testing. For example, "sudden death" Weibull tests are completed when the first failure occurs in each group of components, (say, groups of four bearings). If all the bearings are tested to failure, the cost and time required is much greater.

Another advantage of Weibull analysis is that it provides a simple and useful graphical plot of the failure data. The data plot is extremely important to the engineer and to the manager. The Weibull data plot is particularly informative as Weibull pointed out in his 1951 paper. Figure 1-1 is a typical Weibull plot. The horizontal scale is a measure of life or aging. Start/stop cycles, mileage, operating time, landings or mission cycles are examples of aging parameters. The vertical scale is the cumulative percentage failed. The two defining parameters of the Weibull line are the slope, beta, and the characteristic life, eta. The slope of the line, β, is particularly significant and may provide a clue to the physics of the failure. The relationship between the slope and generic failure classes is discussed in Section 1.7 and Chapter 2. The characteristic life, η, is the typical time to failure in Weibull analysis. It is related to the mean time to failure.

1.6 Data, Discrete Versus Life Data

Discrete or counted data was originally used to measure reliability. Tests would be categorized as success or failure. Receiving inspection data would count good parts versus defective parts. This data is modeled with the binomial and Poisson distributions described in Chapter 8. The results are imprecise unless enormous sample sizes are employed. The author has developed a new method using the Crow-AMSAA model that is much more precise and accurate for success-failure and go-no go data when the sample size is large. (See Chapter 9.)

Measured data like age-to-failure data is much more precise because there is more information in each data point. Measured data provides much better precision so smaller sample sizes are acceptable.

Ideally, each Weibull plot depicts a single failure mode. Data requirements are described by D.R. Cox [1984]: "To determine failure time precisely, there are three requirements:

- A time origin must be unambiguously defined,
- A scale for measuring the passage of time must be agreed to and finally,
- The meaning of failure must be entirely clear."

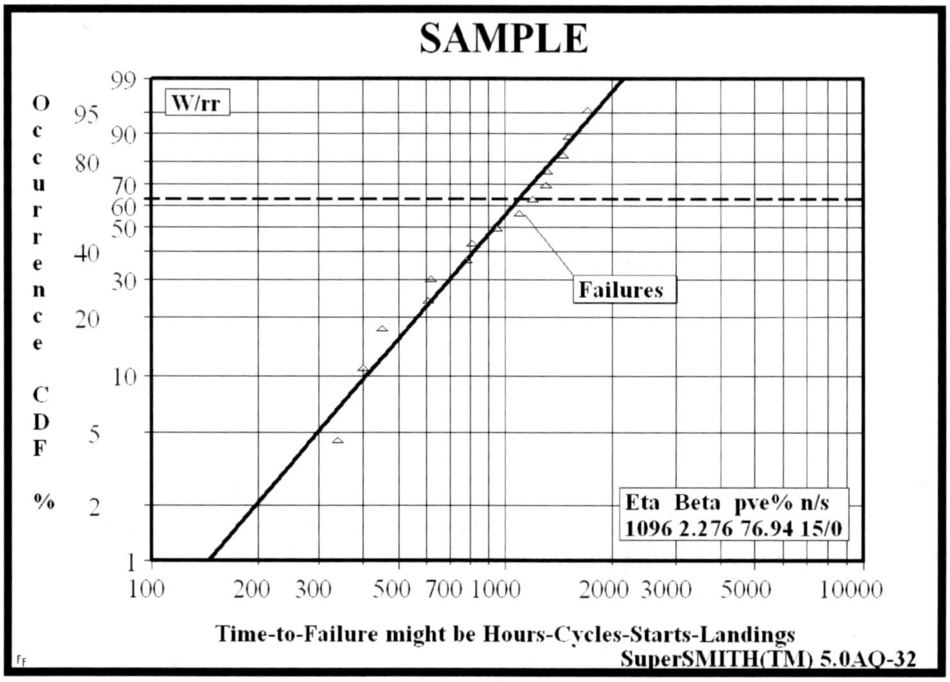

Figure 1-1. The Weibull Data Plot

The age of each part is required, both failed and unfailed. The units of age depend on the part usage and the failure mode. For example, low cycle and high cycle fatigue may produce cracks leading to rupture. The age units would be fatigue cycles. The age of a starter may be the number of starts. Burner and turbine parts may fail as a function of time at high temperature or as the number of cold to hot to cold cycles. Usually, knowledge of the physics-of-failure will provide the age scale. When there is uncertainty, several age scales are tried to determine the best fit. This is not difficult with good software. The "best" aging parameter data may not exist and substitutes are tried. For example, the only data on air conditioner compressors may be the date shipped and the date returned. The "best" data, operating time or cycles, is unobtainable, so based on the dates above, a calendar interval is used as a substitute. These inferior data will increase the uncertainty, but the resulting Weibull plot may still be accurate enough to provide valuable analysis. The data fit will tell us if the Weibull is good enough.

1.7 Failure Distribution

The slope of the Weibull plot, beta, (β), determines which member of the family of Weibull failure distributions best fits or describes the data. The slope, β, also indicates which class of failures is present:

- $\beta < 1.0$ indicates infant mortality
- $\beta = 1.0$ indicates random failures (independent of age)
- $\beta > 1.0$ indicates wear out failures

These classes will be discussed in Chapter 2. The Weibull plot shows the onset of the failure. For example, it may be of interest to determine the time at which 1% of the population will have failed. Weibull called this the "B1" life. For more serious or catastrophic failures, a lower risk may be required, B.1 (age at which 0.1% of the population fail) or even B.01 life (0.01% of the population). Six-sigma quality program goals often equate to 3.4 parts per million (PPM) allowable failure proportion. That would be B.00034! These values are read from the Weibull plot. For example, on Figure 1-1, the B1 life is approximately 160 and the B5 life is 300.

The horizontal scale is the age to failure. The vertical scale is the Cumulative Distribution Function (CDF), describing the percentage that will fail at any age. The complement of the CDF scale, (100 - CDF) is reliability. The characteristic life η is defined as the age at which 63.2% of the units will have failed, the B63.2 life, (indicated on the plot with a horizontal dashed line). For $\beta = 1$, the mean-time-to-failure and η are equal. For $\beta > 1.0$, MTTF and η are approximately equal. The relationship will be given in the next chapter.

1.8 Failure Forecasts and Predictions

When failures occur in service, a prediction of the number of failures that will occur in the fleet in the next period of time is desirable, (say six months, a year, or two years). To accomplish this, the author developed a risk analysis procedure for forecasting future failures. A typical failure forecast is shown in Figure 1-2. Cumulative future failures are plotted against future months. This process provides information on whether the failure mode applies to the entire population or fleet, or to only one portion of the fleet, called a batch. After alternative plans for corrective action are developed, the failure forecasts are repeated. The decision-maker will require these failure forecasts to select the best course of action, the plan with the minimum failure forecast or the minimum cost. If failed parts are replaced as they fail; the failure forecast is higher than without replacement. Prediction intervals, analogous to confidence intervals, may be added to the plot. Chapter 4 is devoted to failure forecasting.

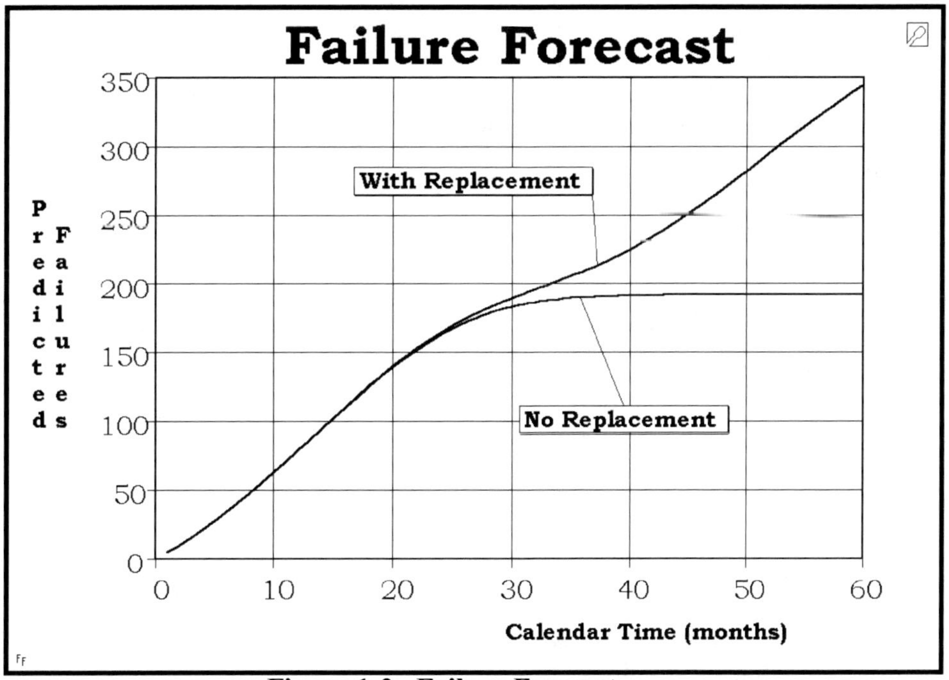

Figure 1-2. Failure Forecast

A further advantage of Weibull analysis is that it may be useful even with inadequacies in the data. Even bad Weibull plots are usually informative to engineers trained to read them. Methods will be described for:

- Identifying mixtures of failure modes,
- Problems with the origin not located at zero,
- Investigation of alternate aging parameters,

- Handling data where some part ages are unknown,
- Construction of a Weibull curve when no failures have occurred,
- Identifying "batch" problems where the failure mode only affects a subset of the fleet,
- Identifying grouped or inspection data,
- Identifying suspected outliers.

The Weibull distribution usually provides the best fit of life data. This is due in part to the broad range of distribution shapes that are included in the Weibull family. Many other distributions are included in the Weibull family either exactly or approximately, including the normal, the exponential, the Rayleigh, and sometimes the Poisson and the Binomial. If the Weibull fit is poor, other distributions should be considered unless the sample size is small, less than 21 failures. The data may be plotted on other probability papers to determine which distribution best fits the data. The lognormal is not a member of the Weibull family and it is by far the most significant competitor. The lognormal is the best choice for some material characteristics, for crack growth rate, and for non-linear, accelerating system deterioration. Weibull-lognormal comparisons are much easier with appropriate software. If there is engineering evidence supporting another distribution, this should be weighed heavily against the Weibull. **Moderate size samples, twenty or more failures, are needed to accurately discriminate between the Weibull and other distributions.** With less than 20 failures the Weibull is the best choice, and therefore, **best practice**.

1.9 Engineering Change Test Substantiation

When a redesign is made to correct an existing failure mode, tests are made to show the new design is an improvement. The tests are required as not all redesigns are improvements. How many units must be tested without failure, for how long, to substantiate that the new design is significantly better than the existing design? Alternatively, the test objective may be to demonstrate a design requirement expressed as a reliability or probability of failure at some design life.

The success data from the test may be used to determine a lower confidence bound for the Weibull line for the new design, called a "Weibayes" line. The test design criteria may allow zero failures, or zero or one failure, etc. as alternatives. "Sudden Death" testing is another useful technique. Zero failure test plans have the absolute minimum test time. Trade studies are used to find the optimum test plan to minimize cost and time. See Chapter 6.

1.10 Maintenance Planning

The Weibull plot is extremely useful for maintenance planning, particularly reliability centered maintenance. Beta, (β) tells the analyst whether or not scheduled inspections and overhauls are needed. If β is less than or equal to one, overhauls are not cost effective. With βs greater than one, the overhaul period or scheduled inspection interval is read directly from the plot at an acceptable probability of failure. For wear out failure modes, if the cost of an unplanned failure is much greater than the cost of a planned replacement, there is an optimum replacement interval for minimum cost. Using Weibull failure forecasting, quantitative trades are made between:

- Scheduled and unscheduled maintenance,
- Forced retrofit and convenience retrofit,
- Non-destructive inspections versus parts replacement,
- Corrective action versus "do nothing,"
- Different times-between-overhauls,
- Optimal replacement intervals.

Planned maintenance induces cyclic or rhythmic changes in failure rates. The rhythm is affected by the interactions between characteristic lives of the failure modes of the system, β, the inspection periods, and parts replacements. This phenomenon is illustrated in Chapter 4.

1.11 System Analysis and Math Models

Mathematical models of components and entire systems like a sports car, a big truck, or a nuclear power system may be produced by combining (statisticians say convoluting) tens or hundreds of failure modes. Most of these modes are represented by Weibulls but some may be lognormal, or even binomial. The combination may be done by Monte Carlo simulation or by analytical methods. If the data cannot be segregated into individual failure modes or if the early data is missing, the Crow-AMSAA or the Kaplan-Meier models may still be applied to provide trending and failure forecasting. System models are useful for predicting spare parts usage, availability, module returns to depot, and maintainability support costs. These models are frequently updated with the latest Weibulls. Predictions may be compared with actual results to estimate the model uncertainties and fine-tune the model.

If the data from a system, such as a diesel engine, are not adequate to plot individual failure modes, it is tempting to plot a single Weibull for the system. This plot will show a β close to one. This is roughly equivalent to using mean-time-to-failure (MTTF) and exponential reliability. It masks infant mortality and wear out modes. This approach is not recommended, as the results are not meaningful for the individual failure modes. This method was common in the fifties and sixties and is still used by those unaware of the advantages of Weibull Analysis and the newer methods for system models. Better methods are provided for these data deficiencies. See Chapter 8.

1.12 Weibulls with Curved Data

The Weibull plot is inspected to determine how well the failure data fit a straight line. Sometimes failure points do not fall along a straight line on the Weibull plot, and modification of the simple Weibull approach is required. The data are trying to tell us something in these cases. Weibull illustrated this concept in his 1951 paper. The bad fit may relate to the physics of the failure or to the quality of the data. If the points fall on gentle curves, it may be that the origin of the age scale is not located at zero. See Figure 1-3.

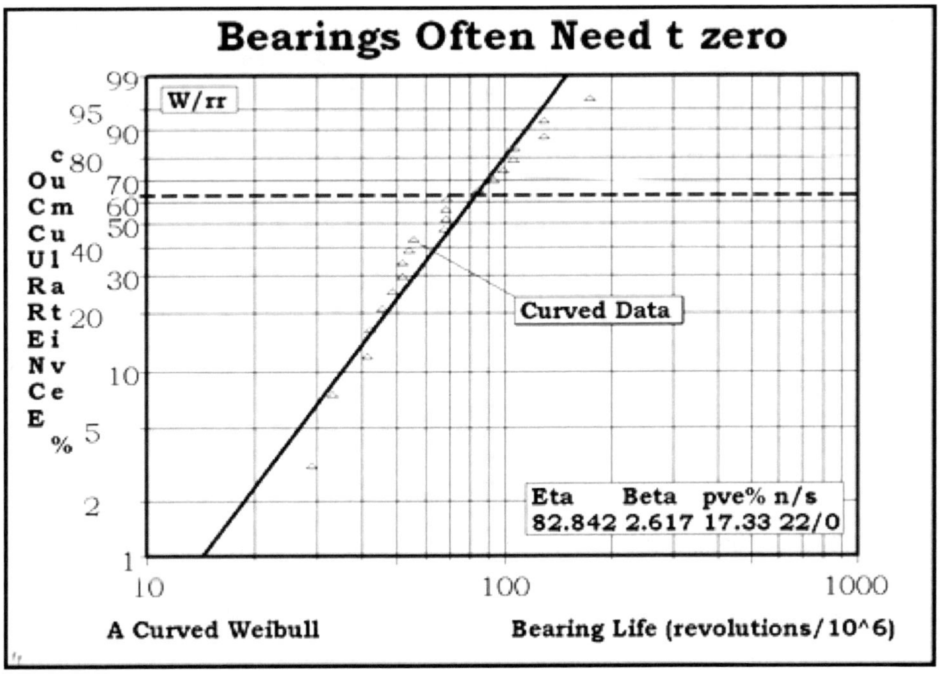

Figure 1-3. Bearings Rarely Fail Early in Life With Curved Data

There are usually physical reasons for this. The manufacturer may not have reported early failures that occurred in production acceptance. With roller bearing unbalance, it takes many rotations for the wobbling roller to destroy the cage. The bearing cannot fail instantaneously. This leads to an origin correction equal to the minimum time necessary to fail. The correction is called t_0. If t_0 is positive, it provides a guaranteed failure free period from zero to time t_0. If the t_0 correction is negative, "old" parts, perhaps with shelf life aging, may be the cause. The t_0 correction is treated in Chapter 3.

Lognormal data plotted on Weibull probability paper will appear curved, concave downward like Figure 1-3. Therefore, the Weibull with the t_0 correction and the lognormal distribution are candidates for distributional analysis to tell us which is the best distribution. The critical correlation coefficient provides an easy solution. This capability is in SuperSMITH Weibull (SSW). Moderate size samples are required to discriminate between these distributions, **at least 20 failures**. Otherwise, the standard two-parameter Weibull is the best choice [Liu].

1.13 Weibulls With Corners and Doglegs

Another common cause of bad Weibulls is a mixture of failure modes. The plot will often show doglegs and sharp corners. This indicates there is more than one failure mode, a mixture of modes. See Figure 1-4. The preferred procedure is to use engineering physical analysis to classify the failure data into different modes. Examination of the broken parts, or failure reports, or laboratory tests is often helpful. Separate Weibull plots are made for each failure mode. If the classification is correct, the separate Weibull plots will show good fits, with their points falling along the plot` line. On each plot, the failure points from the other failure modes are treated as successful (censored, suspended or non-failure) units. The mixture shown in Figure 1-4 is frequently observed with hydromechanical components, a shallow slope followed by a steep slope. For this classic mixture, if additional failure analysis is impossible, a statistical solution based on the likelihood ratio test may provide the separate Weibull failure modes. Appendixes F and J provide a discussion of more complex mixtures including batch problems.

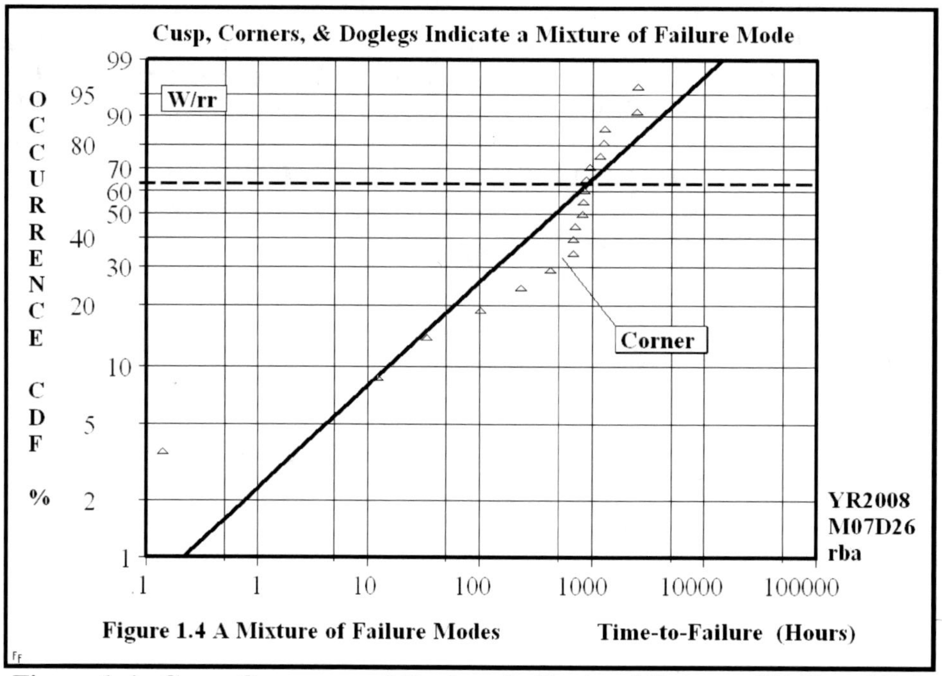

Figure 1-4. Cusp, Corners, and Doglegs Indicate a Mixture of Failure Mode

As the number of failure modes mixed together increases to say five or more, the doglegs and corners tend to disappear and β will tend toward one (unless all the modes have the same β and similar η). See Section 3.7.

1.14 Weibayes

A Weibull plot may be needed even when no failures have occurred. For example, an engineering redesign is made to correct a failure mode experienced in service. Redesigned units are tested to prove that the problem is corrected. How many units must be tested for how long to have statistical confidence that the problem is solved?

When parts exceed their design life, it may be possible to extend their life by constructing a Weibull with no failures. The method is called Weibayes analysis. It has many applications and is presented in Chapter 6. Methods to design experiments to substantiate new designs using Weibayes theory is also presented in Chapter 6. In preliminary design, when there is little or no data on the new designs, Weibayes may be used to estimate the Weibulls for the new design. If the number of failures is extremely small and there is good prior knowledge of the slope, β, Weibayes will be more accurate than the Weibull.

1.15 Small Sample Weibulls

Statisticians always prefer large samples of data, but engineers are forced to do Weibull or Weibayes analysis with very small samples, even as few as one to three failures. When the result of a failure involves safety or extreme costs, it is inappropriate to request more failures. The primary advantage of Weibull analysis is that engineering objectives can often be met even with extremely small samples. Of course, small samples always imply large uncertainties. To evaluate these small sample uncertainties, confidence bounds are available in SSW. Chapter 7 is devoted to confidence methods.

1.16 Updating Weibulls

After the initial Weibull, later plots may be based on larger failure samples and more time on successful units. Each updated plot will be slightly different, but gradually the Weibull parameters, β and η, will stabilize and approach the true Weibull as the data sample increases. With the appropriate fit method, the important engineering inferences about B.1 life and the failure forecasts will not change significantly as the sample size increases. With complete samples (no suspensions) β and η oscillate around the true unknown value.

1.17 Deficient (Dirty) Data

Special methods have been developed for many forms of "dirty" data including:

- Censored-suspended data
- Mixtures of failure modes
- Failed units not identified
- Inspection data & coarse data
- Suspension times or ages missing
- No failure data
- Nonzero time origin
- Extremely small samples
- Early data missing
- Data plots with curves and doglegs

1.18 Establishing the Weibull Line, Choosing the Fit Method

The standard engineering method for establishing the Weibull line is to plot the time to failure data on Weibull probability graphs using median rank plotting positions and regression analysis to fit the line as described in Chapter 2. This method is recommended as **best practice.** Many statisticians prefer an analytical method called maximum likelihood. The likelihood calculations require a computer. Both methods have advantages and disadvantages. The pros and cons will be discussed in Chapter 5. There are also special methods for different types of inspection data and warranty data that are treated in Chapters 5 and 8.

1.19 Related Methods and Problems

Weibull analysis is the main theme of this text, but there are some types of data and some types of problems that can be analyzed better with other math models that are described later and included in the software. For example, the Weibull distribution is also the **Extreme Value** Type III minimum distribution. The Extreme Value Type I is called the **Gumbel distribution** and both the minimum and maximum forms have useful applications. For example gust loads on aircraft are modeled with Gumbel maximum distribution. See Figure 1-5.

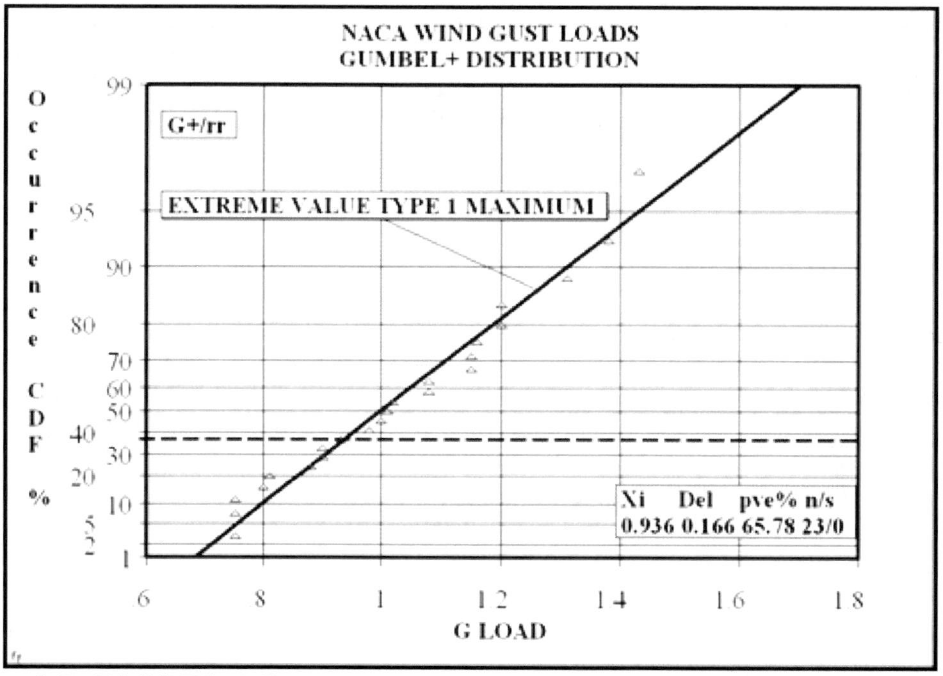

Figure 1-5. NACA Wind Gust Loads Gumbel + Distribution (Also shown in Chapter 8)

Some organizations feel the **Crow-AMSAA** (C-A) model is more important than Weibull analysis. It is extremely useful. It is more robust than Weibull, that is, it provides reasonable accurate results when the data has serious deficiencies. It works well even with mixtures of failure modes and missing portions of data. Weibull plots allow only one failure mode at a time. It will track changes in reliability which Weibull cannot. C-A is the best tool for trending significant events for management, such as outages, production cutbacks and interruptions, accidents, and in-flight loss of power. The original C-A objective was to track the growth of reliability in R&D testing. It is still **best practice** for that application. It is also **best practice** for predicting warranty claims by calendar time and tracking systems in-service.

Kaplan-Meier (K-M) survival analysis is good for "snapshot" data, (a large portion of initial data is not available). This tool was developed in the medical industry and has good statistical properties. We have borrowed K-M for our reliability work. K-M is also used for warranty claim predictions as a function of age, and for some forms of inspection data.

Wayne Nelson's **Graphical Repair Analysis** provides a rigorous approach for predicting warranty claims by age of the units when repeated warranty repairs of the same type are expected.

1.20 Summary

Chapter 1 has provided an overview of Weibull analysis. There are many methods because there are many kinds of data and applications. The real effort involved with this Weibull analysis is obtaining and validating good data. The analysis, given good software, is almost trivial by comparison.

- The next chapter (2) will discuss good Weibull plots and good data and the standard method of analysis.
- Chapter 3 will treat bad Weibulls and "dirty" data. Case studies will be employed to illustrate the methods and solutions.
- Chapter 4 is devoted to failure forecasting, Monte Carlo, and optimal parts replacement intervals.
- Chapter 5 presents alternative data formats and special solution methods.
- The Weibayes concept, substantiation tests and Weibull Libraries are treated in Chapter 6.
- Chapter 7 discusses confidence intervals and testing two or more data sets to see if they are significantly different.

- Chapter 8 treats a number of related math models.
- Chapter 9 presents Crow-AMSAA modeling and warranty analysis.
- Chapter 10 summarizes the Handbook methods, analysis, and interpretations of the plots. Bob Rock's logic diagram in Chapter 10 will lead you step-by-step though the **best practices** for your particular data.
- Chapter 11 is a collection of interesting case studies contributed by industry experts

There is little herein on **qualitative methods**, as the emphasis is on **quantitative methods**. However, qualitative methods are extremely important. **Failure Reporting, Analysis, and Corrective Action System** (FRACAS) is required for accurate life data analysis. **Failure Mode and Effect Analysis** (FMEA) and **Fault Tree Analysis** (FTA) can significantly improve the design process and are needed to identify the root cause of failure modes. **Design review** is prerequisite to achieving high reliability and is significantly improved by use of a Weibull library to avoid previous design mistakes. Pat [O'Connor] provides a good introduction to these methods. The Weibull library is discussed in Chapter 6.

A comment on **software** is appropriate here. As the primary purpose of this Handbook is to be the workbook for the Weibull Workshops, it is necessary to present the software capabilities with the methodology. Further, for readers not attending the workshop, good software is needed to eliminate the drudgery of hand calculations and in general, increases the productivity of the analyst many times over. The author has been deeply involved in the development of SuperSMITH® Weibull (SSW) & SuperSMITH Visual (SSV) software and therefore, recommends their features and friendly characteristics. Further, no other software does **all the methods** presented in the Handbook. Almost all of the plots herein are from SuperSMITH, a bundled set of Windows software including SSW and SSV as well as YBath™ (SSY). This software is used in our Weibull Workshops. Free DEMO versions of the software are available on our web sites. Wes Fulton is the creator of the software, and Carl Tarum oversees new software research. When in the office during office hours (and not travelling), they are available to answer questions concerning potential applications and software use. The initial title pages of this handbook contain their contact information. They would like to hear from you.

This page is intentionally blank.

CHAPTER 2: PLOTTING THE DATA AND INTERPRETING THE PLOT

2.1 Foreword

Basic Weibull analysis consists of plotting the data on Weibull probability paper and then interpreting the plot. There is significant information available on the Weibull plot. It is so informative that even bad data is worth plotting.

The first step is to try to get good data. The times-to-failure (or cycles–to-failure or mileage-to-failure) are needed plus the current age of units that have not failed. Data acquisition is the most difficult part of Weibull analysis assuming there is good friendly software available. Next, these data are ranked (i.e., rearranged so the earliest failure is listed first and the oldest failure is last), plotted on Weibull probability paper, and then the plot is interpreted. Each of these steps will be discussed in this chapter.

2.2 Weibull Data

Life data is special because we must know the "age" of the parts. Weibull failure data is life data. There are **two** kinds: (1) **standard life data** consists of the exact "age" of the parts that have failed and those that have not failed and (2) **grouped** or **interval data** for which exact ages are unknown so the data is grouped into intervals of time. Inspection data is an example, grouped into inspection intervals. Inspection data is “dirty” or deficient data as we do not know the exact time to failure. If data is collected monthly or weekly, it is also grouped data and dirty for the same reason. This is coarse data as the intervals are too long compared to exact times.

"Age" may be operating time, starts and stops, landings, takeoffs, low cycle fatigue cycles, mileage, shelf or storage time, cycles or time at high stress or high temperature, or many other parameters (Figure 2-1). Usually the appropriate aging parameter is obvious from the physics of the failure mode, but sometimes there may be confusion. For example, does a turbine blade erode due to operating time, or time at high temperature, or hot-cold cycles? Without engineering knowledge to tell us which aging parameter is best, the choice is determined from Weibull plots of the alternatives. The best aging parameter is the one with the best fit compared to a straight line.

Sometimes life data for the best aging parameter is not available. For example, air conditioning compressors probably age as a function of operating time or start-stop cycles. However, calendar time between production acceptance and return from warranty claim may be the only data available. This is "dirty data" because it has deficiencies. The Weibull plot against calendar time will have a poorer fit and more uncertainty than the (unobtainable) plot against the best aging parameter. A measure of goodness of fit (r^2 or pve%) will determine if the calendar time Weibull plot may be useful. This problem is common. Automotive data rarely contains starts and stops or operating time, so mileage is used. Electric utilities may not have operating cycles, so they are forced to use operating time or calendar time. As Weibull analysis becomes more widely accepted, data quality usually improves because management recognizes the cost effectiveness of good data.

Materials characteristics such as strength, creep, stress rupture and fatigue often are plotted on Weibull probability paper. Here the horizontal scale may be stress, load or temperature. Although these plots may use forcing parameters instead of aging parameters, the plot is interpreted the same way.

2.3 The Weibull Plot Scales

The horizontal scale is the age or time parameter (t). This scale is logarithmic as shown in Figure 2-1. For automobiles and trucks mileage is used. Operating time or cycles are used for in-service systems and components. The vertical scale is the Cumulative Distribution Function (CDF), that defines the proportion of the parts that will fail up to age (t) in percent. The statistical symbol for the CDF is F(t), the probability of failure up to time t. The complement of F(t) is reliability, the probability of not failing up to time t. R(t) = 1-F(t).

Weibull coined a name for the time that a given percentage will fail. He called these "B lives". In Figure 2-1, the B1 life for the new design is about 80 units. This means 1% of the new design parts will fail by the time they age 80 units. It also means the reliability at 80 units is 99%. The block in the lower right corner provides η; β; pve% (a measure of goodness of fit); and n, (the fleet size of 8), and s, the number of suspensions, zero in this case. The scales are derived so the data conforms to a straight line if it is a Weibull distribution. Other notation on Figure 2-1 will be explained later. Weibull analysis applies to only <u>one failure mode at a time</u>. A single part may have several failure modes and each mode has its own Weibull plot.

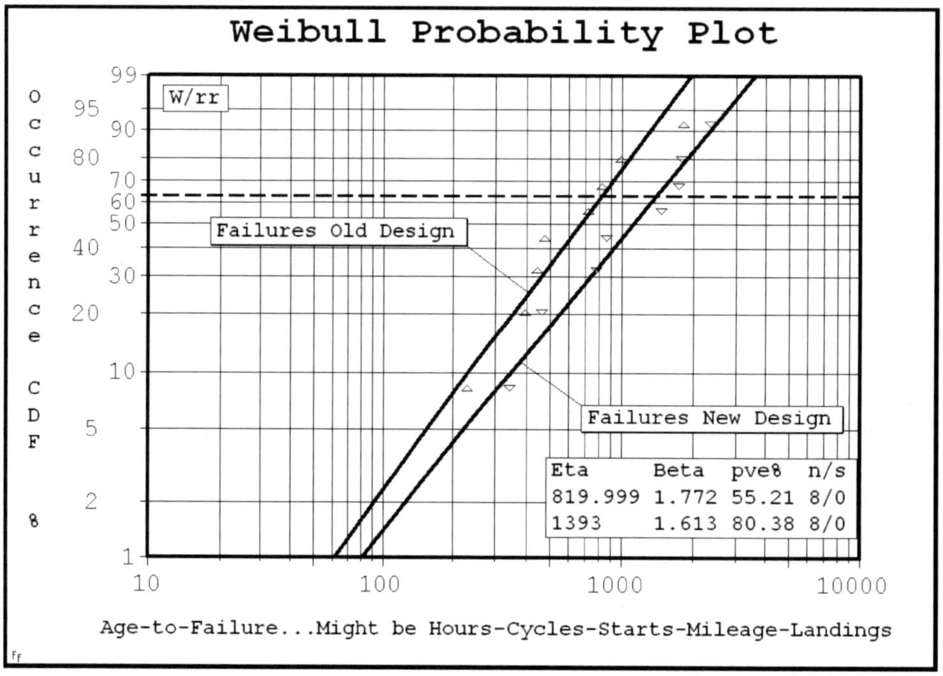

Figure 2-1. A Typical Weibull Plot

2.4 η (Eta) and β (Beta)

The two-parameter Weibull distribution is by far the most widely used distribution for life data analysis. The Weibull cumulative distribution function (CDF) provides the probability of failure, F(t), up to time (t):

$$F(t) = 1 - e^{-(t/\eta)^{\beta}} \tag{2-1}$$

where:

F(t) = fraction failing up-to-time t (CDF), t=failure time,

e = 2.718281828..., the base for natural logarithms, η =characteristic life or scale parameter,

β =slope or shape parameter

The complement of the CDF is reliability, (R), the probability that failure will not occur up to time (t).

$$R(t)=e^{-(t/\eta)^{\beta}} \tag{2-2}$$

The two parameters are η, the characteristic life, and β, the slope of the Weibull line. β shows the class of failure mode such as infant mortality, random, or wear out. These interpretations are discussed later in this chapter. β is called the shape parameter as it determines which member of the Weibull family of distributions is most appropriate. Different members have different shaped probability density functions (PDF). See Figure 2-2. The Weibull distribution fits a broad range of life data compared to other distributions. The PDF is the derivative of the CDF with respect to time (t). Failure data plotted as a bar chart against time will have the same shape as the PDF.

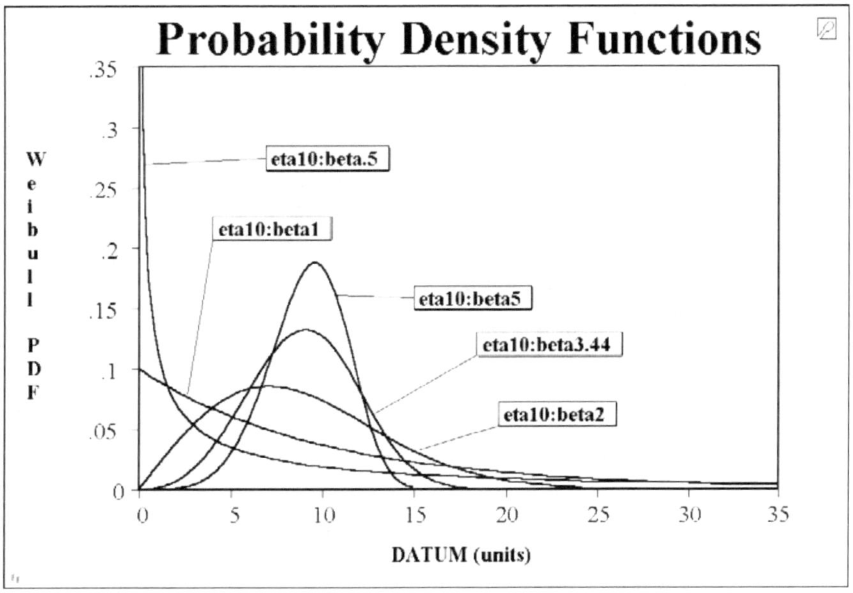

Figure 2-2. The Shapes of the Weibull Family

The Weibull probability density function, PDF, is:

$$f(t) = (\beta/\eta)(t/\eta)^{\beta-1}\, e^{-(t/\eta)^{\beta}} \tag{2-3}$$

Although the PDF shape for $\beta = 3.4$ looks like the normal distribution, normal data plotted on a Weibull plot can have any β. σ, the standard deviation of the normal varies inversely with β. Small sigmas correspond to large betas. Normally distributed data makes good Weibull plots but not usually the other way around. As a result the normal distribution is rarely used in life data analysis.

η is equal to the mean-time-to-failure (MTTF) when the slope, β, equals one. η and MTTF are related by a gamma function of β shown in Equation 2-4, and derived in Appendix G.

$$\text{MTTF} = \eta\, \Gamma\,[1+1/\beta] \tag{2-4}$$

- When $\beta = 1.0$, MTTF = η, the Exponential distribution.
- When $\beta > 1.0$, MTTF is less than η.
- When $\beta < 1.0$, MTTF is greater than η. See Figure 2-3.
- When $\beta = 0.5$, MTTF = 2 x η.

MTBF, the average time *between* failures, should not be confused with MTTF, mean-time-to-failure, *the average life*. MTBF is estimated by dividing total operating time on all units by the number of failures observed. They are different parameters, although equal when there are no suspensions. Otherwise they can be vastly different. MTBF is used with repairable systems. See the Glossary, Appendix A.

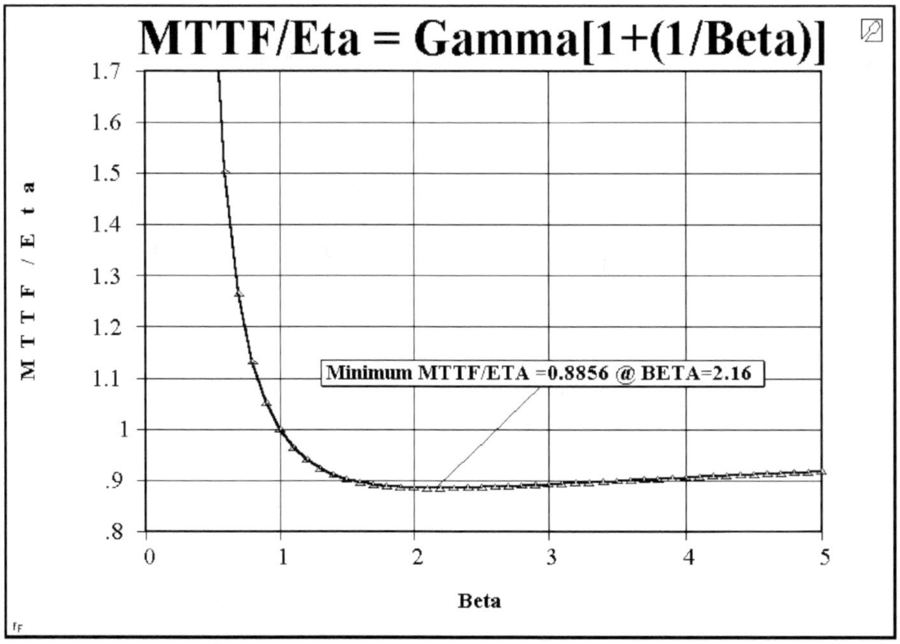

Figure 2-3. MTTF/Eta = Gamma[1 + (1/Beta)]

η is the age at which 63.2% of the units will fail. This is true for all Weibulls. Setting t = eta in Equation 2-1 illustrates the reason.

$$F(t) = 1 - e^{-(\eta/\eta)^{\beta}} = 1 - e^{-(1)^{\beta}} = 1 - (1/e) = 0.632, \text{ regardless of } \boldsymbol{\beta}$$

This means that for all Weibulls 63.2% of the units will fail by the Characteristic Life, η.

2.5 Weibull Analysis - An Example

A benign failure does not affect the performance of the system, and it will go unnoticed until the system is inspected. This was the case for the compressor inlet air seal rivets in the following example. The flare part of the rivet was missing from some of the rivets during inspection. After thorough failure analysis, a redesign was started. An accelerated laboratory test using the old rivets established a baseline. The results are shown in Table 2-1. Rivet serial numbers 3, 6, and 7 are considered non-representative failures as they were produced by two other failure modes. For instructive purposes only, the three suspensions are ignored for the first analysis, but will be considered correctly later. That leaves five failure points for flare failure. (Weibull plots could also be made for flare loosened, 2 failures + 6 suspensions and lug failed, 1 failure + 7 suspensions.)

Table 2-1. Baseline Rivet Data

Serial Number	*Failure Time (Minutes)*	*Remarks*
1	90	Flare Failure
2	96	Flare Failure
3	100	Flare loosened
4	30	Flare Failure
5	49	Flare Failure
6	45	Flare loosened
7	10	Lug failed
8	82	Flare Failure

To make the Weibull plot, rank the data from the low time to the high time failure. See Table 2-2. This ranking will establish the plotting positions for the time, t, axis and the ordinate, F(*t*), in % values. Each failure is plotted at its time-to-failure (*t*) and an estimate of F(*t*), the percentage of the total population failing before it. The true percentage values are unknown. In the next section a Weibull plot of this data is produced ignoring suspensions. **This is wrong.** Always include suspensions. It is done here to illustrate the effect of suspensions. The correct plot with suspensions will be produced in Section 2.9.

2.6 Median Ranks

The median rank is the most popular approach of estimating the Y-axis plotting positions. Median ranks are recommended as most accurate and therefore, **best practice**, but there are other plotting positions discussed in Chapter 5. Weibull employed *mean ranks* in his paper $[Weibull 1951] but later recognized $[Benard] and [Johnson 1964] suggestions to use median ranks. In Section 5.4 we will compare the accuracy of several plotting positions. With reasonably sized samples there is little difference in results with different plotting positions. NOTE: The "$" symbol preceding a reference means the document may be downloaded from Paul Barringer's website (see References).

Median rank tables are provided in Appendix I. Enter the tables for a sample of five and find the median ranks shown in Table 2-2 for the five flare failure times shown in the middle column. These same median rank plotting positions are used with all types of probability paper i.e., Weibull, lognormal, normal, and extreme value. Note that if two data points have the same time to failure on the X-axis, they are plotted at different median rank values on the Y-axis, each point gets its own individual vertical location.

Table 2-2. Ranked Flare Failure Rivet Data

Order Number (i)	*Failure Time t (Minutes) (X)*	*Median Rank % (Y)*
1	30	12.94
2	49	31.38
3	82	50.00
4	90	68.62
5	96	87.06

The median rank table is calculated using the incomplete beta distribution. The median rank values of F(t), probability of failure, are found by setting the beta probability equal to 0.5 (50%) for each value of sequence number. (See page I-1 for the Excel function). For example, for a set including 3 failure points (N = 3):

1. For the lowest data point the median rank estimate, the probability plotting position is (20.63%).
2. For the middle data point the probability plotting position is (50.00%).
3. For the highest data point the probability plotting position is (79.37%).

The median estimate is preferred to the mean or average value for non-symmetrical distributions. The median is more typical in these cases. For example, the majority of the population makes far less than the average (mean) income, but half the population makes more (or less) than the median income. Most life data distributions are skewed and therefore, the median plays an important role. [$Mischke 1979] is recommended for readers that would like more detailed explanation. NOTE: The "$" symbol preceding a reference means the document may be downloaded from Paul Barringer's website (see References).

The rivet failure data from Table 2-2 is plotted below in Figure 2-4.

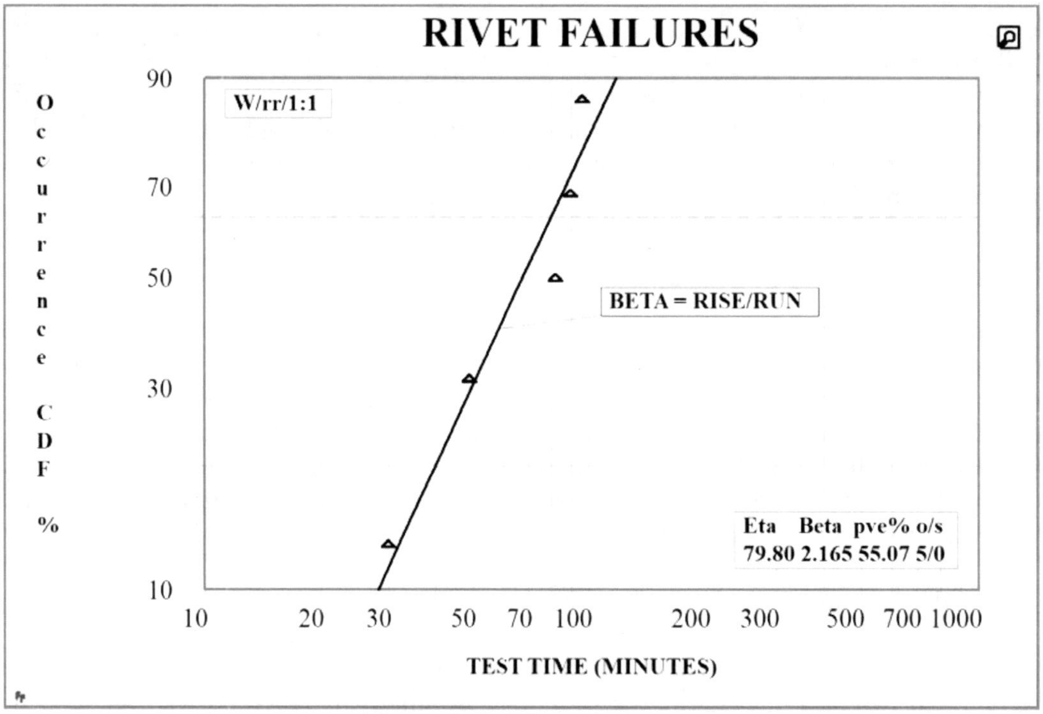

Figure 2-4. Rivet Failures

2.7 The Weibull Plot

Plot the points from Table 2-2 on one-to-one ratio Weibull probability paper and "eyeball" (draw) a best fit line through the points. Measure the slope of the line with a ruler by taking the ratio of rise over run measured with a ruler to estimate beta. Note, make sure you are using 1:1 scale Weibull probability paper. Weibull plots generated by computers are rarely 1:1 scale. Most of the illustrations in this book are not 1:1. See the problem section at the end of this chapter for 1:1 Weibull paper. Select a starting point and measure one inch in the horizontal direction (run). Then, measure vertically, (rise), to the line intersect. In Figure 2-4 the slope is an estimate of β = rise/run = 2/1 = 2. Analytical methods for establishing the line using regression analysis and maximum likelihood estimates will be discussed in Chapter 5.

Two parameters define the Weibull distribution. The first is β, the slope or shape parameter, and the other is the characteristic life, eta (denoted by η). η is the 63.2 percentile of the distribution. In Figure (2-4), the 63.2 percentile intersects the line at 79.8 minutes; therefore, η = 79.8 minutes.

2.8 "B" Life

Waloddi Weibull suggested that bearings should be compared by their "B10" life, the age at which 10% of the bearings will fail. Weibull's suggestion has become the world standard for quoting bearing life. It is not clear where the B notation came from. The author believes it may be from the German "Brucheinleitzeit" (fracture initiation time). Others believe "B" stands for bearing. Some bearing manufacturers now use "L10" (L for life) instead of the "B10" notation. Aircraft industries use much lower B lives for design requirements. In aerospace, B1 life is used for benign failures, B0.1 for serious failures and B0.01 for catastrophic failures. Some organizations quote reliability with a confidence level, such as "99% reliability with 95% confidence for catastrophic failures." The author prefers the simple B life without the confidence interval. B lives may be read directly from the Weibull plot or determined more accurately from the Weibull equation. For example, the B1 life for the rivet data in Figure 2-4 would be 9.5 minutes. The age at which 50% of the units fail, the B50 life, is the median time-to-failure, about 67 minutes in Figure (2-4).

2.9 Suspended Test Items

Non-failed units or units that fail by a different failure mode are "censored" or "suspended" units. These data cannot be ignored even though the suspensions are never plotted. Times on suspended units must be included in the analysis.

Suspensions are not plotted but the rank order and the median ranks must be adjusted for suspensions. Leonard Johnson is credited for the method of adjustment employed herein [Johnson 1964]. The Johnson equation given in the original USAF Handbook represented a significant obstacle for students trying to hand plot Weibulls. Fortunately, Drew Auth of AVCO/LYCOMING, simplified Leonard Johnson's formula for the author. Auth's simple formula gives exactly the same adjustment to the rank order (i). It is easier to apply. The Auth adjusted rank value then is used in Bernard's equation to calculate an *adjusted median rank*. The procedure is to rank the Table 2-1 data with the suspensions and use the Auth equation below, to determine the adjusted ranks, accounting for the presence of the suspensions. (This corrects the previous Weibull, Table (2-2), and Figure (2-4), which excluded the three suspensions.)

$$\textbf{i = Adjusted Rank} = \frac{\textbf{(Reverse Rank)*(Previous Adjusted Rank)+(N+1)}}{\textbf{(Reverse Rank+1)}} \qquad (2\text{-}5)$$

The rank order numbers are adjusted for the effect of the three suspended items in Table 2-3.

Table 2-3. Adjusted Ranks for Suspended or Censored Data

Rank	*Time*	*Reverse Rank*	*Adjusted Rank (i)*	*Median Rank*
1	10S	8	...Suspended...	
2	30F	7	[7 X 0 +(8+1)]/ (7+1) = 1.125	9.8%
3	45S	6	...Suspended...	
4	49F	5	[5 X 1.125 +(8+1)]/ (5+1) = 2.438	25.5%
5	82F	4	[4 X 2.438 +(8+1)]/ (4+1) = 3.750	41.1%
6	90F	3	[3 X 3.750 +(8+1)]/ (3+1) = 5.063	56.7%
7	96F	2	[2 X 5.063 +(8+1)]/ (2+1) = 6.375	72.3%
8	100S	1	...Suspended...	

The adjusted ranks are used in Benard's approximation to calculate the adjusted median ranks, as it is easier than interpolating in the table. Benard's approximation is given in the next section. The results in Table (2-3) are plotted in Figure (2-5) below. The major effect of the suspensions is to increase η. Beta is minimally affected. Note: suspended items do not affect rank numbers *until after they occur* (earlier failure times have unadjusted rank numbers). Also, if two items fail at the same age, they are assigned sequential rank order numbers.

2.10 Benard's Approximation

Benard's approximation for the median rank is sufficiently accurate for plotting Weibulls and estimating the parameters. (Benard's approximation is accurate to 1% for N = 5 and 0.1% for N = 50.) It is also easier than interpolating in the tables for the adjusted median ranks that are not an integer value. $[Benard].

$$\text{Benard's Median Rank} = (i - 0.3) / (N + 0.4) \qquad (2\text{-}6)$$

Here "i" is the adjusted rank and "N" is the sum of failures and suspensions. The median ranks are converted to percentages to plot on Weibull paper. For example, considering the first failure in Table 2-3 with an adjusted rank of **1.125**: Median Rank (%) = (**1.125** - 0.3) x 100 / (8 + 0.4) = 9.82%

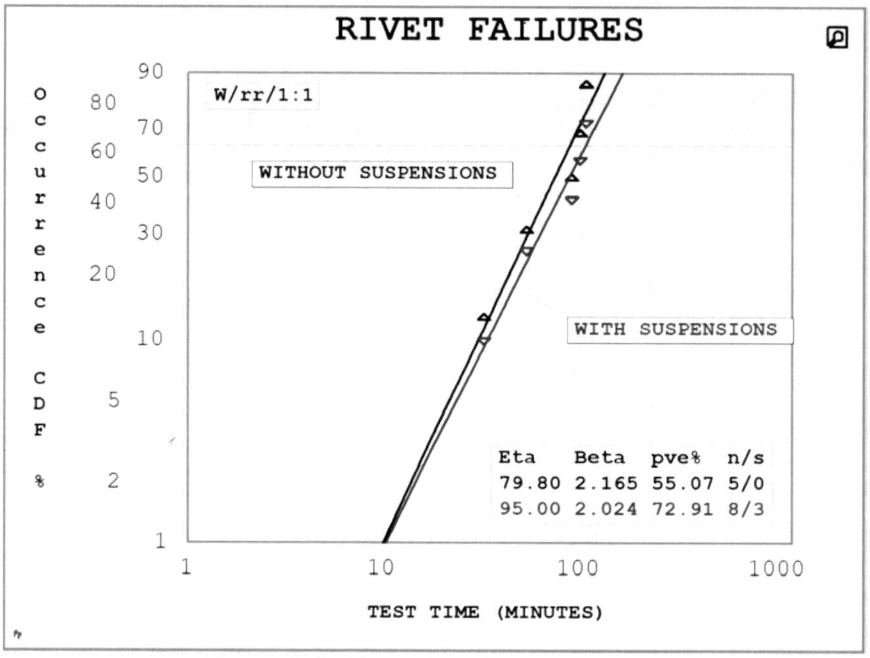

Figure 2-5. Total Test Time (minutes)

2.11 Suspensions Increase Eta

What effect did the suspensions have on the Weibull plot? The slope, beta, hardly changed but the characteristic life, eta, increased. This effect generally is true: adding suspensions has little effect on beta, but increases eta. Figure 2-5 is the *correct* Weibull plot as it includes the suspensions. Thus if suspensions are ignored, results will be pessimistic. These are the **steps to plot data sets** with suspensions:

1. First rank the times, both failures and suspensions, from earliest to latest.
2. Calculate the adjusted ranks for the failures (suspensions are not plotted).
3. Use Benard's approximation to calculate the adjusted median ranks.
4. Plot the failure times (x) versus the adjusted median ranks (y) on standard 1 x 1 Weibull paper. (Incidentally, the same plotting positions may be used to plot the data on other probability paper such as lognormal.)
5. Estimate eta by reading the B63.2 life from the plot.
6. Estimate beta as the ratio of the rise over run measured with a ruler.
7. Interpret the plot.

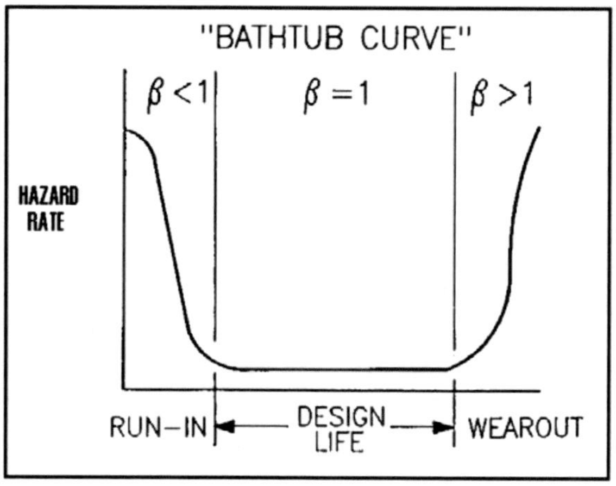

Figure 2-6. The Bathtub Curve for a Good Component

2.12 Interpreting the Weibull Plot

The first look at the Weibull plot answers two questions; how good is the fit and what is the beta, the slope? The Weibull plot provides clues about the failure mechanism, since different slopes, betas, imply different classes of failure modes. The bathtub curve, Figure 2-6, shows the relationship between beta and failures throughout the life of a component. The *"hazard rate"* is the instantaneous failure rate.
Figure 2-7 shows a bathtub curve for human beings. Personally we hope the "failure rate" during the design life is zero from all causes.

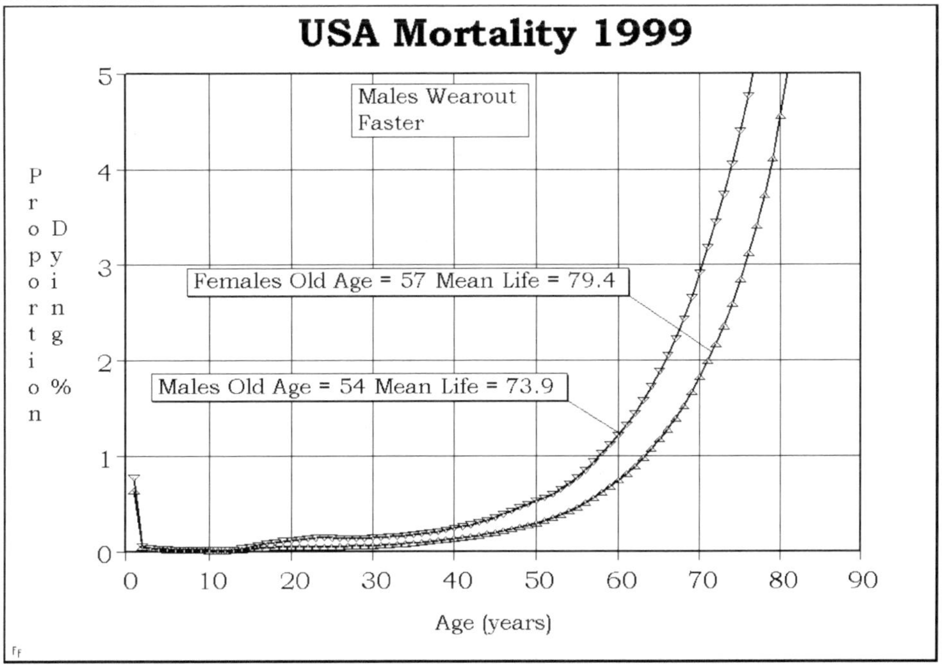

Figure 2-7. The Bathtub Curve for Human Mortality

2.13 Beta < 1 Implies Infant Mortality

The term infant mortality stems from the high mortality of infants. Both electronic and mechanical systems can initially have high failure rates. Manufacturers provide production acceptance tests, "burn-in," and environmental stress screening, to end infant mortality before delivery to the customer. Therefore, betas less than one lead us to suspect:

- Inadequate burn-in or stress screening,
- Production problems, misassembly, quality control problems,
- Overhaul problems,
- Solid state electronic failures.

Until 1980 most experts thought electronics had a beta of one, the exponential distribution. At the 1980 RAMS Conference three papers were presented indicating the correct beta for *solid state electronics is close to 0.7-0.8* if proper environmental stress screening had been done. If *stress screening or burn-in is inadequate* the slope will be shallower, *0.2-0.3*. Were the experts wrong or did the physics of failure change during the transition from radio tubes to solid state components? The author believes the latter is the answer. Note in these sections 2.13-2.16, *italics* relate to a particular range of betas, in this section betas less than one.

If the dominant failure mode for a component is beta <1 and the component survives infant mortality, it will improve with age as its hazard rate declines with age. With survival the hazard rate decreases and reliability increases. See Figure 2-8 below. Overhauling such a component is not appropriate, as old parts are better than new. Components with infant mortality are like red wine: the longer you leave it alone, the better it gets.

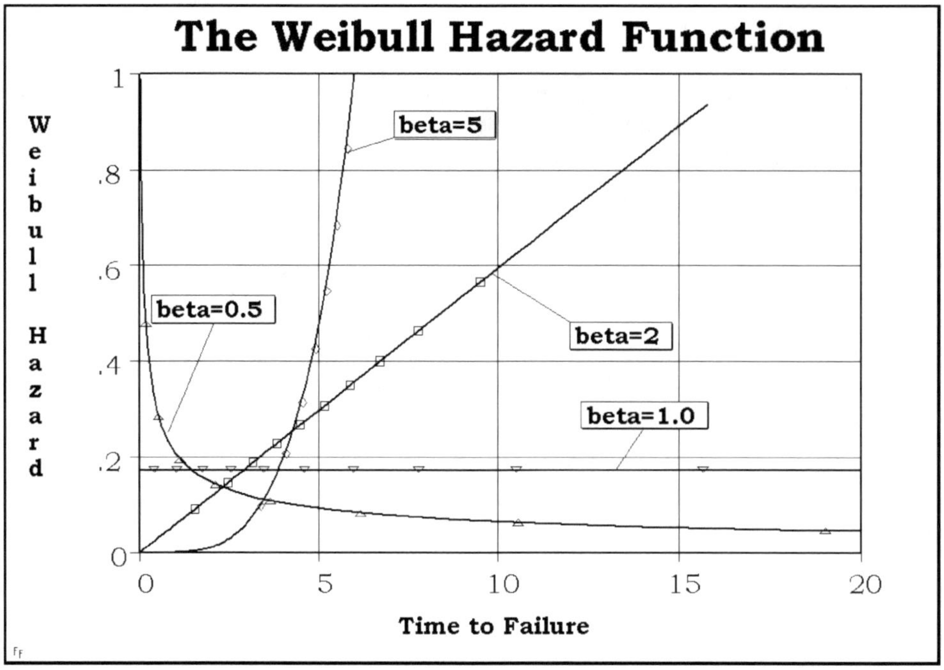

Figure 2-8. The Weibull Hazard Function

The Weibull hazard function, h(t), is the instantaneous failure rate:

$$h(t) = (\beta / \eta)(t / \eta)^{\beta - 1} \tag{2-7}$$

2.14 *Beta = 1.0* Implies Random Failures

By random we mean the failures are independent of time. These failure modes are ageless. An old part is as good as a new one if its failure mode is random. Therefore, we might suspect:

- Maintenance errors, human errors, abusive events
- Failures due to nature, foreign object damage, lightning strikes on transformers, woodpecker attacks on power poles
- Mixtures of data from 3 or more failure modes (assuming they have different betas)
- Systems or component Weibulls (mixtures)
- Intervals between failures.

Here again, overhauls are not appropriate. The Weibull distribution with a beta = 1 is identical to the exponential distribution which is also the first term of the Poisson distribution. Of those that survive to time (t), a constant percentage fails in the next unit of time. This is known as a constant hazard rate, the instantaneous failure rate.

2.15 *1.0 < Beta < 4.0* Implies Early Wear Out

If these failures occur within the design life they are unpleasant surprises. There are many mechanical failure modes in this class and *beta is predictable* for generic failure modes.

- *Low cycle fatigue – beta varies from 2.5 – 4.0*
- *Ball bearing failures – beta = 2.0, roller bearings – beta = 1.5*
- *Corrosion, erosion- beta = 2- 3.5,*
- *however stress corrosion will be 5.0 or greater*
- *V-belts - beta = 2.5*

Overhauls or parts replacement at low B lives can be cost effective. The period for overhaul is read off the Weibull plot at the appropriate B life. If the failure produces a safety hazard, the recommended useful life should be very low, B.1-B.01. If the failure mode is benign, the recommended age for overhaul or part replacement may be much higher B1-B10. If the failure mode is wear out, betas greater than one, and the cost of an unplanned failure is higher than the cost of planned replacement, there will be an optimal replacement interval. Replacing the part at the optimal replacement time will minimize cost. See Section 4.8.

2.16 *Beta > 4.0* Implies Old Age (Rapid) Wear Out

"But our machines have now been running seventy or eighty years, and we must expect that, worn as they are, here a pivot, there a wheel, now a pinion, next a spring, will be giving way; and however we may tinker them up for a while, all will at length surcease motions."
...Thomas Jefferson at (71) years of age writing to John Adams (78) on July 5, 1814. Both died on the Fourth of July 1826, exactly fifty years after the signing of the Declaration of Independence.

Steep betas within the design life are a source of concern. There is risk that the entire fleet will fail quickly as they age into the steep Weibull. On the other hand, if the Weibull characteristic life is well beyond the design life, there is a negligible probability of failure before overhaul or retirement. In this case, steep betas are a source of happiness. Most steep Weibulls have a safe period before the onset of failure within which the probability of failure is negligible. The steeper the slope, beta, the smaller variation in the times to failure and the more predictable the results. A vertical Weibull, with a beta of infinity implies perfect design, quality control and production. For example, the latest powder metallurgy may provide cleaner metal with fewer inclusions. Turbine blades made from purer metal should have steeper betas than blades made with dirtier metals. Some manufacturers now use steep betas to specify and control the quality of vendor parts.

Typical failure modes with old age, rapid wear out include:

- stress corrosion,
- material properties,
- brittle materials like ceramics,
- some forms of erosion.

All wear out modes have increasing failure rates with age and therefore, decreasing reliability. Overhauls and inspections may be cost effective to replace parts that produce significant failures. Airlines remove compressor and turbine disks at their B.01 life to reduce the risk of non-contained turbine failures in flight. Nuclear steam generator tubes are inspected remotely for the presence of cracks from stress corrosion, fatigue and flaws. Pumps have bearing and seal failures. Buried electric cables short-out with age. Railroads have roller bearing failures that cause derailments. Pacemaker batteries wear out. Dental implants fall out. All of these are Weibull wear out failure modes that require inspections and corrective action.

2.17 Weibull Modes May Be "Covered"

Suppose a part has two failure modes. If one is well in advance of the other, failures from the second mode will never occur unless the first failure mode is eliminated. The part will always fail from the first mode if the lines are widely separated. The first mode is said to "cover" the second. In humans, cancer may precede and therefore, "cover" heart disease. See Figure 2-9 below. All parts and components have multiple failure modes but often one mode covers the others to the extent we are unaware of them.

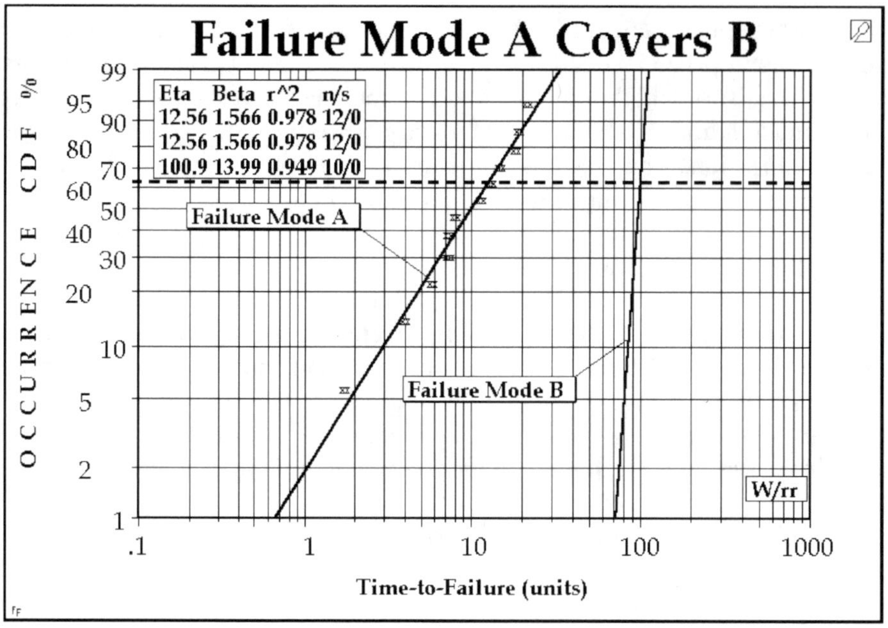

Figure 2-9. Weibull Failure Mode A "Covers" B

For complex systems, there are **always failure modes beyond the known modes**. Existence of these unknown failure modes makes warranties and guarantees risky because development tests may not uncover the hidden modes. In human mortality the same concept applies. If a fatal illness is eliminated by a breakthrough in medicine, it may have little effect on the maximum life span because other fatal illnesses are "uncovered". During the last century the average life span has more than doubled but the probability of surviving for 100 years has hardly changed because of uncovering other failure modes. For old systems that have operated many multiples of their design life, there is no sign that all the failure modes are ever solved or even found. For this reason, systems that involve safety must be exposed to accelerated testing well beyond their design life to uncover unknown failure modes. The manufacturer must test well beyond the age of the customer's fleet leaders to assure that catastrophic modes are eliminated. Accelerated life tests are often used. However, these tests may be too costly except for extreme safety requirements like aircraft, automobiles, surgical implants, and nuclear systems.

The author has reviewed data on complex systems that shows the incidence of the first failure for each mode plotted against age using the Crow-AMSAA model. This is a linear function on log-log paper and *never reaches a plateau*. These results imply that there are always unknown failure modes that will occur in the future. There are always Weibulls beyond the Weibulls we know about! See Figure 9-12.

2.18 Weibull Paper and Its Construction

Few readers will ever need to construct Weibull probability paper. However, understanding the construction provides a better understanding of its use. All probability papers have scales that transform the cumulative probability distribution into a linear scale. If data is plotted on the transformed scale and it conforms to a straight line that supports the supposition that the distribution is appropriate. If the data is plotted on several probability papers, say normal, lognormal and Weibull; the plot with the best fit is usually the most appropriate distribution. Of course, whenever possible, knowledge about the physics of failure should support the choice of distribution. The Weibull distribution is defined mathematically as follows by equation (2-1):

$$F(t) = 1 - e^{-(t/\eta)^{\beta}}$$

F(t) defines the cumulative fraction of parts that will fail by a time t. The fraction of parts that have not failed up to time t is 1 - F(t). This is the reliability at time t, denoted by R(t). By rearranging the distribution function, a linear relation is obtained:

$$1/(1 - F(t)) = e^{(t/\eta)^{\beta}}$$

$$\ln(1/(1 - F(t)) = (t/\eta)^{\beta}$$

$$\ln\ln(1/(1 - F(t))) = \beta \ln(t) - \beta \ln(\eta)$$

$$y = \beta x + \alpha$$

The expression ($y = \beta x + \alpha$) is the equation for a straight line. By choosing ln(t) as x, the scale on the abscissa, and lnln(1/(1 - F(t))) as y, the scale on the ordinate, the cumulative Weibull distribution is represented as a straight line. As shown in Tables 2-4, and 2-5, Weibull paper can be constructed as follows.

Table 2-4. Construction of Ordinate (Y)

F(t)	lnln(1/(1 - F(t)))	Col 2 Value +(6.91)
0.001	-6.91	0 units
0.010	-4.60	2.31
0.1	-2.25	4.66
0.5	-0.37	6.54
0.9	0.83	7.74
0.99	1.53	8.44
0.999	1.93	8.84

Table 2-5. Construction of Abscissa (t)

t (hr)	ln(t)
1	0 units
2	0.69
3	1.10
4	1.39
5	1.61
10	2.30
15	2.71
20	3.00
.	.
100	4.61
1000	6.91

The paper is scaled in inches. It will have a one-to-one relationship for establishing the slope of the Weibull. The Weibull parameter β is estimated by simply measuring the slope of the line on 1-to-1 Weibull paper. Of course, the scales can be made in any relationship. For example, assume the range of the failure data, the X parameter, is less than one log cycle. Multiply the X scale by three, to produce 1-to-3 paper. On 1-to-3 paper the slope is multiplied by three to estimate β. These considerations were important in past decades when most Weibulls were hand plotted. Today with good software available, hand plotting is rare. SSW automatically provides optimal scales for each data set, but these may be changed with the "Zoom" option. SSW also provides 1-to-1 scales as an option. Sample 1-to-1 scale Weibull paper is included at the end of this chapter. Appendix H has additional 1-to-1 scale paper you may copy and use.

2.19 Weibull Analysis - The Standard Method

When data is entered in a computer, seven methods are in popular use for estimating β and η; all are available in the software: The first method is described in this chapter, the others in the next chapter except for (MLE) which is treated in Chapter 5.

1. The standard method and **best practice** for small and moderate size samples (2-100), median rank regression (X onto Y) curve fitting using the times-to-failure as the dependent variable,
2. Median rank regression curve fitting using the median ranks as the dependent variable (Y onto X), (not recommended),
3. Maximum likelihood estimation (MLE) for very large samples, over 500 failures,
4. Maximum likelihood estimation with Reduced Bias Adjustment for small and large samples, 500 or less failures,
5. Grouped or interval data analysis for coarse and inspection data may use the "inspection option",
6. Interval analysis of destructive inspection or NDE (nondestructive evaluation) data, (regression is Y on X) may use the "Probit method",
7. Interval MLE may be used for all kinds of interval data (5 and 6, above)and appears to be slightly more accurate than the "inspection option" or the "Probit method" based on new research by $[Marquart 2004].

For standard data (not interval data), the author recommends median rank regression [1] as the standard. This method is used for most engineering Weibull analysis throughout the world with heritage directly from Waloddi Weibull. If you could only use one method this would be the best choice. The last three methods are alternative methods used for special situations. All methods will be compared in detail in Chapter 5. This chapter was devoted to good Weibulls based on good data using the standard method. In Chapter 3, we will attack dirty interval data, uncertainties, and some of the methods needed to analyze these deficient data sets. Chapters 5 and 8 include additional methods for dirty data.

If beta is known from prior experience, a method called Weibayes may be employed. For small samples, say 20 or less, Weibayes is much more accurate than Weibull and is **best practice** <u>if</u> beta is known. Weibayes is treated in Chapter 6.

2.20 Problems

In prior decades, thousands of Weibull data sets were hand calculated and hand plotted. Today, personal computers and friendly software have eliminated these cumbersome chores and provide solutions that were difficult by hand such as maximum likelihood solutions. Therefore, it may seem strange that we presented hand calculation methods for Weibull analysis here. The author recommends every reader plot at least one or two Weibulls by hand to really understand the method. Two such problems are provided in this section. There is standard 1:1 scale Weibull Probability Paper following the problems for your use.

After those problems, the student is encouraged to use good friendly software for as many of the remaining problems as possible, herein and elsewhere. The PlayTIME Computer Tutorial supplied with the SuperSMITH software has dozens of problems laid out step-by-step.

Problem 2-1: Ten fatigue specimens were put on test. They were all tested to failure. The failure times in hours were as follows: 150, 85, 250, 240, 135, 200, 240, 150, 200, and 190. Rank the data and note that identical failure times are assigned sequential rank order numbers. Look up the median ranks in Appendix I.

Rank Order Number	Failure Time	Median Rank

a. Plot the failure times and the median ranks on 1 x 1 scale Weibull paper.
b. What is the slope and what class of failure mode is it?
c. What is the characteristic life?
d. What is the B1 life?
e. What percent of the samples are expected to fail before 200 hours?
f. What is the reliability at 300 hours?

Problem 2-2: Plot a Weibull with Suspensions. There have been five failures out of eight parts in service. The time in hours on the parts is as follows:

Serial #	831	832	833	834	835	836	837	838
Time	9.0	6.0	14.6	1.1	20.0	7.0	65.0	8.0
	Fail	Fail	Susp	Fail	Fail	Susp	Fail	Susp

a. Construct a Weibull plot.
b. What are β & η?
c. What class of failure mode does this β indicate? Hints: Rank the data from the first failure or suspension to the last. Enter in the table. Remember suspensions have no effect on adjusted ranks or median ranks <u>until after they occur</u>. N is total sample size (occurrences + suspensions) and i is Adjusted Rank.

$$\mathbf{i = Adjusted\ Rank = \frac{(Reverse\ Rank)*(Previous\ Adjusted\ Rank)+(N+1)}{(Reverse\ Rank+1)}}$$

To convert adjusted ranks into median ranks use Benard's Approximation:

$$\mathbf{Median\ rank\ (\%) = [(i - 0.3)/(N + 0.4)]\ x\ 100}$$

Order	Time	F/S	Reverse Rank	Adjusted Rank	Median Rank

Where F/S indicates failed or suspended for use of the adjusted rank.

For Problem 2-1

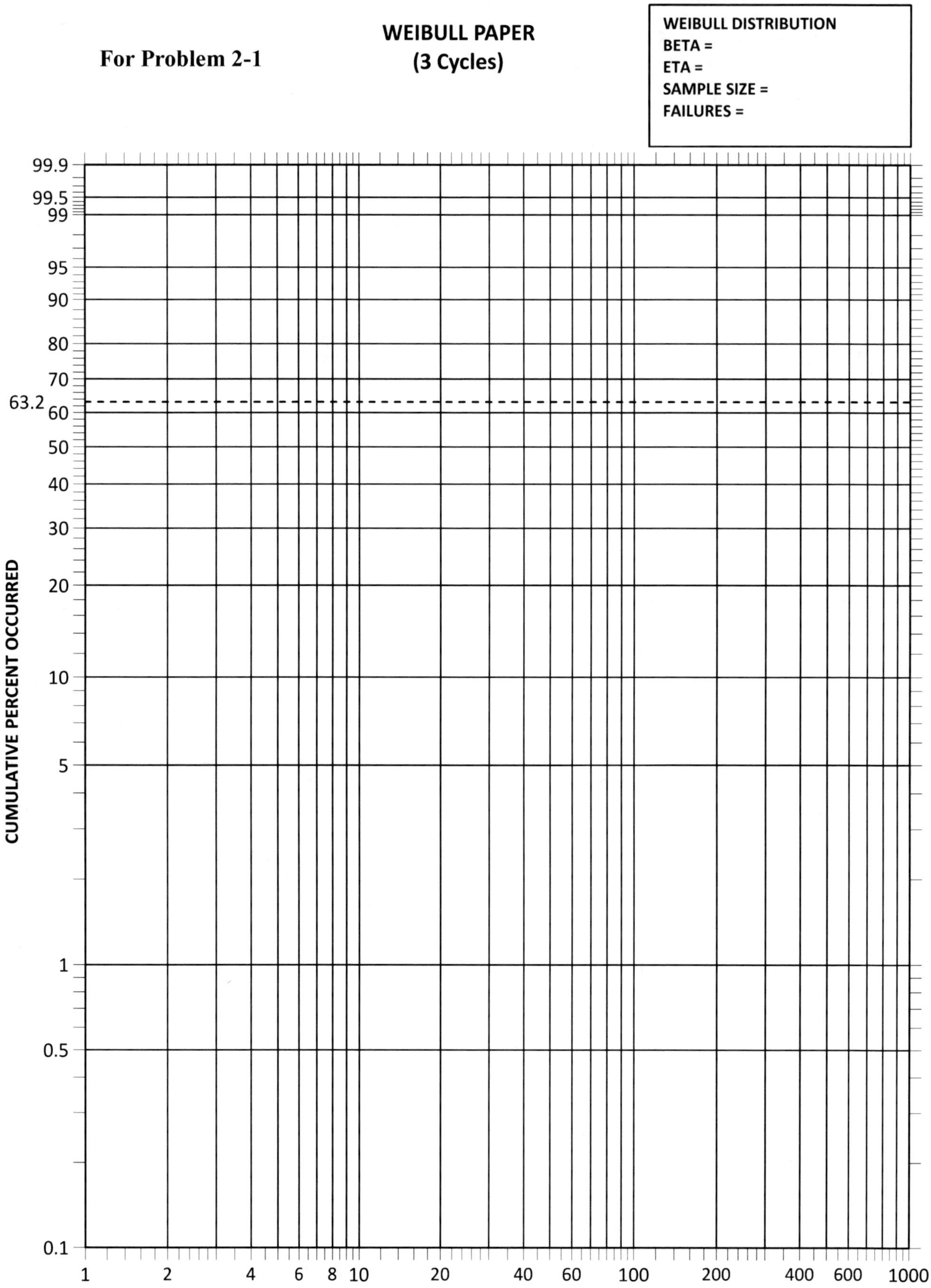

Customize the X-axis to your data. Start at the far left with a scale less than your actual data. If your smallest data is 85 then start your left horizontal scale at 10. Remember the vertical axis is % of cumulative failures.

WEIBULL PAPER
(3 Cycles)

For Problem 2-2

WEIBULL DISTRIBUTION
BETA =
ETA =
SAMPLE SIZE =
FAILURES =

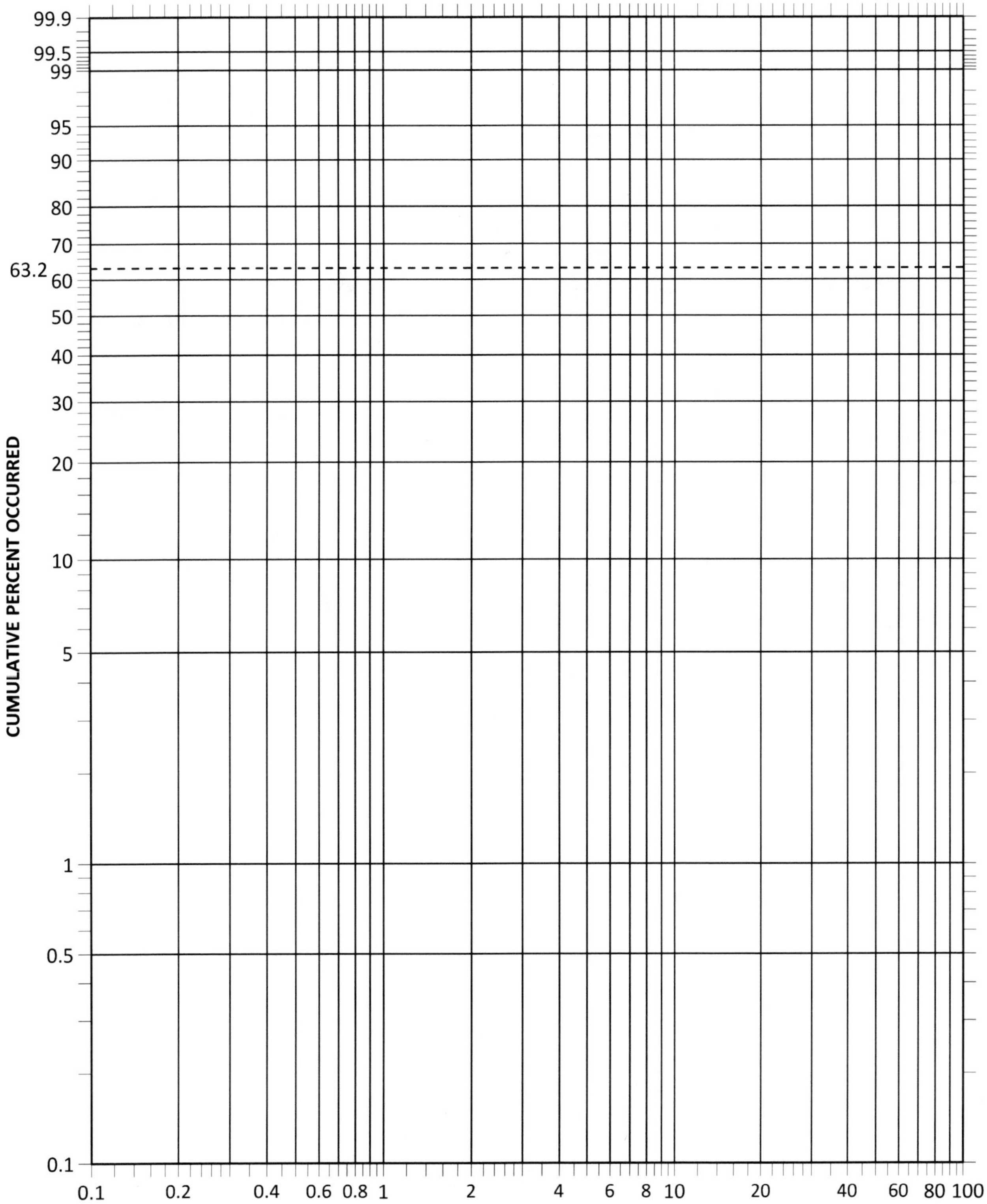

Problem 2-3: Understanding Good Weibulls

a. What is the hardest part of Weibull analysis?
b. Air conditioning compressor failure data is perhaps as dirty as dirty can be. In most cases the only data on failed compressors is the date shipped from the compressor manufacturer to the air conditioner manufacturer and the date returned under warranty. Operating time is highly seasonal and is not reported. How would you determine if this data is good enough to do Weibull analysis?
c. The slope parameter is β. It is sometimes called what other name?
d. What is another name for a Weibull distribution with $\beta = 1.0$ and η = MTTF?
e. What is the standard method for doing Weibull analysis?
f. What class of failure, [infant mortality, random, or wear out], and what range of β would you expect for the following data sets?
    1. Failures from inadequate stress screening or burn-in of electronic components?
    2. Woodpecker attacks on wooden electric power poles?
    3. Alzheimer's disease? Chicken Pox or Whooping cough?
    4. Bearings from spalling or skidding balls? Fatigue cracking?
    5. Turbine vanes may fail early from hot spots from plugged fuel nozzles, a quality problem, or much later from stress corrosion cracking? What two classes?

Solutions to these problems are in Appendix K. (...No peeking!!!)

Personal Computer Exercises: "PlayTIME with SuperSMITH, " is recommended.

Problem 2-4: Weibull Closure Property Parts Versus Assemblies

Should you analyze blade failures or disk/blade assembly failures? The answer is it does not make any difference if you do it properly as the two Weibulls are consistent. For example, let the blade Weibull have a $\beta = 3.0$ and $\eta = 1000$ hours. For simplicity, assume the disk contains 4 blades. The Weibull closure property [Tobias] is:

$$\eta_{system} = \left(\sum_{i=1}^{N} \left(\frac{1}{\eta_i^{\beta}} \right) \right)^{-(1/\beta)}$$

If $\beta_i = 3.0$ and $\eta_i = 1000$, $N = 4$, then $\eta_{system} = (4/1000^3)^{(-1/3)} = 630.0$ hours. Plot the two Weibull lines i.e., component line and system line, on Weibull probability paper.

Check on solution: Compare the first time-to-failure for one of four blades to the first failure of the system. The median rank (Appendix I) for first of four failures is 15.91%. The B15.91 life for the blade Weibull is 557.5 hours from the plot or your calculator. The median rank for one failure is 50% and the B50 life for the system Weibull is 557.5 hours.

CHAPTER 3: DIRTY DATA, "BAD" WEIBULLS, AND UNCERTAINTIES

3.1 Foreword

Chapter 2 was devoted to good Weibull plots. Chapter 3 is devoted to "bad" Weibull plots. Bad Weibull plots are often informative. In contrast to other distributions, the bad plot can provide significant information for the engineer who learns how to interpret results. Therefore, always plot life data on Weibull probability paper as the first step in the analysis.

The greatest benefit of the Weibull method is that it can be useful for engineering with extremely small samples and data sets that have major deficiencies which we call "dirty data." Every industry has dirty data, although the situation is improving as management realizes the cost effectiveness of acquiring high quality data. The objective of this chapter is to present the methods for interpreting bad Weibull plots, the analysis of dirty data, small sample risks, goodness of fit, and measures of uncertainty.

"It is a capital mistake to theorize before one has data. Insensibly one begins to twist facts to suit theories, instead of theories to suit facts." Sir Arthur Conan Doyle (1892).

3.2 Small Sample Uncertainties

Most statisticians balk at the use of the extremely small sample Weibulls illustrated herein. However, in cases of safety and extraordinary financial loss, there may be no alternative to employing Weibulls with only 1, 2 or 3 failures. The analyst cannot request more "crash and burn" type failures. There is an urgent need to provide a source of direction rather than waiting for more failures. It is, therefore, prudent to consider the associated uncertainties of these small sample applications.

It is remarkable that Weibull engineering analysis may be useful with an order of magnitude fewer samples than other commonly used statistical distributions. By comparison, if success-failure data is analyzed with the binomial distribution, a minimum sample size might be three hundred. If dimensions and performance measurements are normally distributed, a minimum sample would be thirty. It is fair to ask why Weibull engineering analysis is possible with only a few failures.

- First, with service data, failures tend plot in the lower left hand corner of the Weibull because of the multitude of suspensions. See Figure 3-15. This is the area of engineering interest, precisely where the data appears. This is the neighborhood of B0.1 to B1 life.
- The second reason is that the data set includes both failures and successes (suspensions). Although there may be only three failures, there may be thousands of suspensions. Suspensions are not weighted as much as failures but they do count.
- With regression the failure data in the lower left corner is over weighted. See Chapter 5. Although this degrades mathematical rigor, it increases the accuracy in the area of interest.
- Weibull probability paper is unique in that the lower area is expanded. In other words the area of small probabilities is magnified for engineering purposes.

Therefore, for engineering predictions of B1 life or B0.1 life and for failure forecasts, small sample plots may be useful, if the risks and uncertainties are known and acceptable. For statistical purposes, it is true that larger samples are needed to accurately estimate the Weibull parameters, eta and beta, and to determine which distribution is most appropriate, lognormal or Weibull. For these statistical purposes a minimum of twenty one failures are needed, but thirty would be better, particularly for estimating beta. The Weibull analyst can quantify the statistical uncertainty of his small sample by using SSW software.

Although the author discourages using confidence intervals for presentations, this usage for the Weibull analyst to gauge his uncertainties is highly recommended. The improvement in uncertainty with increasing the sample size is illustrated in Figures 3-1, 3-2 and 3-3. The hour-glass curves on each side of the Weibull line, bound 90% confidence intervals. Rigorously these intervals contain the true unknown B life with frequency 90%. Approximately, they contain 90% of the expected variation in B lives or reliability. For example, in Figure 3-1 for sample size three, the B5 life varies from about 150 to 916 in repeated samples from the same parent Weibull just due to statistical uncertainty.

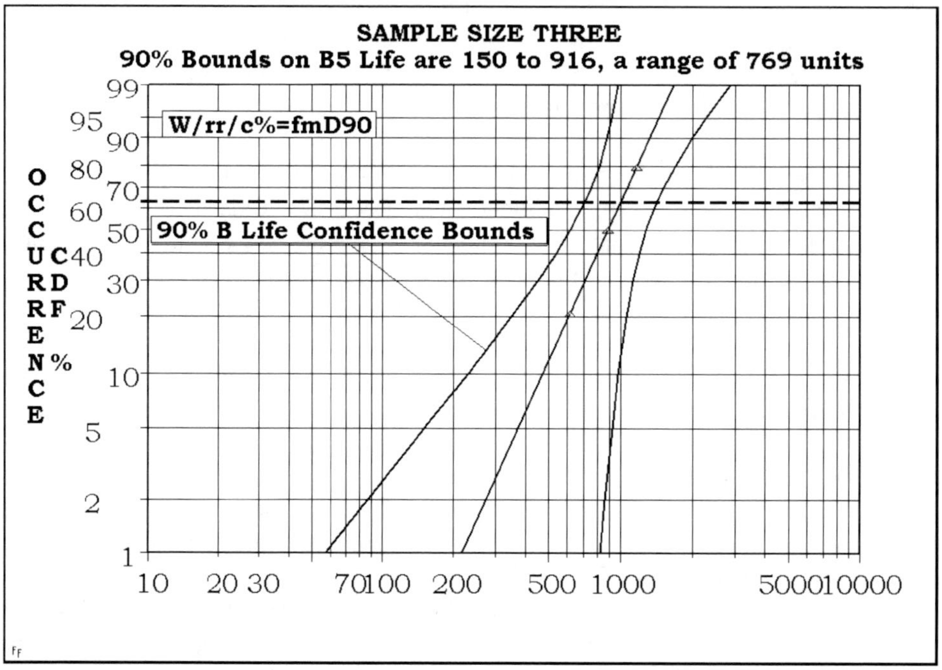

Figure 3-1. Sample Size Three

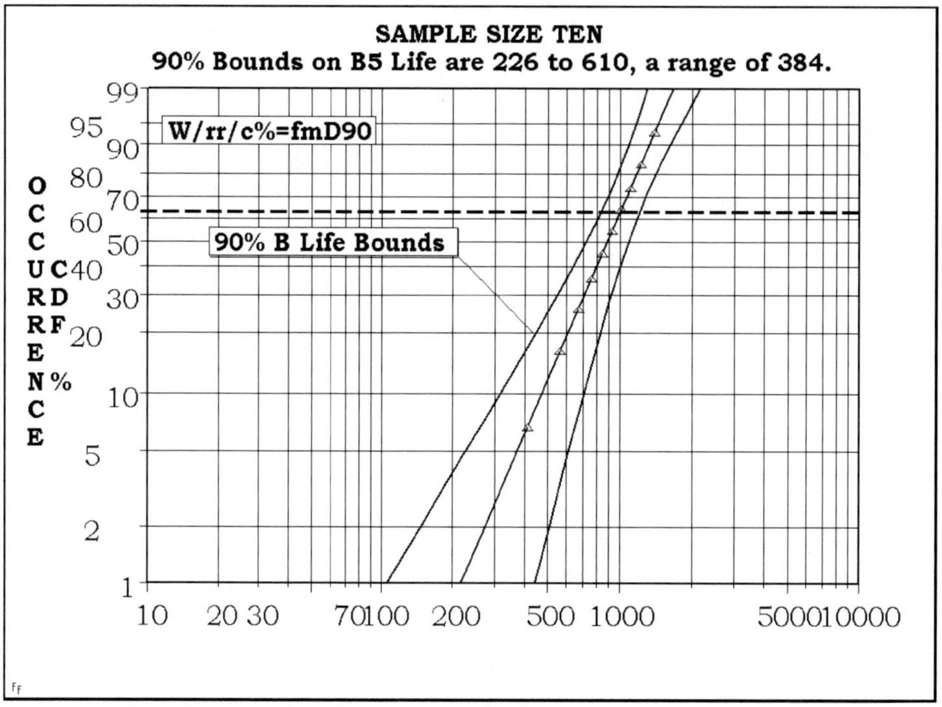

Figure 3-2. Sample Size 10

Figures 3-2 and 3-3 show the improved uncertainties for sample sizes of 10 and 100. Chapter 7 provides a detailed discussion of confidence intervals. There are methods for reducing the small sample uncertainty. Obviously, the sample size may be increased, if this is not too expensive. A more cost effective method is to employ prior experience with the subject failure mode. This method, called "Weibayes," is described in Chapter 6 and depends on assuming the slope, β. The approach is based on engineering experience and a Weibull library to select β. It will reduce uncertainty by factors of two or three for small samples.

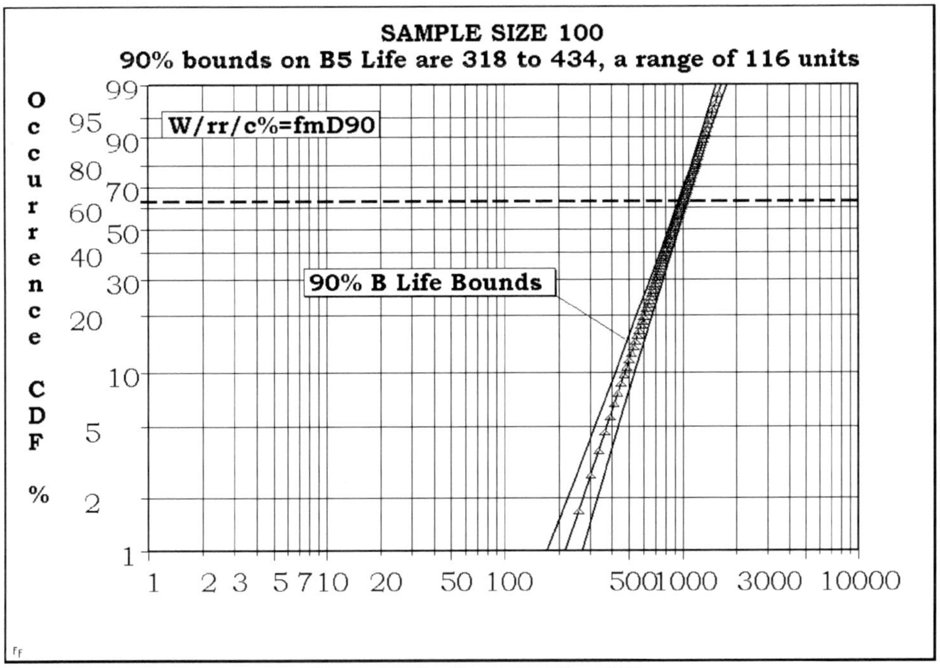

Figure 3-3. Sample Size 100

3.2.1 Goodness of Fit

If the data clusters around a straight line on probability paper that is evidence the data comes from the subject distribution. The experienced analyst will have a sense of fit by studying the plot. However, small samples make it difficult to gauge goodness of fit. There are complex statistical measures of goodness of fit such as Chi-Square, Kolmogorov-Smirnov, Anderson-Darling, Cramer-von Mises, and Nancy Mann's tests. The author prefers the simple correlation coefficient. The correlation coefficient, "r," measures the strength of a linear relationship between two variables so it is ideal for our purpose. "r" is a number between -1 and +1, depending on the slope. As life data probability plots always have positive slopes, they always have positive correlation coefficients. The closer r is to one the better the fit. Calculation of the correlation coefficient is illustrated in Appendix B. This section will introduce goodness of fit based on "r." The likelihood function may also be used for goodness of fit which will be discussed in Chapter 5. New research and other measures of goodness of fit will be described in Appendix D.

Mathematicians object to using the correlation coefficient on probability plots because using median ranks artificially increases the observed correlation. To overcome this objection, the author employed Monte Carlo simulation programmed by Wes Fulton to approximate the distribution of the correlation coefficient from ideal Weibulls based on median rank plotting positions. The 90% critical correlation coefficient (CCC) provides a measure of goodness of fit. If your correlation coefficient, "r", is <u>larger</u> than the CCC, you have a good fit. See Figure 3-4. If your "r" is <u>smaller</u> than the CCC, your data is significantly different from a Weibull with 90% confidence and you have a bad fit.

The CCC is found by ranking the "r" values for the correlation coefficient from 10,000 Monte Carlo simulation trials and choosing the 1000th value. Thus the 90% CCC is the tenth percentile, the "10 p" of the 1000 values. If your "r" is larger than the CCC, the "10 p", you have a good fit. That is, the upper 90% of the r values are "acceptable." Coefficients less than the CCC indicate a bad fit. See [Edgeman 1990], and [Gan and Koehler 1992] for related studies with other plotting positions.

The correlation coefficient squared is the coefficient of determination, "r^2". Statisticians prefer r^2 to r as a goodness of fit measure. r^2 equals the percentage of the variation in the data explained by the fit to the distribution. For example, if the observed r^2 is 93%, it implies that 93% of the variation in the data is "explained" by the fit of the trend line to the distribution. **CCC^2 is the tenth percentile of r^2**. SuperSMITH provides r, r^2, and CCC^2.

Figure 3-4 provides critical values (CCC^2) of r^2 based on ten million trials and up to a million failures for the 2-parameter and 3-parameter Weibull and lognormal distributions. Notice the distributions have different CCC^2. In other words, the three parameter Weibull r^2 must be much higher than the r^2 for the other two distributions to be an equally good fit. For example, consider a sample of 10 failures whose r^2 is 0.90. On the plot this would be a good fit for the 2-parameter Weibull and lognormal, but a bad fit for the 3-parameter Weibull. The 10p CCC^2 values are provided by SSW. (The lognormal CCC^2 line is also the normal CCC^2 line.)

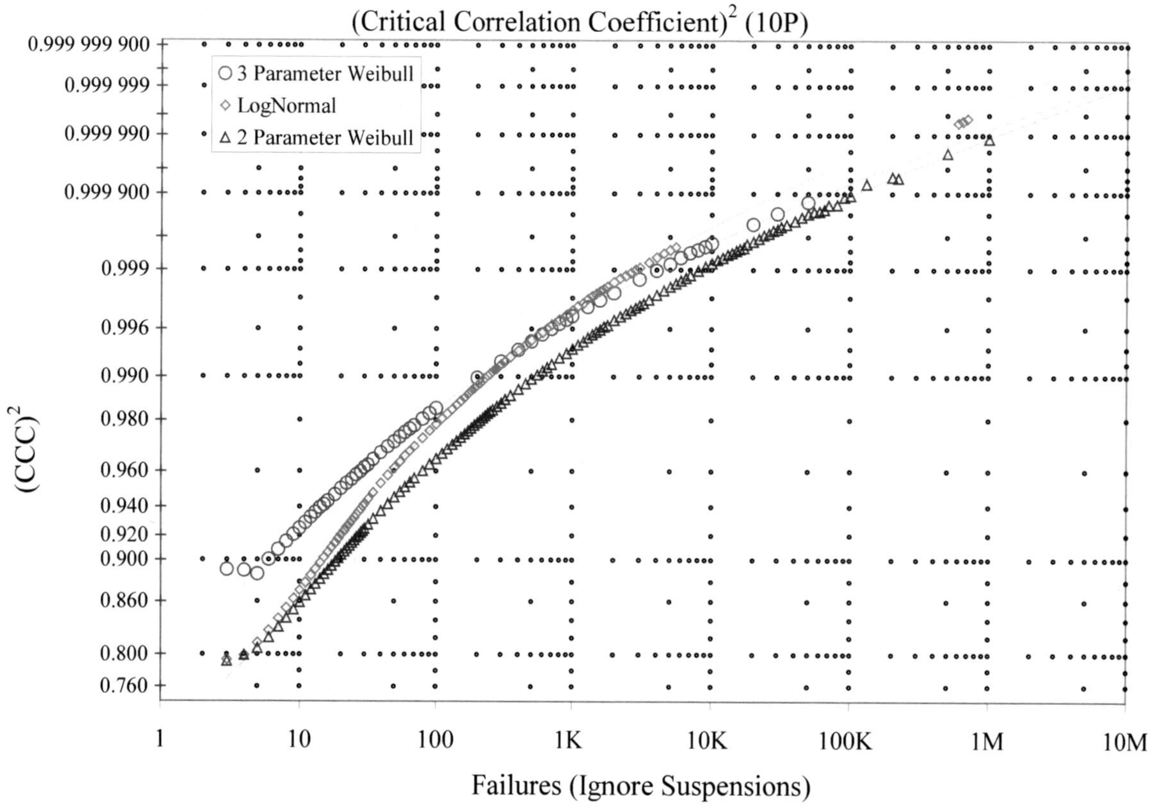

Figure 3-4. CCC^2 Versus Failures

SuperSMITH also provides the percentile, the p-value, of the observed r and r^2. It is called "pve%". This is the most accurate indicator of goodness of fit. See $[Liu] for a comparison of this approach with other measures of goodness of fit. NOTE: The "$" symbol preceding a reference means the document may be downloaded from Paul Barringer's website (see References).

To compare the fit of one distribution with another, say a Weibull with a lognormal, we need a moderate size sample, say 21 or more failures, and the pve%. The distribution with the highest pve% among competing distributions is the best statistical choice. (r^2 - CCC^2) has been used in the past as a guide but pve% is more accurate and now **best practice**.

Wes Fulton explains goodness-of-fit p-value "pve%" as follows: The pve% for a particular data set is the rank of the actual correlation coefficient among all possible correlation coefficient values for the sample size and the model selected. The pve% values for probability plotting must be between 0 and 100, the higher the better. The SSW software provides the pve% for the Weibull 2-parameter, Weibull 3-parameter, lognormal, normal, and extreme value (Gumbel) distributions.

The **pve%** is an estimate of true p-value and is displayed on the software plot and the report output. A pve% less than 10 indicates a bad fit. If the pve% is above 10 then the model is not rejected at a 10% significance level, indicating a good fit.

The pve% offers many benefits as a goodness of fit indicator. Chi Chao Liu [1997] concluded for regression that the p-value provides an excellent indication of goodness-of-fit. Distribution analysis determines the model with the best fit using the highest value of pve% as the indicator of the best selection. The pve% goodness of fit measurement for regression is now the default for the SSW software. However, there is capability to show r^2 or (r^2 - CCC^2).

Prior knowledge of the physics of failure and past experience should weigh heavily in choosing the best distribution. There can be real concerns about the validity of the data. Often **the suspicion will be that the data is non-homogenous;** several heats of metal were used, several vendors produced the units, the service environment was not uniform, the testing procedure was poorly defined, the data collection was poor, there is a mixture of failure modes, and on and on. These concerns may be sufficient to negate the analysis. The solution to these worries is in the data. What does the data say? **If the data fit the distribution well it implies these concerns are negligible, but if the fit is poor they may be significant**. Remember the basic premise of statistics is that no two things, parts, tests, twins or data points are exactly alike. Statistical methods allow us to deal with uncertainty. In other words a good fit may be a cure for insomnia.

Carl [Tarum] of Delphi using the author's CCC values discovered that these critical correlation coefficients are approximately Weibull distributed and generated the results in 1995 shown in previous editions of the Handbook, a Weibull plot showing the CCC up to 1000 failures. By extending the coverage to a million failures it is now obvious that the straight line Weibull is inadequate as curvature is now obvious. The more recent simulations (2006-2007) by Carl, Todd Marquart, and Paul Barringer have produced an updated plot shown in Figure 3-4 and in Appendix D and incorporated in SSW.

3.3 Suspensions

A unit that has not failed by the failure mode in question is a suspension or censored unit. It may have failed by a different failure mode or not failed at all. A bolt that fails in the bolt head is a suspension in a pull test for thread failures. An "early suspension" is one that was suspended before the first failure time. A "late suspension" is suspended after the last failure. Suspensions between failures are called "internal" or "random" suspensions.

As a rule, *suspensions increase the characteristic life*, eta, but have little effect on the slope, beta. Early suspensions have negligible effect on the Weibull plot. Late suspensions have more significant effects, and may reduce the slope, beta. The presence of many *late suspensions usually indicates a batch problem*; newer units are failing, older units are not failing. Internal suspensions are more difficult statistically, particularly for confidence interval estimates. Figure 3-5 shows the effect of 90 late and early suspensions with 10 failures. Internal suspensions will tend to shift the Weibull line somewhere between the early and late suspensions shifts. Monte Carlo simulation is the recommended method for generating confidence intervals for data sets with internal suspensions and/or very small sets. The SSW software provides this capability. Confidence bounds are discussed in Chapter 7.

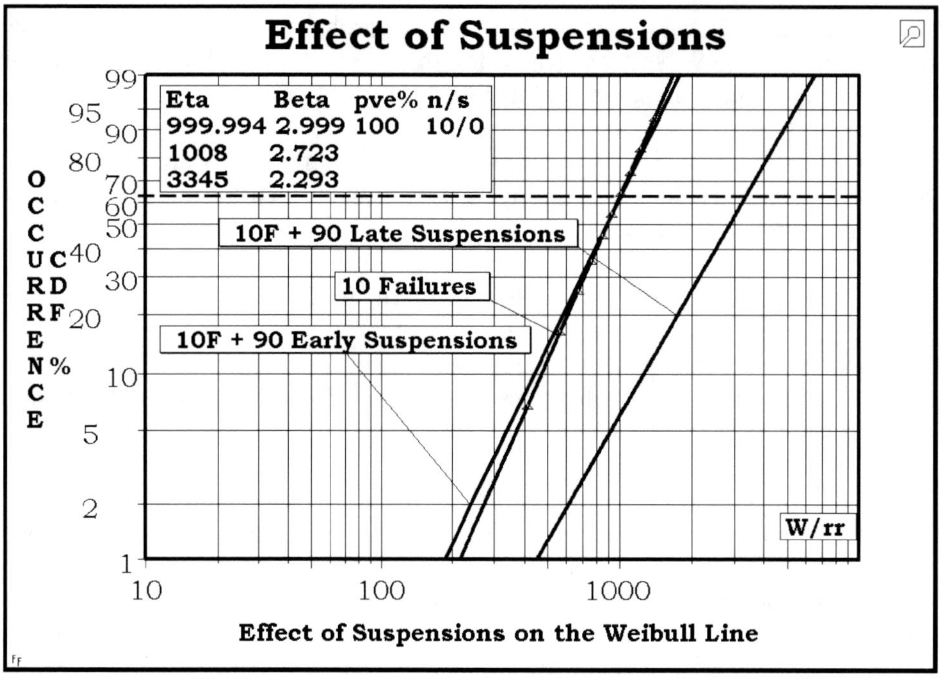

Figure 3-5. Effect of Suspensions on the Weibull Line

Statisticians always prefer complete samples (i.e., no suspensions), but in industry suspensions are very common. For example, all in-service and warranty data sets have suspensions. Fortunately, the Weibull and lognormal distributions handle suspensions well.

3.4 Suspect Outliers

From time-to-time you may suspect that the first or last point in a data set is a wild point, not a member of the data set for some reason. For unknown reasons materials curves for low cycle fatigue and creep often have a suspect first point. Deleting this point is very attractive and may reduce the weight of the design by a significant amount. This will be a critical, perhaps dangerous, decision. Lloyd Nelson [Journal of Quality Technology, July 2003] has an excellent comment on outliers, "the first question is...does the presence of this value make any significant difference in the results of the analysis? ...If not, concern about the presence of this value will disappear." Otherwise these points are important to your life data analysis.

With a suspect point you should investigate the engineering aspects of data recording, test records, instrument calibrations, etc., to identify the cause of extreme scatter of the point. This is the proper thing to do. Statistics can be of some help in this investigation but should not be used to justify rejecting an outlier without supporting engineering.

Using [Gumbel's] extreme value theory the author developed the distributions of the first and last data points for complete data sets, i.e., no suspensions. Wes Fulton then modeled the test for outliers using simulation. The outlier test is conducted as follows: The null hypothesis is that the suspect point is not an outlier. However, if it is an outlier it can have a large effect on the parameter estimates. Thus, the suspect point is deleted from the set, and new parameters and an outlier p-value calculated. This p-value, the percentile of the wild point compared to statistical expectations for complete sets is calculated. If it is very close to 100% or 0%, the point should be labeled as a "suspect" outlier. The critical p value is set to provide 90% confidence that the point is *suspect* if its p-value exceeds the critical value. Note this means that randomly sampled data sets with no outliers, will indicate an outlier about 10% of the time. See Figure 3-6.

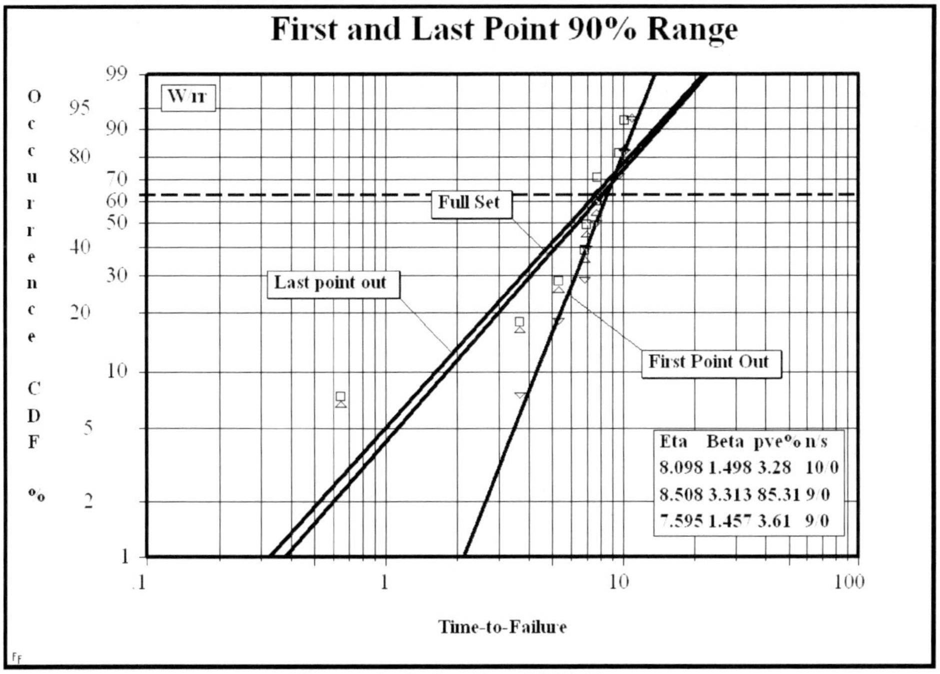

Figure 3-6. Outlier 90% Range

The author <u>does not recommend</u> deleting or rejecting points <u>based on statistics only;</u> there should always be engineering reasons supporting the statistics. Data is precious and should not be rejected without sufficient evidence. In the SuperSmith Weibull software select the calculator option for an outlier test. See Chapter 11 for Case Study 11.11 with an outlier.

3.5 Curved Weibulls and the t_0 Correction

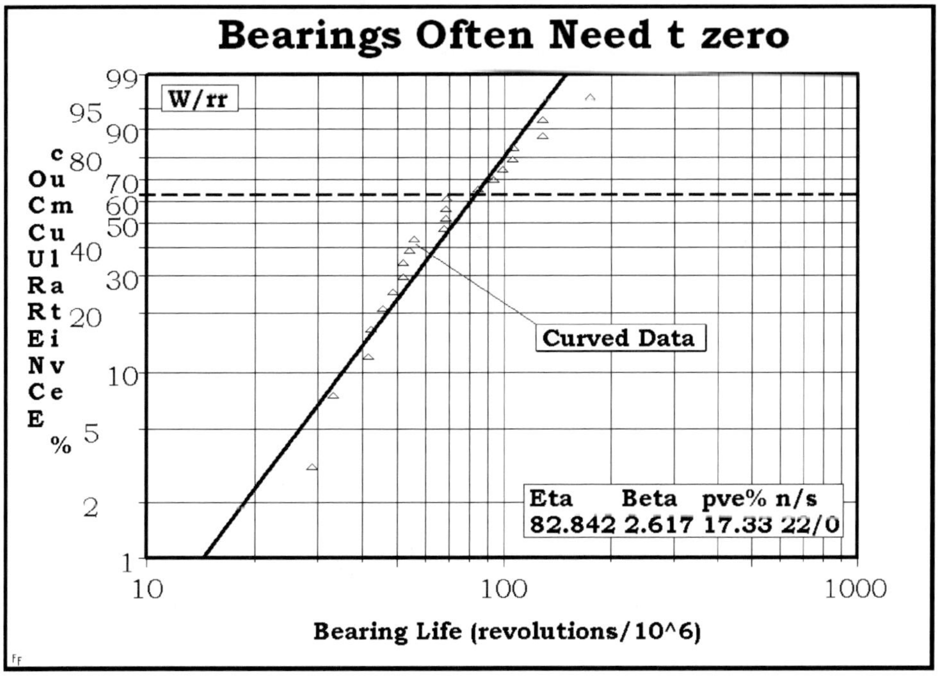

Figure 3-7. Bearings Often Need t_0 Age-Shift Adjustment

The data on a Weibull plot may appear curved as illustrated in Figure 3-7. The message from the data is the origin may be in the wrong place. For a curved plot time may not start at zero. For example, it may be physically impossible for the failure mode to occur early in life. A bearing failure due to spalling or unbalance cannot occur without bearing rotation inducing enough damage to fail the bearing. Time starts when failures are possible. Figure 3-8 shows the same data with the origin shifted to 24 million revolutions. There is a guaranteed failure free period (the first 24 million revolutions), within which the probability of failure is zero. Notice the significant improvement in the goodness of fit. The pve% value for r^2 improves from 17 to 77. The mathematics consists of subtracting t_0 = 24.06 from each data point in Figure 3-7 to obtain Figure 3-8.

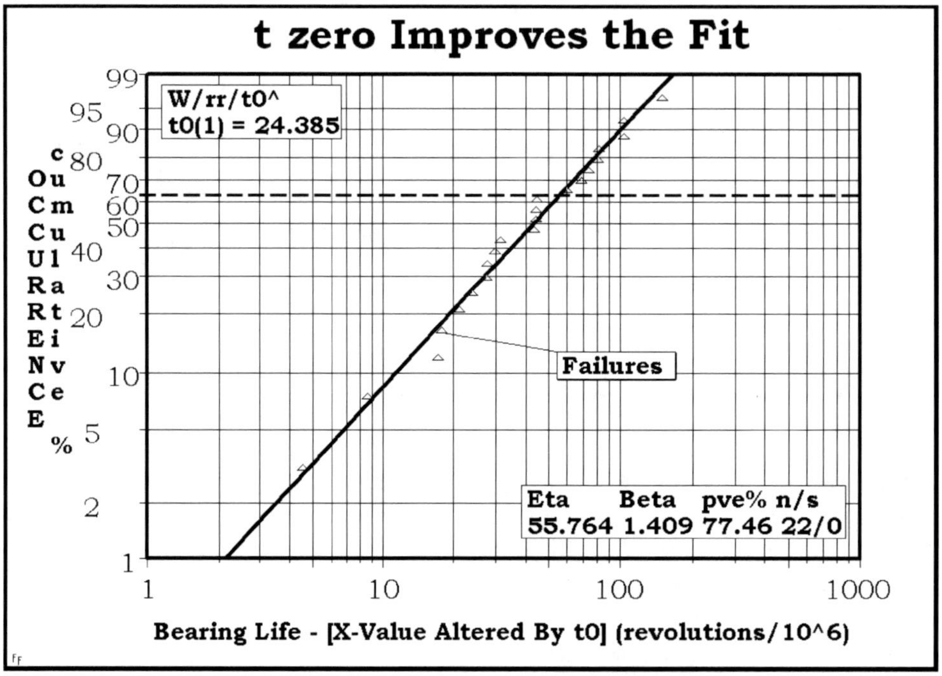

Figure 3-8. The t_0 Age Adjustment Improves the Fit

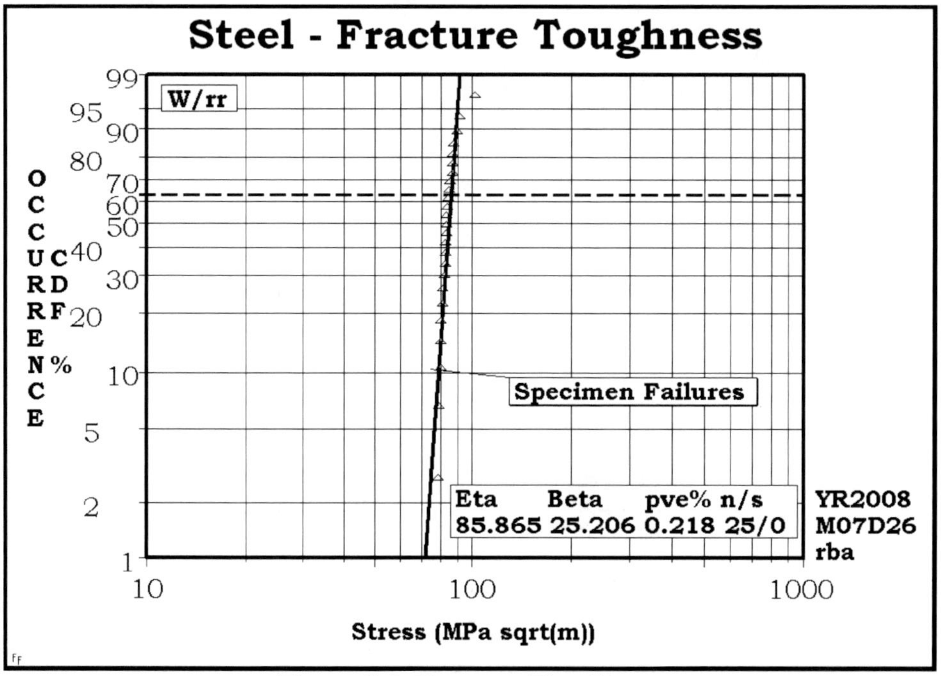

Figure 3-9. Fracture Toughness

Figure 3-9 is another typical example, fracture toughness of steel plate. Note that because beta is so steep the plot looks good even though there is extreme curvature. Steep betas hide patterns that are disclosed when the X axis is expanded. The model indicates it is physically impossible to fail the plate at a low level of stress. Figure 3-10 shows the effect of the t_0 shift.

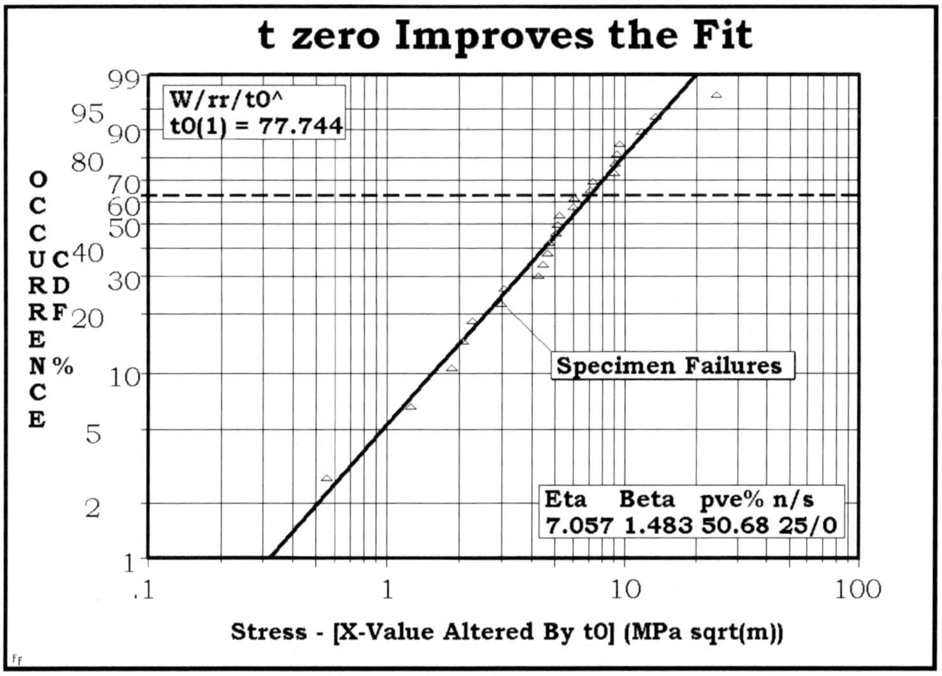

Figure 3-10. t_0 Improves the Fit

There are many possible reasons for an origin shift. The manufacturer may have put time or mileage on the system as part of production acceptance, but reported that the units are "zero time" at delivery. The purpose of production acceptance is to eliminate the infant mortality failures. Electronic components often are subjected to burn-in or environmental stress screening for the same purpose. In these cases the units have aged before being delivered as "zero time" systems.

For material properties, where the Weibull ordinate is stress or strain, it may be impossible for fracture or creep or other properties to produce failure near the origin on the scale. Spare parts like rubber, chemicals and ball bearings may age in storage and use part of their life on the shelf, requiring a negative t_0.

For these reasons and others, the Weibull plot may be curved and needs an origin shift, from zero to t_0. Three parameters, t_0, β, & η, are included in the Weibull cumulative distribution function as follows:

$$F(t) = 1 - e^{-((t - t_0)/\eta)^{\beta}} \qquad (3\text{-}1)$$

where: **t = failure time and t_0= starting point or origin of the distribution**

Warning! The high beta in Figure 3-9 gives the false impression of a reasonable fit, but pve% = 0.218, a really bad fit. Anytime beta is above 5 or 6 the data should be scrutinized carefully as it will appear to be a good fit but can have curvature or outliers. When the t_0 correction is applied to the data, the resulting plot will follow more of a straight line if the correction is appropriate. Figure 3-10 shows the fracture data in Figure 3-9 with the t_0 correction. Note that the Weibull ordinate scale and the characteristic life are now in the t_0 *domain*. To convert back to real time, add t_0 back. For example, in Figure 3-10, η = 7.057 and t_0

= 77.8. To convert η to real time, $\eta + t_0 = 7.057 + 77.8 = 84.9$. As this addition is tiresome and often leads to errors, a new software option has been added to produce the t_0 plot with the scale in real time ("as recorded"). Of course this produces a curved t_0 Weibull line when plotted in the real time domain but the real time domain is easier to explain without adding or subtracting data. "As recorded" is recommended.

The three parameter correlation coefficient with t_0 will always show a higher correlation coefficient better fit than the two parameter simply because it is a more complex model. Similarly, a quadratic curve fit will have a higher correlation coefficient than a linear fit. The following **four criteria should always be met** before using the three parameter Weibull:

1. The Weibull plot should show **curvature.**
2. There should be a **physical explanation** of why failures cannot occur before t_0. For example, bearings cannot spall instantaneously. Many rotations are required to produce the damage.
3. **A larger sample size, at least 21 failures** should be available. It takes much more data to estimate three parameters than two parameters. If there is prior knowledge from earlier Weibulls that the third parameter is appropriate, a smaller sample size, say eight to ten may be acceptable.
4. The correlation coefficient **p-value should be greater than the p-value for the next best distribution.** See Figure 3-4.

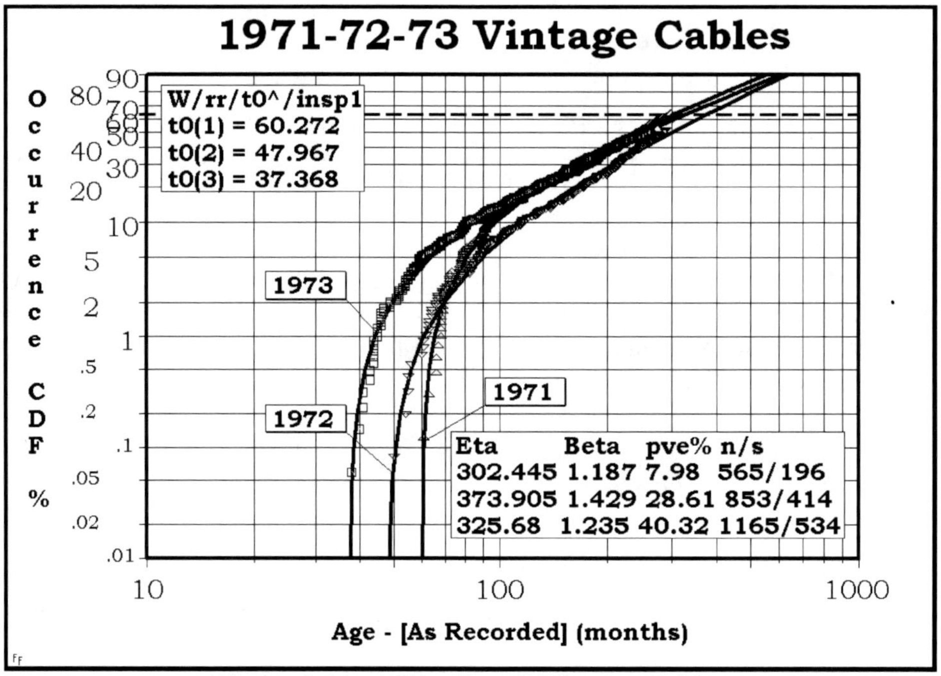

Figure 3-11. 1971 - 72 - 73 Vintage Cables

Concave downward plots occur much more often than concave upward. Concave upward suggests a negative t_0, and is more difficult to explain physically. Some parts have been aged before installation. Shelf life, burn-in for electronics, and production are examples. Another possibility is the classic mixture of two failure modes, the BiWeibull. There are several ways to estimate t_0. A curve may be "eyeballed" through the data and extrapolated down to the horizontal time scale. The intersection will be an approximate t_0. If the earliest portion of the data is missing, t_0 may compensate for the missing data, although this may not always be successful. The author has worked with Charlie Williams of Florida Power on buried cable failures in Florida. The cable data was grouped by vintage year and the aging scale was in months. There was no data recorded prior to 1976. Note the t_0 corrections in Figure 3-11. 1976-1971 is five years or 60 months, 1972 is missing 48 months, 1973 is missing 36 months! (Please do not expect this kind of close agreement, but it shows the value of good data in large quantities.) We thank Florida Power for permission to use their data. Note this SuperSMITH scale is in real or as recorded time, not the t_0 time domain. The computer will iterate on t_0 until the correlation coefficient is maximized. t_0 will always be less than the first failure time. (Left suspensions are earlier than the first failure and should be deleted before calculating t_0; otherwise t_0 cannot exceed the earliest left suspension. There is an option in SSW to allow this deletion; select $t_0 <$ minimum occurrence value instead of $t_0 <$ minimum data value to allow t_0 to exceed the left- early suspensions. Note that the deletion of left suspensions will have a small effect on the plot line.)

In summary, concave downward plots indicate the origin needs to be shifted to the right, subtracting t_0 from each time-to-failure to get a straight line fit in the t_0 time domain. Concave upward plots indicate the origin has to be shifted to the left and t_0 must be added to each time-to-failure to get a straight line fit. The plot in "as recorded" time scale is easier to understand. Case Study 11.9 is an example.

3.6 Curved Weibulls and The Lognormal Distribution

The lognormal distribution is not a member of the Weibull family of distributions but has application to life data analysis. Lognormal data plotted on Weibull paper is concave downward, very much like the three parameter Weibull. See Figure 3-12. The same data plotted on lognormal probability paper follows more of a straight-line trend. See Figure 3-13. If x is lognormally distributed, the distribution of x will be skewed to the right and log x will have the familiar bell-shaped normal distribution. Figure 3-14 shows this effect with the probability density function plotted.

The three-parameter Weibull plot with positive t-zero correction looks just like the lognormal plot with small samples, so these are competitive distributions. The author has witnessed many arguments over which distribution best fits the data. With bearing failures, the arguments went on for two decades until the industry arbitrarily selected the three parameter Weibull as the standard. Technically however, the lognormal is sometimes a better choice even with bearing data and this is true with the data in Figure 3-7. *If the distribution of log x is normal, x is lognormally distributed.* All the standard normal statistics apply in the logarithmic domain. The lognormal parameters are the mean of log x, (mu), and the standard deviation of log x, (sigma). Both mu and sigma are *expressed in logarithms*. Plotted on lognormal probability paper, x should follow a straight line. With lognormal data the antilog of the mean of log x will be less than the mean of x. The antilog of the mean of log x is the median value of x ("Med"). Sigma is also a logarithm and therefore, its antilog is a multiplying factor. The equations are given in Appendix G.

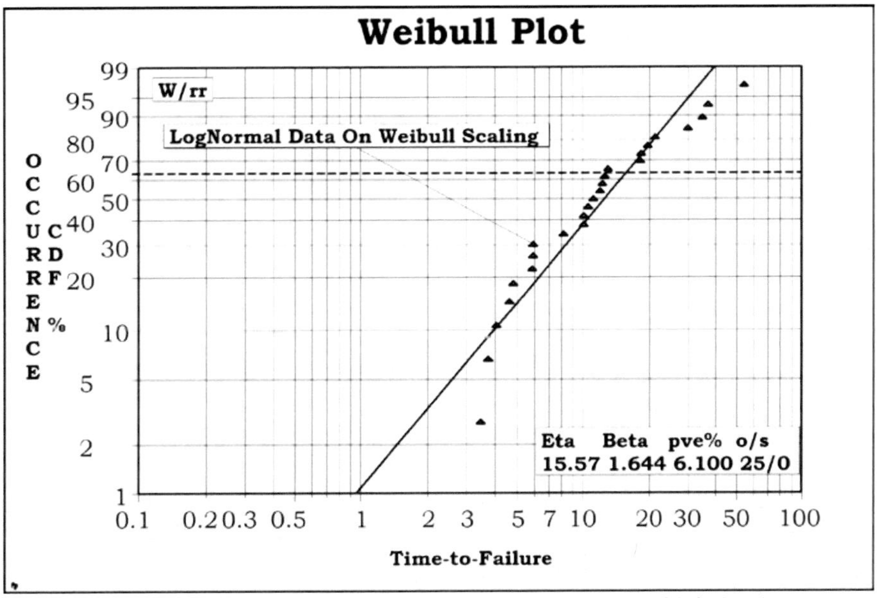

Figure 3-12. Weibull Plot of Lognormal Data (pve% is low)

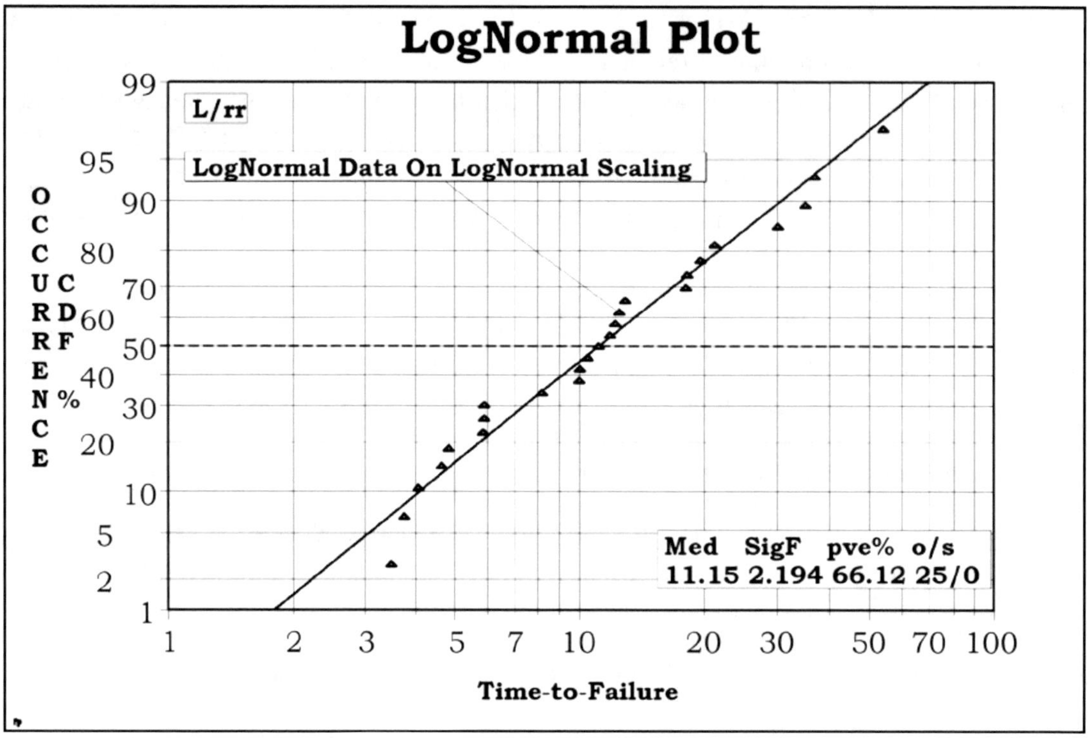

Figure 3-13. Lognormal Plot of Lognormal Data (pve% is much better)

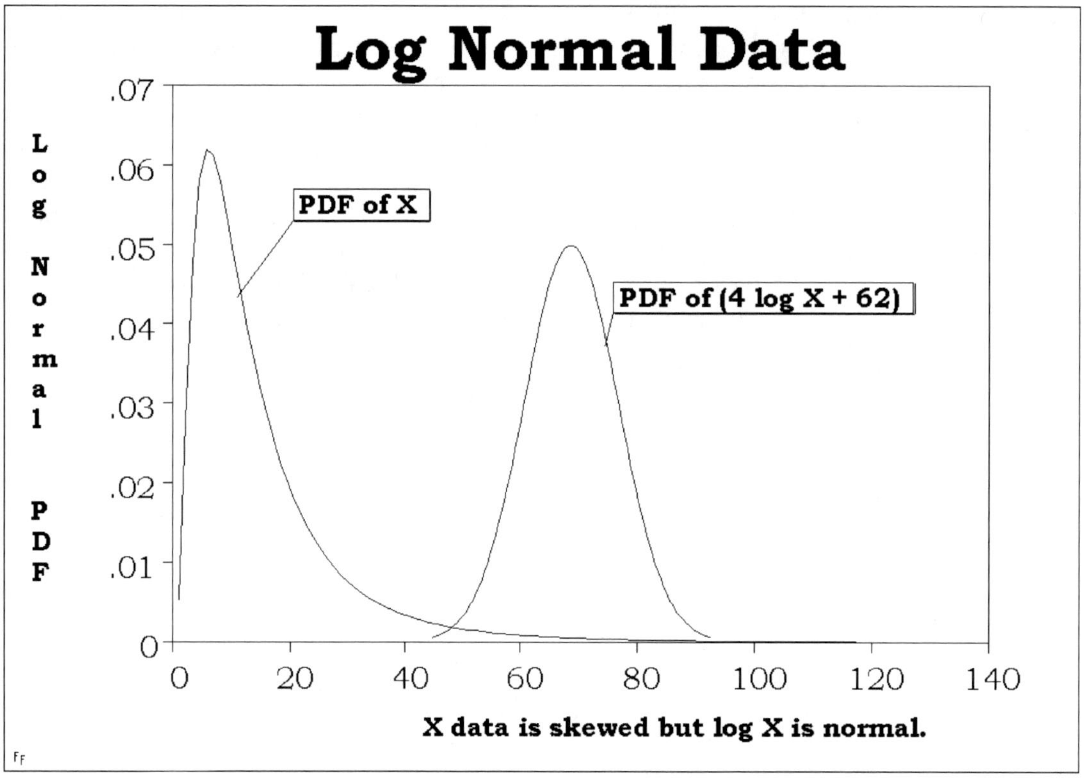

Figure 3-14. Lognormal Probability Density Functions

Physically, the lognormal models a process where the time-to-failure results from a multiplication of effects. Progressive deterioration will be lognormal. For example, a crack grows rapidly with high stress because the stress increases progressively as the crack grows. If so, the growth rate will be lognormal. On the other hand, if the growth rate is linear with time as it may be in a low stress area the Weibull will be more appropriate. Vibration tends to loosen bolts and fasteners which increases the vibration. In a gas turbine, loss of blade tip clearance will cause an increase in fuel consumption. If the rate of blade tip loss accelerates, the increase in fuel consumption will be lognormal. The lognormal has many applications such as materials properties, personal incomes, bank deposits, growth rate of cracks, and the distribution of flaw sizes.

Some metallurgists use the following model to predict in-service times-to-failure (T_s) from laboratory specimen times-to-failure (T_{lab}):

$$T_s = K_1 * K_2 * K_3 \ldots * K_n * T_{lab}$$

The K factors account for the differences between the laboratory and service for temperature, stress, acceleration, corrosion, etc., that affect the time-to-failure in-service. Taking the logarithms:

$$\ln T_s = \ln K_1 + \ln K_2 + \ln K_3 \ldots \ln K_n + \ln T_{lab}$$

Therefore, if the K factors are insignificant, that is, close to one (the log of one is zero), the in-service times-to-failure (T_s) will equal the laboratory specimen times-to-failure (T_{lab}). If (T_{lab}) has a Weibull distribution, (T_s) will also have a Weibull distribution. However, if the K factors are significant, the in-service times will be lognormally distributed because of the Central Limit Theorem: sums of samples from any shape distributions tend to be normally distributed and the Ln of (T_s) is such a sum. Note the multiplication in this model produces the lognormal times-to-failure.

Data should be plotted on Weibull and lognormal probability paper to compare the goodness of fit. The three parameter Weibull and the 2-parameter lognormal will usually provide correlation coefficients within plus or minus one per cent of each other. Do not be deceived by this into thinking the two distributions are equivalent. Use pve% to determine the best fit. The software provides solutions for two and three parameter Weibulls and lognormals, so comparisons are easily made with pve%. Statistically at least twenty one failures are needed to discriminate between the Weibull and the lognormal $[Dumonceaux] and $[Liu]. [Meeker 2001] says, " The sample size needed to distinguish between lognormal and Weibull depends strongly on the shape of the distribution and the kind of censoring. With heavy censoring (e.g. a few percent failing) and decreasing hazard, hundreds of observations are not enough." NOTE: The "$" symbol preceding a reference means the document may be downloaded from Paul Barringer's website (see References).

While there are many statistical distributions other than the Weibull, the lognormal distribution is the second choice for life data analysis. The lognormal distribution should be the first choice if there is good prior information and more than twenty failures. For example, many material characteristics employ the lognormal distribution. Times-to-repair and crack growth-to-rupture are often lognormal. Progressive deterioration cause-of-failure would also indicate lognormal for data variability. Some semiconductor chip failures are lognormal. The distribution of the Weibull slope parameter, beta, is approximately lognormal, while eta is more normally distributed. $[Liu] suggests that if there are less than twenty failures, the Weibull two parameter will be a better *predictor* (more conservative in the lower tail) than the lognormal even if the data is known to be lognormal. This is not always true comparing the Weibull three parameter with the lognormal. See Case Studies 11.11 and 11.13. Considering that distribution analysis is not creditable with small samples, (20 or fewer failures) and that the Weibull 2 parameter distribution is more conservative than the lognormal, **best practice** is always use the Weibull 2 parameter for small samples.

3.7 Data Inconsistencies and Multimode Failures

There are other subtleties in Weibull data that may provide insight. Examples are given illustrating the following:

a. Failures are mostly low-time parts with high time parts unaffected, suggesting a batch problem.

b. Serial numbers of failed parts are close together, also suggesting a batch problem

c. The data has a "dogleg" bend or cusps when plotted on Weibull paper, probably caused by a mixture of failure modes.

d. The first or last point appears suspect as an outlier, indicating data problems or perhaps evidence of a different failure mode.

3.7.1 Low-Time Failures

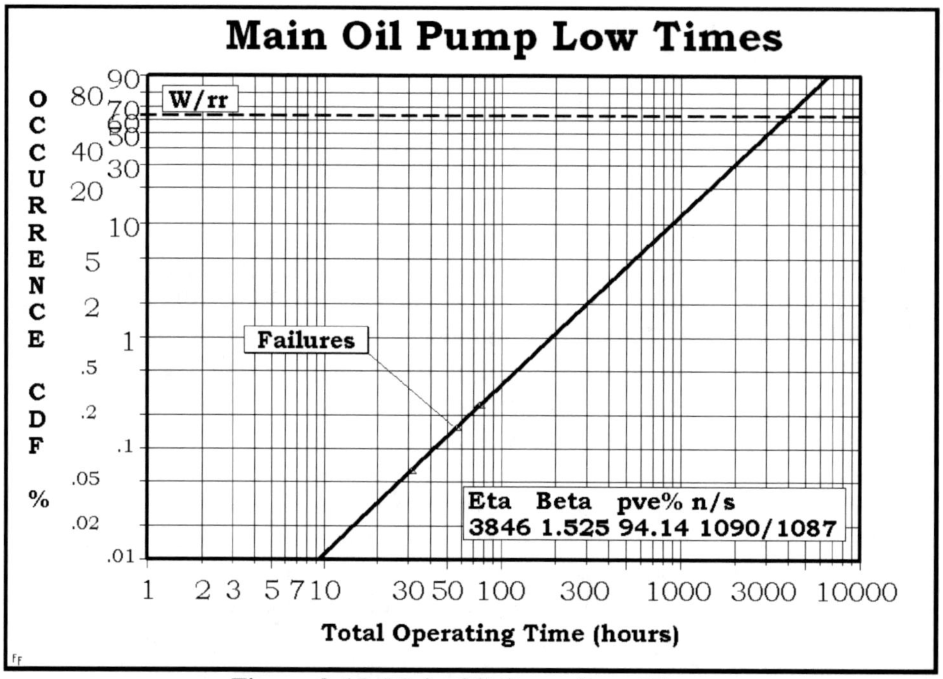

Figure 3-15. Main Oil Pump Low Times

Figure 3-15 is an example of low-time part failures on main oil pumps. Gas turbine engines are tested before shipping to the customer, and with more than 1000 of these engines in the field with no problems, what was going wrong? Examining the failed oil pumps, it was found they contained oversized parts. Something changed in the manufacturing process and created this problem, a batch problem. The oversized parts caused interference with gears in the pump resulting in failure. It was traced to a machining operation and corrected. Low-time failures can suggest wear out with a slope more than one, but more often, they will show infant mortality, with slope less than one. Low-time failures provide a clue to production or assembly change, especially with many successful high-time units in the field. Overhaul and scheduled maintenance also may produce such "batch" effects. Times since overhaul or maintenance may provide a clue. The presence of many late suspensions also points to a batch problem. The next chapter will be present using failure forecasting to confirm presence of a batch problem. Chapter 8 summarizes batch clues and analysis. Appendixes F and J contain advanced material on detecting batch effects.

3.7.2 Close Serial Numbers

Similar reasoning can be extended to other peculiar failure groupings. For example, if low-time units have no failures, mid-time units have failures, and high-time units have no failures, a batch problem is strongly suggested. Something may have changed in the manufacturing process for a short period and then changed back. Closeness of serial numbers of the failed parts suggests a batch problem. Figure 3-16 is a prime example of a process change happening midstream in production. Bearings were failing in new augmentor pumps. The failures occurred at approximately 200 to 400 hours. At least 650 units had more time than the highest time failure. These failures were traced to a process change incorporated as a cost reduction for manufacturing bearing cages. NOTE: Figure 3-16 was edited from original for improved display.

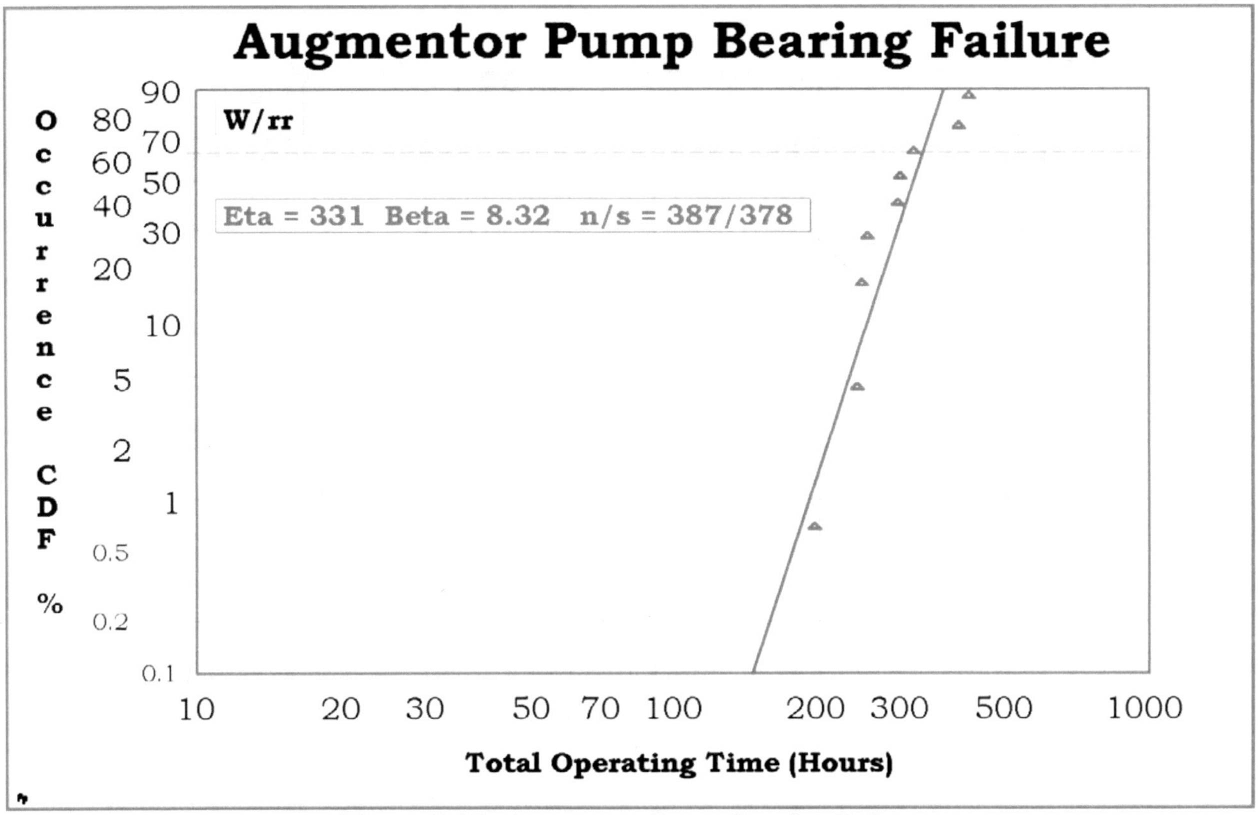

Figure 3-16. Augmentor Pump Bearing Failure

Figure 3-17A. Negative t-zero

3.7.3 Mixtures of Failure Modes A Weibull plot containing a "dogleg bend" is a clue to the potential of multiple competitive failure modes (see Figure 3-18). This was the case for a compressor start bleed system binding problem. Upon examination of the data, 10 out of 19 failures had occurred at one base. It was concluded that the base's location contributed to the problem. The base was located on the ocean and the salt air was the factor. The data were categorized into separate Weibull plots with this engineering knowledge. The first Weibull had a slope of 0.75. This could be considered an infant mortality problem, while the ocean base Weibull had a stress corrosion wear out failure mechanism with β = 11.9. More attention to maintenance resolved the problem.

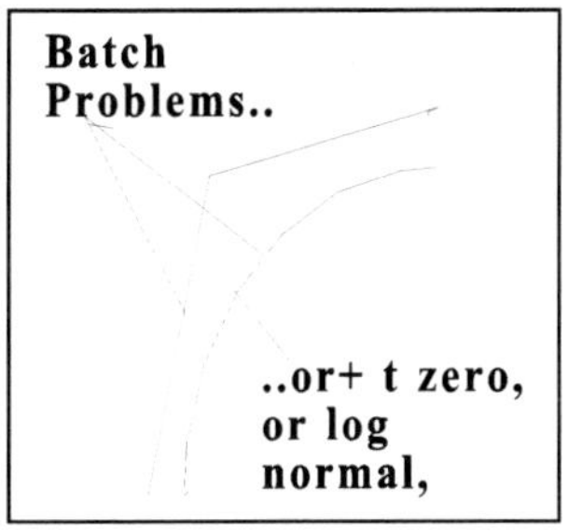

Figure 3-17B. Positive t-zero

Dogleg Weibulls are caused by mixtures of more than one failure mode. These are competitive failure modes, competing to produce failure. For instance, fuel pump failures could be due to bearings, housing cracks, leaks, etc. If these different failure modes are plotted on one Weibull plot, several dogleg bends will result. When this occurs, a close examination of the failed parts is the best way to separate the data into different failure modes. If this is done correctly, separate good Weibulls will result. With small samples it is hard to distinguish curves from cusps. (see Figures 3-17A and 3-17B). There can be mixtures of modes and populations, perhaps batches and competing failure modes. A steep slope followed by a shallow slope usually indicates a batch problem. A steep slope followed by a gentle curve to the right indicates there are some "perpetual survivors" that are not subject to the failure mode. For example, there may be defects in some parts but not all the parts, a batch problem.

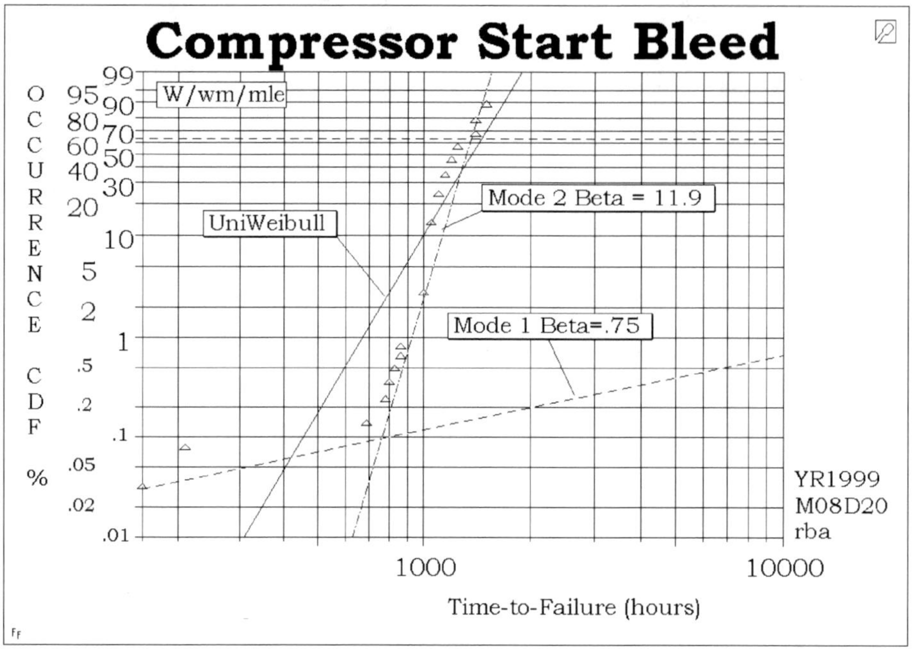

Figure 3-18. The Classic Bi-Weibull

Many hydromechanical components show infant mortality from production and quality problems followed by wear out later in life as competing failure modes. This is called the **"Classic Bi-Weibull,"** a shallow slope followed by a steep slope. The author recommends engineering analysis to separate or categorize the data by failure mode. Many analysts assume that all failures earlier than the corner belong to the first Weibull and all the later failures belong to the second. This is rarely true, but statistically it may be approximately true. This approach will be illustrated with an example.

EXAMPLE: Southern Pacific Railroad schedule overhauls of their locomotive power units 36 months after the previous overhaul. However they experienced early failures after overhaul indicating a quality problem in the overhaul process. See Figure 3-19. Many hydromechanical parts experience this combination of a few failures early in life followed by wear out. The corrective action is to identify and eliminate the quality problem. The failed units were not available for inspection to categorize them into separate failure modes. Therefore a **statistical separation** based on the likelihood ratio test was employed to estimate the parameters of the two Weibulls using SSW. This test will be treated in Chapter 7. The approach is to consider all possible "corners" dividing the data into two sets. The corner showing the strongest evidence of two sets is selected if there is strong evidence supporting two sets versus one set. To estimate the four parameters at least 21 failures are needed for creditable results. See Case Study 11.4.

More complex solutions for mixtures of two and three failure modes including batch effects are provided by the **"YBath"** software created by Carl Tarum and included with the SuperSMITH software. For mixtures of three failure modes, at least 100 failures should be available. However, engineering separation of data mixtures is always preferred to statistical methods.

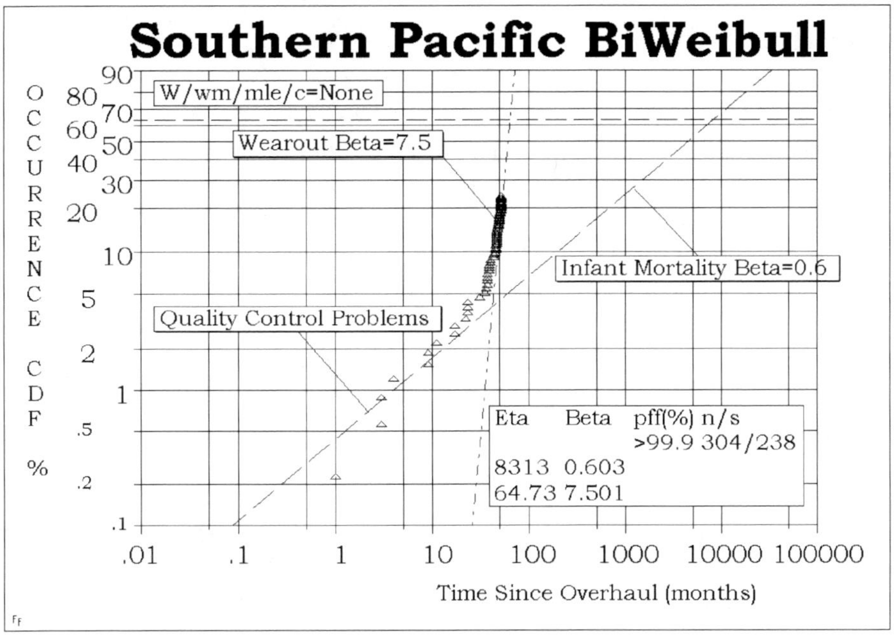

Figure 3-19. Southern Pacific BiWeibull

Many failure modes mixed together will have a better fit, as the doglegs will disappear and β will tend toward one. Thus Weibulls for a system or component with many modes mixed together will tend toward a β of one. These Weibulls should not be employed if there is any way to categorize the data into separate, more accurate failure modes. Using a Weibull plot with mixtures of many failure modes is equivalent to assuming the exponential distribution applies. The exponential results are often misleading and yet this is common practice. MIL-HDBK-217 uses the exponential assumption for modeling electronic systems whereas field failure data will indicate infant mortality, or wear out failures. See case Study 11.7.

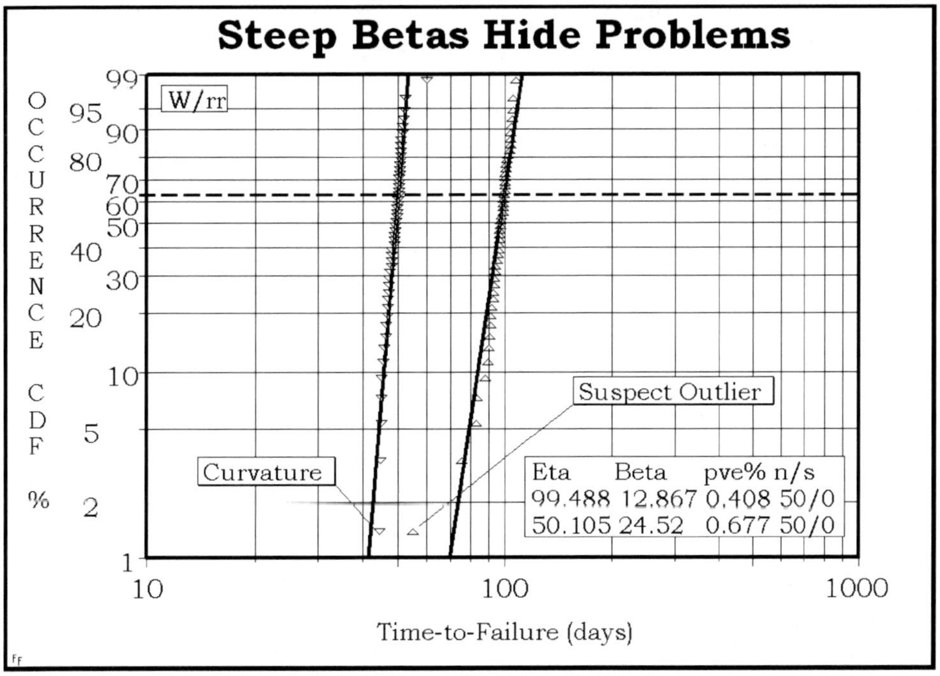

Figure 3-20. Steep Betas Hide Problems

3.8 Steep Slopes May Hide Problems

A word of caution is suggested for steep slopes, β. The steep plot often hides bad Weibull data. All the messages from the data such as curves, outliers, doglegs tend to disappear. Apparently good Weibulls may have poor fits. An example is Figure 3-20. Here, at first glance the plots appear to be good fits, but there is curvature and perhaps an outlier. The pve% values are well below 10. If SSW is used, a simple solution to amplify the data problems is to use the "zoom" option to magnify the curvature portion. This will make the problems with the data more obvious. One expert suggests a distribution analysis is needed for all Weibulls with $\beta > 6$. The author would add "if there are 21 or more failures," but careful scrutiny of the data plot is always a good idea with steep betas with or without a distribution analysis.

3.9 Bad Weibull Patterns

A summary of some bad Weibull patterns:

- Data *curved downward* may be a three parameter Weibull or a lognormal. With large samples a flat spot in the middle of the plot can indicate a lognormal. The Weibull 3P is a continuous curve.
- Data *curved upward* may be a three parameter Weibull with a negative t_0.
- *Cusps and corners* indicate a mixture of failure modes. There may or may not be a batch problem.
- *Steep followed shallow* is almost always a batch problem.
- *Steep followed by a shallow curve* usually indicates "perpetual survivors."
- *Vertical columns of points* indicate interval or inspection data.
- *Intervals between failures and mixtures of many failure modes* will tend to a beta of one.
- With intermediate and large samples, *the lognormal will show a flat spot in the middle while a Weibull 3P is continuously curved.*

3.10 Conclusion

The reader may feel overwhelmed at this point with special methods, data formats, what to look for, etc. It is not the author's intent to confuse the reader. It is my intent to share experience. Waloddi Weibull preached simplicity; plot the data, and look at it, and the author strongly agrees. Waloddi said these methods may have "some utility." He was correct. This handbook may appear to be complex because the "menu" of possible data sets in industry is infinite. Many of you have probably browsed through Chapter 3, leaving some sections under the heading, "if I ever need this material, I know where it is." That is an entirely acceptable attitude, and applies to the remainder of this Handbook. Use what applies in your application; ignore the rest. Chapter 10 provides a summary of the methods for handling the good, the bad and the dirty data. Further, it now includes a useful logic diagram that will take you step-by-step through your analysis.

3.11 Problems

Problem 3.1 t_0:

The following set of failure points will result in a curved Weibull: 90, 130, 165, 220, 275, 370, 525, and 1200 hours. Plot on Weibull paper.

a. What value is needed to straighten the Weibull? Hint: "Eyeball" a curve through the data and read an approximate t_0 where it intersects the bottom scale.
b. Will the value found in "a" be added or subtracted from the failure values?
c. Repeat with SSW using the t_0 options to find a more accurate answer.

Problem 3.2 Interpreting the Plot:

In the first three chapters comments and hints were given on interpreting the data plot. Study Figures 3-21 through 3-32. Assume you have no prior knowledge. If you are in a Weibull Workshop talk it over with the "experts" sitting near you. Use the information provided on the plot. List your comments and the actions you would take in analyzing each data set.

3-21 ______________________________

3-22 ______________________________

3-23 ______________________________

3-24 ______________________________

3-25 ______________________________

3-26 ______________________________

3-27 ______________________________

3-28 ______________________________

3-29 ______________________________

3-30 ______________________________

3-31 ______________________________

3-32 ______________________________

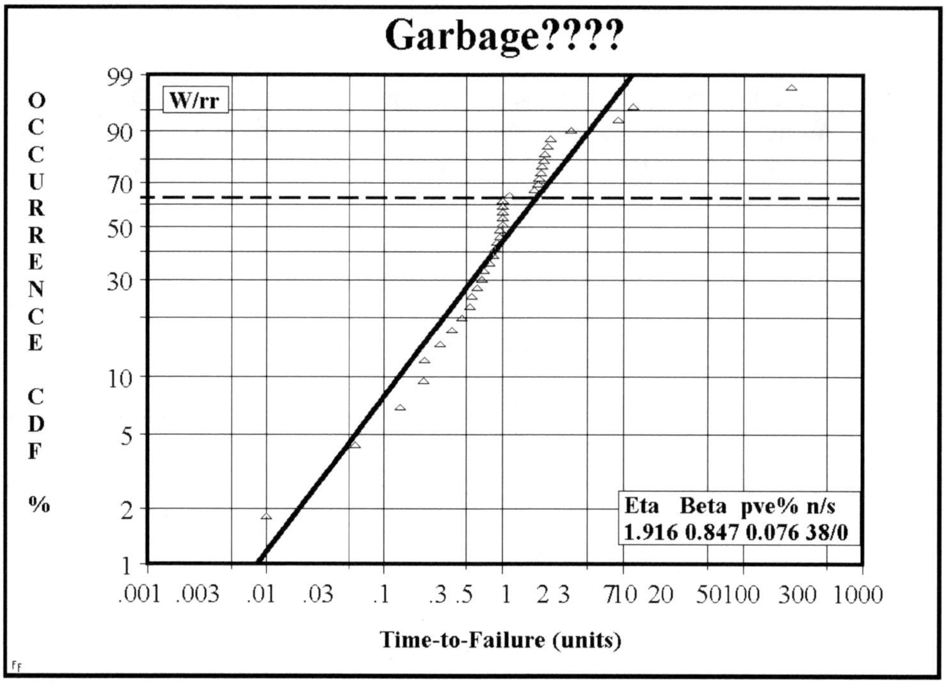

Figure 3-21. Garbage

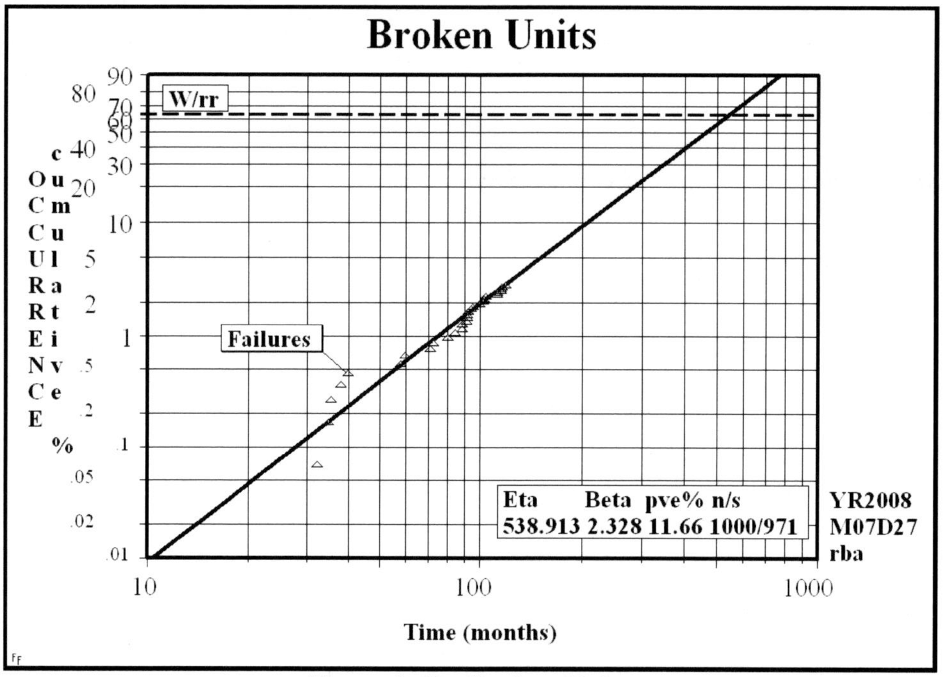

Figure 3-22. Broken Units

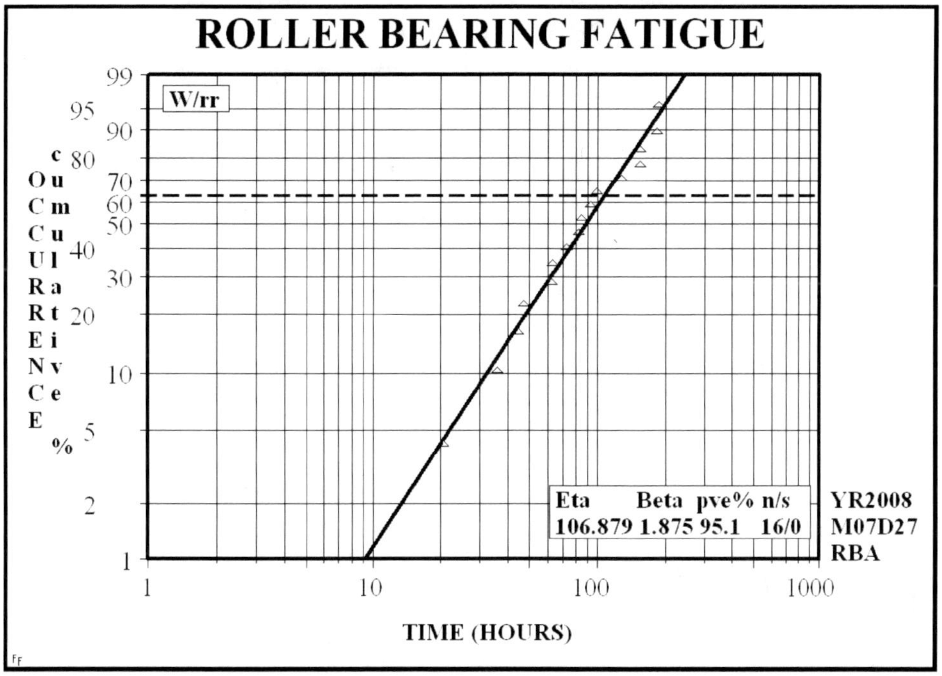

Figure 3-23. Roller Bearing Fatigue

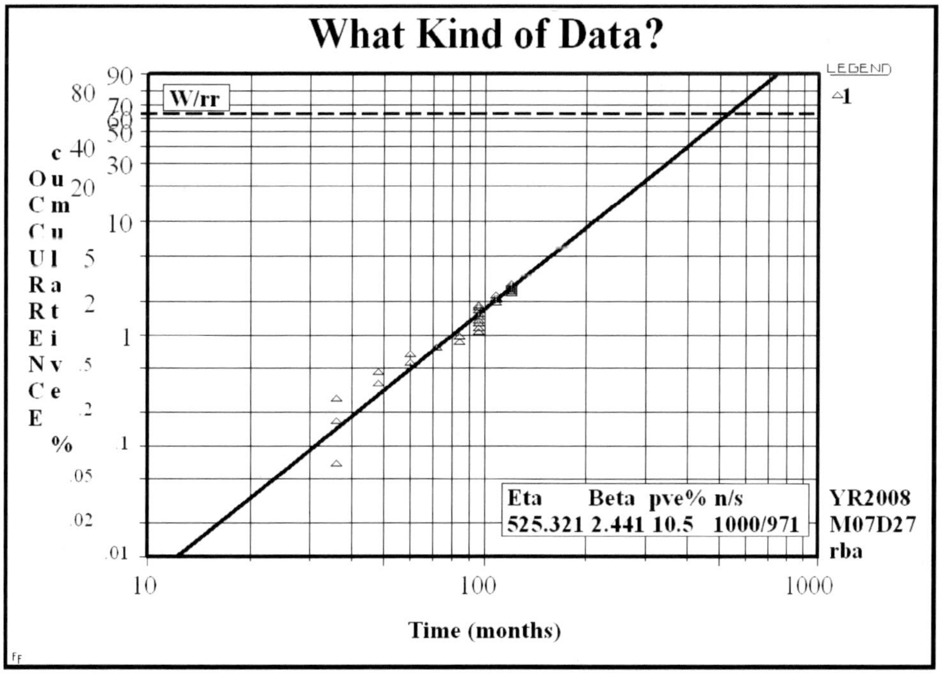

Figure 3-24. What Kind of Data?

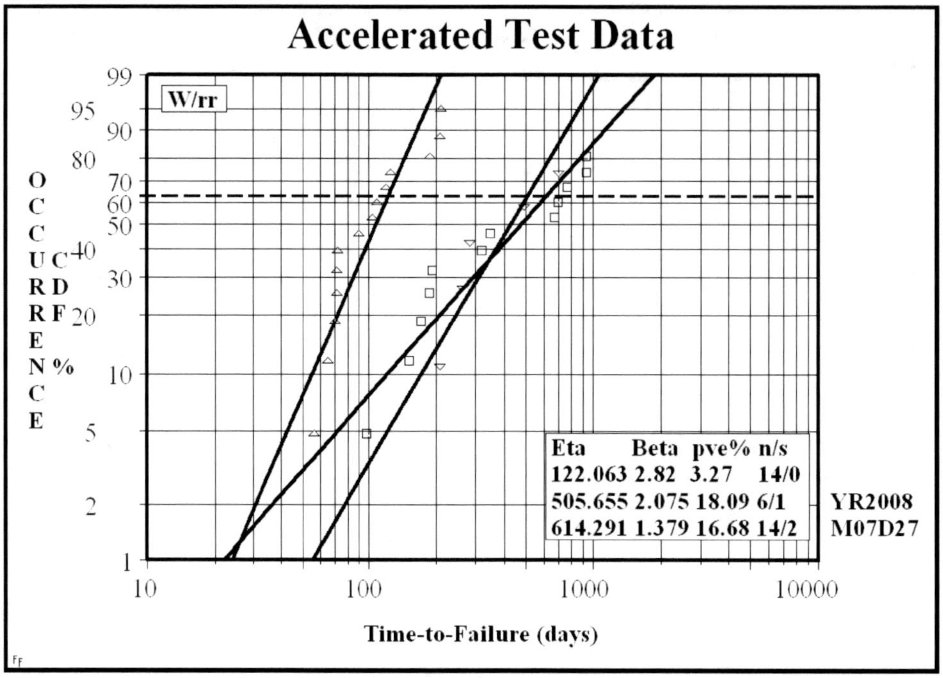

Figure 3-25. Accelerated Test Data

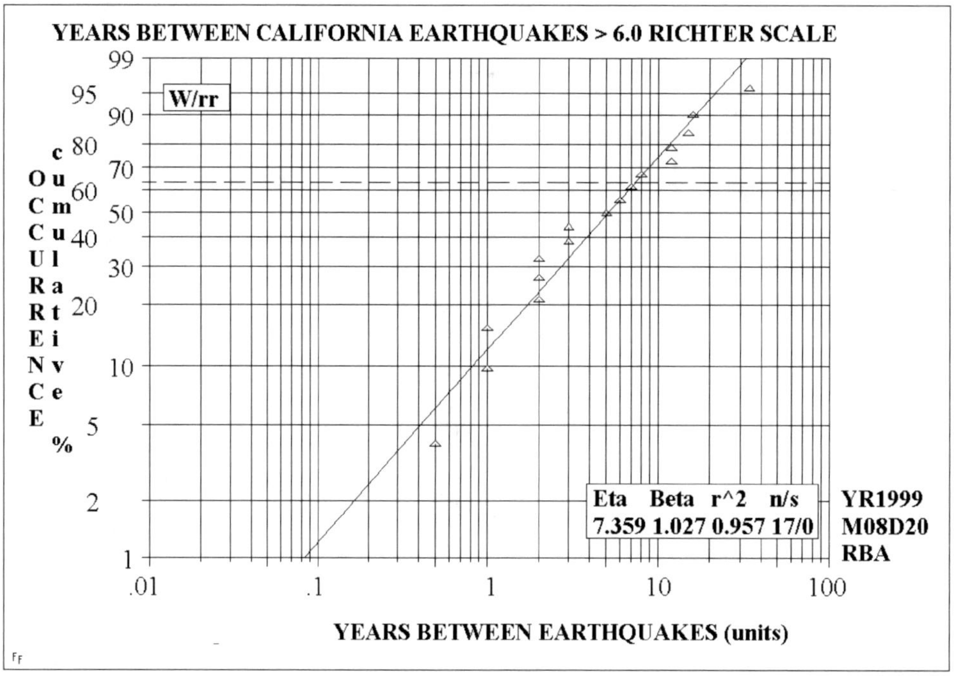

Figure 3-26. Earthquakes > 6.0 Richter Scale pve% =52.23

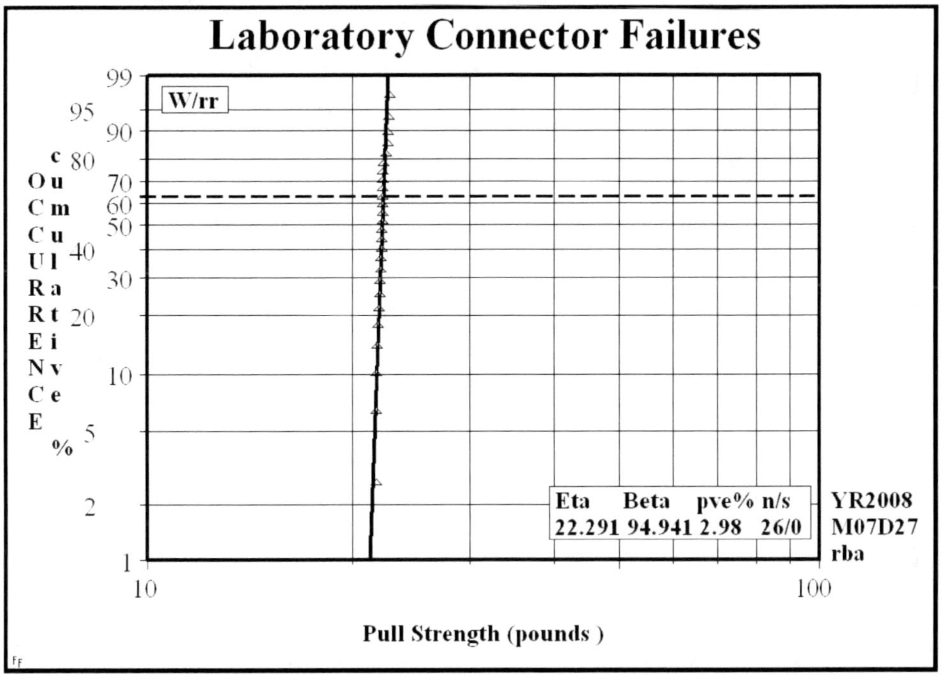

Figure 3-27. Laboratory Connector Failures

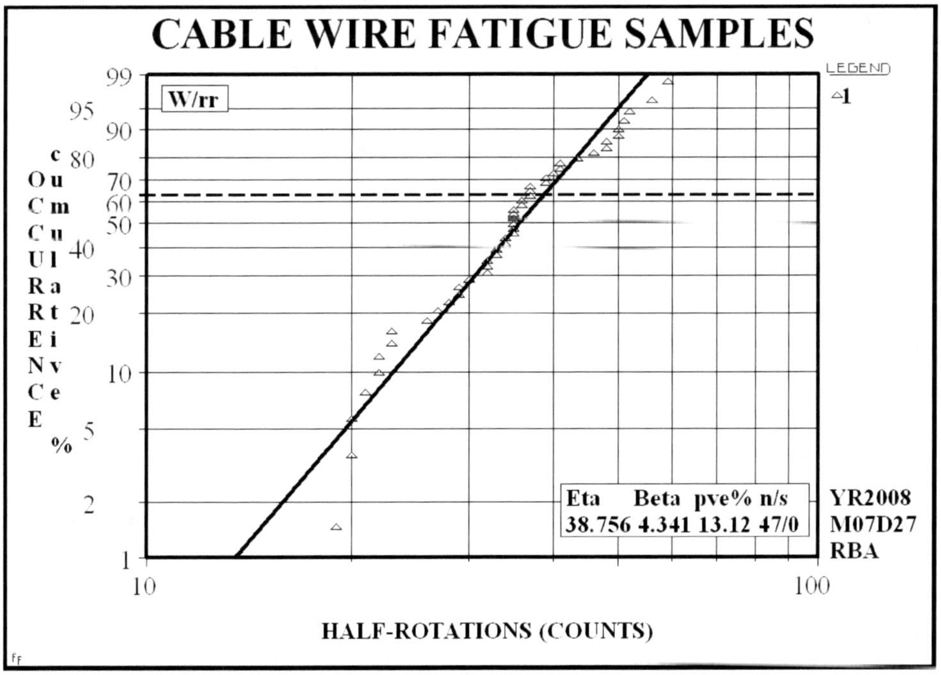

Figure 3-28. Cable Wire Fatigue Samples

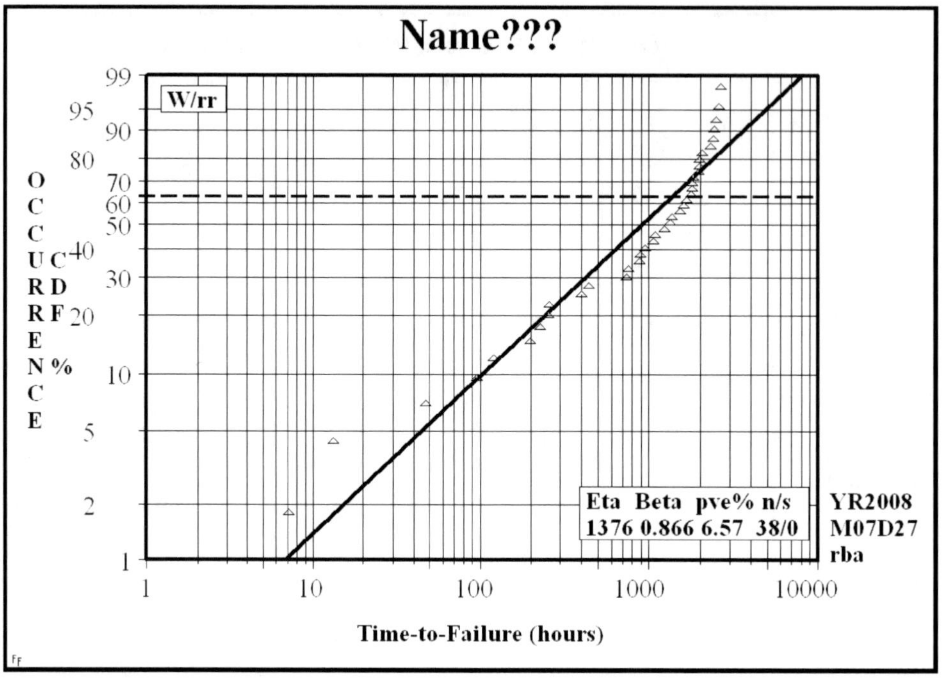

Figure 3-29. Name???

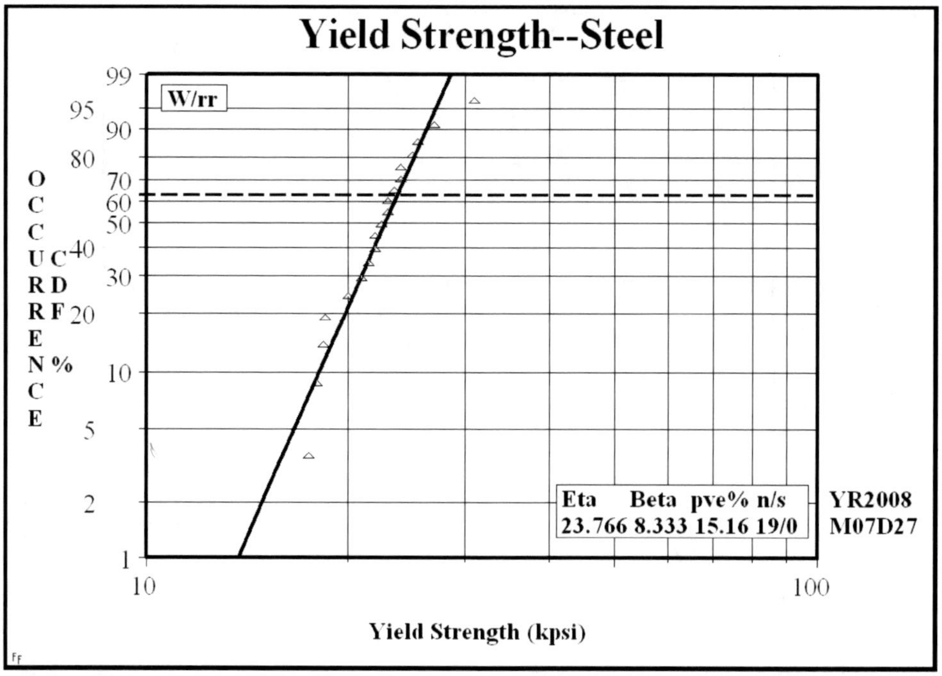

Figure 3-30. Yield Strength – Steel

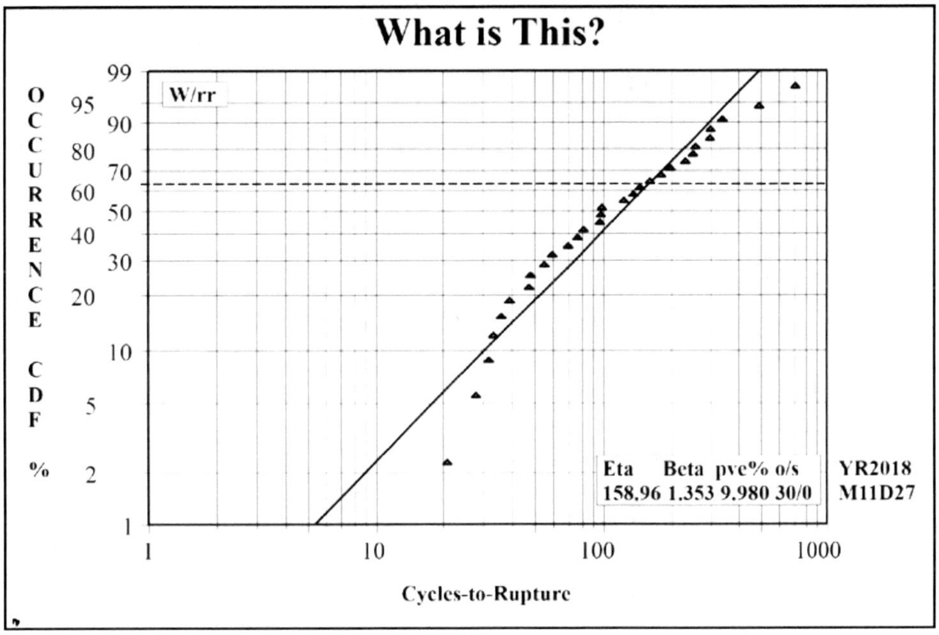

Figure 3-31. What is This?

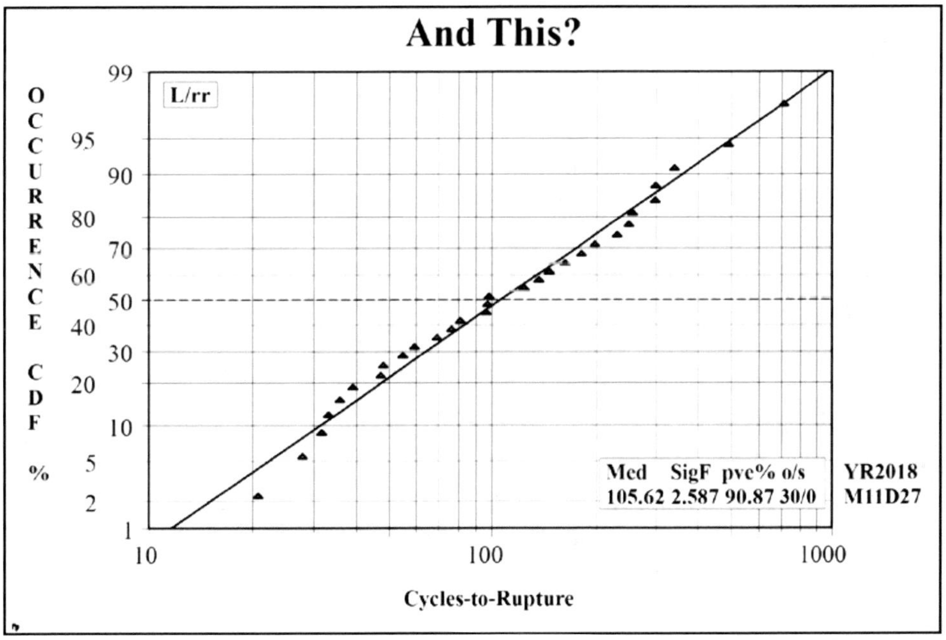

Figure 3-32. And This? (NOTE: pve% is much higher!)

3.12 Dorian Shainin and E.J. Gumbel

Here you will find my comments on two great statisticians, Dorian Shainin and E.J Gumbel. There are similar vignettes of other great statisticians that I admire at the end of chapters that relate to their contributions. These are not required reading.

Dorian Shainin (1914- 2001) was one of the earliest pioneers for Weibull analysis and engineering statistics. He inspired the author to do a doctorate in statistics (although Dorian thought it was a waste of time). He led the team that created "Precontrol," a means of using statistics to actually control a process rather than just test the process to see if it is in statistical control of its own accord. His "Random Balance" preceded Taguchi and was equally controversial. His humor was endless. He claimed he drank carrot juice and vodka; "I can get just as high but I can see better." In his search for the "Red X," the root cause of mysterious problems, he created a very large number of statistical methods such as "Step-Stress," and some very clever investigative strategies. Dorian received every one of ASQ's major awards and was made an honorary member by the Board of Directors. ASQ has now named an award after Dorian. His greatest contribution was getting engineers and managers excited about using statistics. His talent here is unsurpassed, actually never tied.

The consulting firm Dorian founded has continued as Shainin LLC with two of his sons and a grandson. Dorian's son, Peter Shainin, is the manager of the firm and provided this picture.

E.J. Gumbel (1891-1966) spent much of his life studying the statistics of extreme values, (rare events). Born in Germany he resisted the secret rearmament of Germany after World War I and left. He did much of his research in the USA at NBS, now NIST. He and Waloddi Weibull did a sabbatical together at Columbia University and became good friends. Gumbel was fascinated by the fact that the Weibull distribution and Extreme Value Type III minimum are the same distribution. Extreme Value Type I is called the "Gumbel" distribution. It is employed for predicting maximum and minimum values, flood levels, wind gusts, the size of inclusions in metal. The Weibull and the Gumbel minimum are related like the normal and the lognormal through a logarithmic transformation. The SuperSmith software provides the Gumbel distributions as well as the Type III distributions. See Section 8.8 for more on the Gumbel distribution.

CHAPTER 4: FAILURE FORECASTING = RISK ANALYSIS

4.1 Situation

Significant failures have occurred in service, perhaps involving safety or financial loss. Responsible managers demand a forecast of the number of failures expected to occur in the future. How many failures will there be next month, the next six months, the next year? What will the costs be? Managers need failure forecasts to set priorities and allocate resources for corrective action. This is risk analysis, a prediction of the magnitude of the problem, a clear view of the future. "Risk" is employed as a synonym for "failure forecast." The method is applied to the Weibull, normal, and lognormal life distributions for failure and warranty analysis. It was developed by the author with his statisticians at Pratt & Whitney Aircraft and first published in [Abernethy 1983]. Statisticians like to rename things; they call failure forecasting "predictive analysis."

It should be emphasized that a failure forecast is only as good as the quality and quantity of the failure data. Further, the uncertainty will increase as the time span for the forecast increases. Advanced material in this chapter deals with:

- system modeling with and without Monte Carlo simulation and
- determining lowest cost (optimal) replacement intervals for parts that wear out.

4.2 Definition

A risk analysis or failure forecast predicts the expected number of incidents that may occur in a specific period of time.

4.3 Forecasting Techniques

The failure distribution is determined from the failure and suspension data, as discussed in Chapter 2. Forecasting requires additional input:

- the age of the components in service,
- usage rate per unit per month (or year, or day),
- production introduction rate of new units and
- whether failures are catastrophic or failed parts are replaced with zero time parts.

With this information a failure forecast can be produced. The techniques used to produce the failure forecast vary from simple calculations to complicated analysis requiring Monte Carlo simulation. These will be explained in the following sections.

4.4 Calculating Failure Forecasts

The next three sections describe the calculations to predict:

- Expected failures "now", i.e. at the present moment with current data collection and model
- Future failures when failed units are not replaced and
- Future failures when failed units are replaced.

4.4.1 Expected Failures – Now – "Present Risk"

Our objective with the "Present Risk" is to check for a batch problem in our data set; if we have a batch problem, we have the wrong data set. The batch failure mode does not apply to all the units in the population. Detection of batch problems is the principal reason for calculating the Present Risk. To calculate the Present

Risk the first step is to calculate the expected number of failures at the current time which is the Present Risk. Earlier editions of the handbook evaluated the Present Risk as probability of failure by time t_i summed over the number of units, N, including failures, r, and suspensions, s. New research has shown this formula to be slightly biased. More accurately the Present Risk with the Reduced Bias Adjustment (RBA) is:

$$\mathbf{NowRiskRBA} = 2\sum_{i=1}^{r} \mathbf{F}(\mathbf{t}_i) + \sum_{i=1}^{s} \mathbf{F}(\mathbf{t}_i) \tag{4-1}$$

$$\mathbf{Expected\ Failures} = 2\sum_{i=1}^{r}(1 - e^{-(t/\eta)^{\beta}}) + \sum_{I=1}^{s}(1 - e^{-(t/\eta)^{\beta}}) \tag{4-2}$$

Example: Equation 4-2 will be used to calculate expected failures now with present data for a fleet of 20 pumps, one of which has failed at 2000 hours and another at 3000 hours. There are 5 suspensions at ages of 1000 and 4000 hours, and 4 at 2000 and 3000 hours, respectively. SSW software was used to generate the Weibull plot below from the data above where $\beta = 2.63$ and $\eta = 6210$ hours. The first question is, "What is the expected failure quantity from time zero until now for this population?" The value of F(t) can be read directly from the Weibull CDF (Figure 4-1) or calculated from the Weibull Equation 4-1. For example, at time 1000, F(t) = 0.00820, and at time 2000, F(t) = 0.0496. Table 4-1 shows the results.

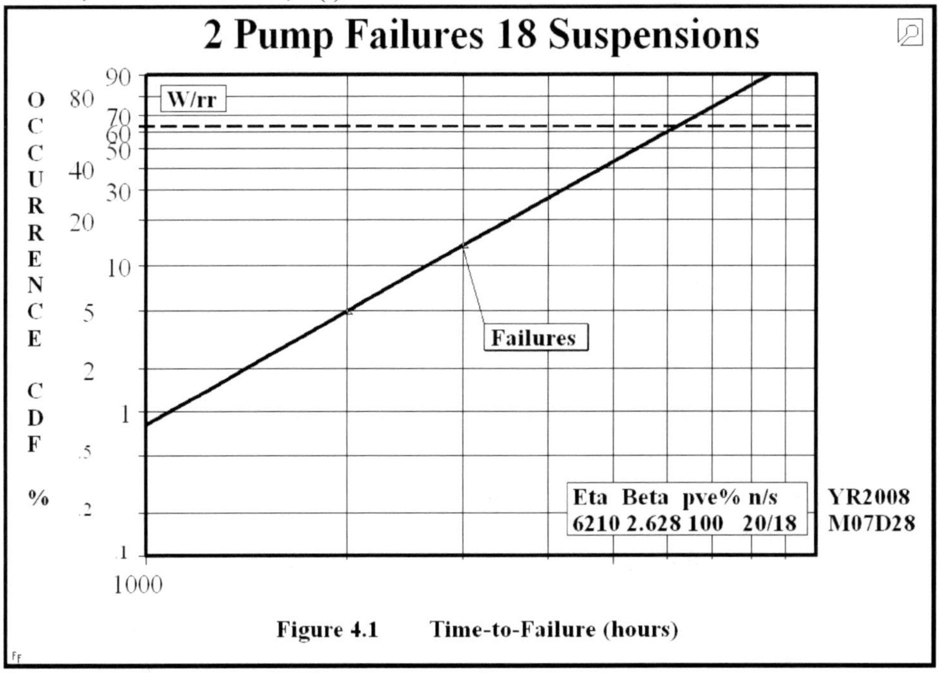

Figure 4-1. 2 Pump Failures 18 Suspensions

Table 4-1. Present Risk Calculations *with the RBA*

Number of Units (N)	*Time (t) on each Unit*	*F(t) From Figure 4-1*	*F(t) * N*
5S	1000	0.0082	0.041
4S + 2(1F)	2000	0.0496	0.298
4S + 2(1F)	3000	0.1374	0.824
5S	4000	0.2700	1.350
18S + 2F = 20		Expected Failures Now = 2.51	

The Present Risk with the RBA is 2.51, which approximates the observed two failures. If the expected-failures-now is much larger than the observed number of failures, the Weibull may not apply to the entire population. This indicates a "batch" problem, i.e., the entire fleet does not have the "illness." Other clues that suggest a batch problem are serial numbers of failed parts bunched together and many late suspensions, i.e., suspensions older than the last failure. Batch problems are very common. If you have a batch problem you have the wrong data set and your first priority is to identify the units in the batch. See Chapter 8 for more details on batch problems and how to analyze them. The failure forecast will be wrong if you use the wrong data set. Appendix F describes the Aggregate Cumulative Hazard (ACH) method for analyzing batch problems. It works for complete samples which the Present Risk method cannot handle.

If a Poisson lower confidence bound on the Present Risk is greater than the observed number of failures, a batch problem is indicated with high confidence. This concept was suggested by Ann Azevedo, Chief Scientist for the FAA, during a Weibull Workshop at the FAA. If the lower 90% bound is greater than or equal to the observed number of failures, there is 90% confidence for a batch problem. If, however, the lower 90% bound on Present Risk is less than actual failure count, there is no evidence of a batch problem. The Present Risk detection capability is available with rank regression, but not with maximum likelihood estimates (MLE). The MLE Present Risk predictions will not reflect or imply a batch problem. SSW can place lower and upper bounds on Present Risk.

4.4.2 Failure Forecast When Failed Units Are Not Replaced

Failure forecasts can be made for any time interval. Let us use the next year as an example. Given the 18 pumps at risk in Table 4-1, the expected number of failures over the next 12 months can be predicted. Yearly usage of each pump will be 1000 hours. The failure forecast for one year sums the predicted failures of the 1000-hour pumps running to 2000 hours, plus the predicted failures of the 2000 hour pumps running to 3000 hours, plus the predicted failures of the 3000 hour pumps running to 4000 hours, etc. The forecast only involves the unfailed pumps, the living suspensions, pumps still "at risk". All future failures will occur on today's survivors.

If pump i has accumulated t_i hours to date <u>without failure</u>, and will accumulate $\boldsymbol{u}$ additional hours in a future period, the failure forecast for the 18 pumps at risk is given by Equation 4-3. $F(t_i)$ is the probability of pump i failing in the first t_i hours of service, assuming it follows a Weibull failure distribution. $\boldsymbol{u}$ is the 12 month usage, 1000 hours. Note that the summation is over the 18 unfailed, suspended pumps. Unfailed pumps are still "at risk" because they have survived up to now. In this example, failed pumps are dead, no longer "at risk."

$$\textbf{Failure Forecast} = \sum \frac{F(t_i + u) - F(t_i)}{1 - F(t_i)} \qquad (4\text{-}3)$$

Table 4-2 summarizes the one year failure forecast calculations for the population of 18 pumps at risk, using Equation 4-3 with a one year usage of 1000 hours. Therefore, the failure prediction is 2.3, the expected number of failures during the next year.

Table 4-2. 12 Month Pump Failure Forecast

Number Of Units (N)	Current Time on Each Pump (Hr.)	Time at Year's End (t + u) (Hr.)	(a) F(t + u)	(b) F(t)	Predicted Failures $\frac{(N)(a-b)}{(1-b)}$
5	1000	2000	0.0496	0.0082	0.2087
4	2000	3000	0.1374	0.0496	0.3695
4	3000	4000	0.2700	0.1374	0.6149
5	4000	5000	0.4321	0.2700	1.1100
N = 18S		12 Month Failure Forecast = Sum = 2.303			

4.4.3 Failure Forecasts When Failed Units are Replaced

Failure forecasts with replacement, called "renewal," are equal to or greater than the failures forecast without replacement. Without replacement the fleet size will decrease with time. With replacement the fleet size is constant and the failure of the replacements increases the number of failures. The replacement unit is usually assumed to be a new, zero-timed, addition to the fleet of the same design, but may be a redesign with an improved Weibull or a repaired part with a worse Weibull. With new units, if the forecast interval is B50 life or less, the forecast with or without replacement is about the same since the replacement units do not age enough to fail. In cases where the failure forecast is greater than the B50 life, the chance of more than one failure over the interval starts to become significant, i.e., the replacement part will age enough to have a chance of failing. In this case, the expected number of failures may be calculated by adding a new zero time unit whenever the failure forecast increases by a unit amount. Numerical accuracy is improved by triggering only the 1st occurrence/replacement at median risk (0.5) and all others at unit risk increase. Renewal capability is included in SSW as well as the following advanced capabilities:

- Continued **production** may add units to the fleet in the months ahead. Production will increase the fleet size and the failure prediction. The production rate may be constant, or it may be seasonal varying by the month of the year, or it may be sampled from a distribution.
- Similarly the **usage rate** may be constant or seasonal varying by the month of the year, or sampled from a distribution.
- **Warranty** analysis predicts warranty claims by the age of the system. The above failure prediction methods are used with one modification; as systems move out of the warranty period, expressed in months and/or mileage, they are removed from consideration as they will no longer generate warranty claims.
- If there is **mixture** of two failure modes and the mixture analysis is used in SSW to estimate the two Weibulls, the Abernethy Risk failure forecast is available including the joint effect of both Weibulls. The calculations for all these variables are complex and require software like SSW.

4.5 Failure Forecast Analysis-Summary

The Expected Failures Now or **Present Risk** sums over the entire fleet, failures plus suspensions.

$$\mathbf{NowRiskRBA} = 2\sum_{i=1}^{r} \mathbf{F(t_i)} + \sum_{i=1}^{s} \mathbf{F(t_i)}$$

The future **Failure Forecast** is summed over only the units "at risk," units that are still operating.

These equations assume failed units are not replaced. F(t) may be calculated from the Weibull, normal lognormal, or whatever appropriate distribution.

$$\mathbf{Failure\ Forecast} = \sum \frac{\mathbf{F(t_i + u) - F(t_i)}}{\mathbf{1 - F(t_i)}}$$

When failed units are replaced, called renewal, the future risk is higher because the fleet size is constant. Without renewal the future risk relates to a decreasing fleet size. The future risk with renewal may be estimated by adding a new production unit to the fleet whenever the future risk increases by a unit amount. SSW provides upper and lower bounds on the future risk with and without renewal.

Several case studies will illustrate the ideas developed in the previous sections. The first two examples, Sections 4.5.1 and 4.5.2, illustrate the failure forecast calculation, with and without renewal. *Section 4.6.1 is a case study using the software for system failure forecasting. (The asterisk indicates advance material which may be skipped on first reading.) The case study in *Section 4.7.1 uses Monte Carlo simulation for system failure forecasting. The last *Section, 4.8, illustrates the methods for determining optimal replacement intervals for parts that have wear out failure modes. In Chapter 11 case study 11.3 illustrates optimal component replacement. Case studies 5.3.1, 11.2, 11.13 & 11.14 illustrate the accuracy of failure forecasting.

* Advanced material may be omitted on first reading.

4.5.1 Case Study 1: Bearing Cage Fracture

Bearing cage fracture times were 230, 334, 423, 990, 1009, and 1510 hours. Suspension times are shown in Table 4-3 and plotted in Figure 4-2 with SSV. The fleet originally contained 1703 bearings in service with ages up to 2050 hours. This "suspension histogram" is the surviving population of bearing cages within which the failures can occur. Figure 4-3 shows the bearing cage Weibull failure distribution. From this Weibull plot we can see that the B10 life (time at which 10% of the population will have failed) is approximately 2590 hours. This was much less than the B10 design life of 8000 hours, so a redesign was undertaken immediately. Additionally, management wanted to know how many failures would be observed before this redesign entered the field.

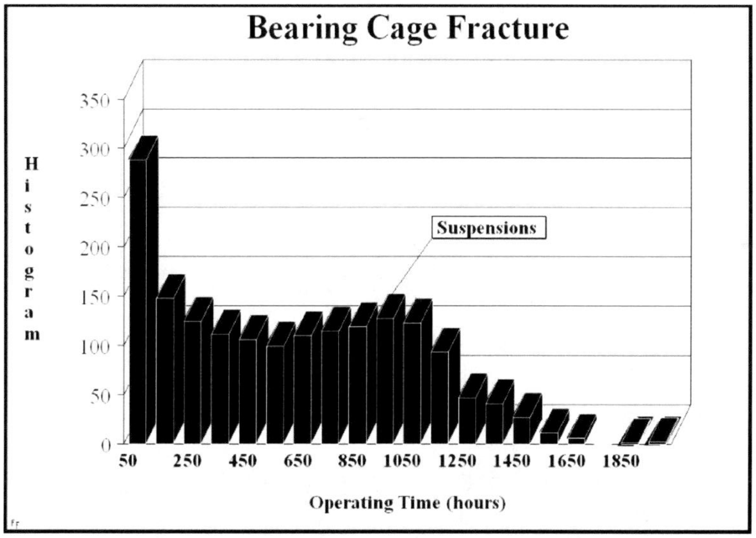

Figure 4-2. Bearing Cage Data from Table 4-3

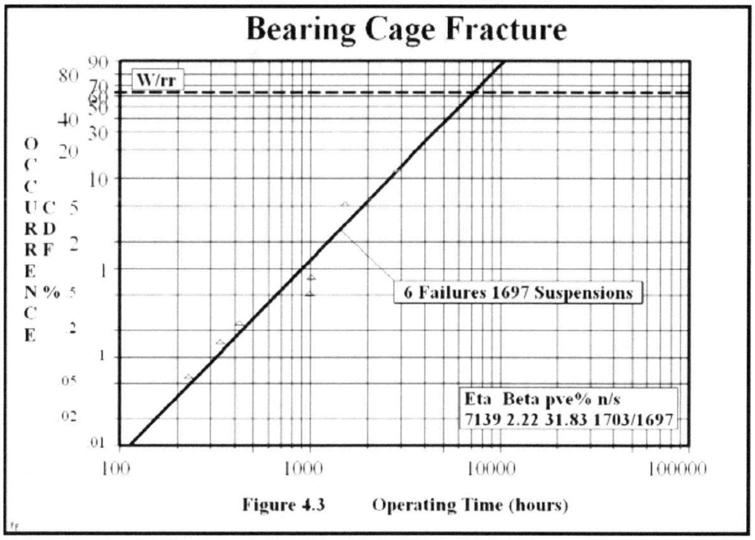

Figure 4-3. Bearing Cage Fracture

The risk questions are:

1. How many failures could be expected before the units reach 1000 hours if all the suspensions were new, zero time parts?

   Calculate the number of units that will fail by 1000 hours, assuming failed units are not replaced. Enter the x-axis of the Weibull plot (Figure 4-3) at 1000 hours and read the y-axis. Approximately 1.3% of the population will fail before 1000 hours. That is, after the entire population of 1703 bearings reach 1000 hours each, 1703(0.013) = 22 bearings would be expected to have failed. (SSW more precisely calculates 21.5 failures).

2. How many failures could be expected in the next year? How many in the next five years?

   Using the methodology explained in Section 4.4 and applying Equation 4-3 with a monthly utilization of 25 hours, or 12(25) = 300 hours in one year, results in the calculations shown in Table 4-3. Thus, about 10.5 more failures can be expected in the next 12 months. Using the same methods SSW calculated and SSV plotted the five-year risk. See Figure 4-4. 102 failures are expected in the next five years. Note that the 12 month forecast is 10.5 failures

3. How many failures could be expected over a 4000 hour period if we replace every bearing at 1000 hours? 2000 hours? *From the answer to Question 1, the probability of a bearing failure by 1000 hours is 0.013. Therefore, if it is assumed that each 1000 hour replacement makes the bearing "good as new" relative to cage fracture, there is a total expectation of failure for each bearing by 4000 hours of approximately 0.013 + 0.013 + 0.013 + 0.013 = 0.052. So, if all 1703 bearings ran to 4000 hours with 1000 hour replacements, 0.052(1703) = 89 failures can be expected.*

Table 4-3. 12 Month Failure Forecast

Number Units (N)	*Current Time on Each (t)*	*Time at End (t + u)*	*(a) F(t)*	*(b) F(t + u)*	*(c) (b - a)/(1 - a)*	*Total of (c*N)*
288	50	350	0.0000	0.0012	0.0012	0.3480
148	150	450	0.0002	0.0022	0.0020	0.2963
124	250	550	0.0006	0.0034	0.0029	0.3607
111	350	650	0.0012	0.0049	0.0037	0.4107
106	450	750	0.0022	0.0067	0.0045	0.4770
99	550	850	0.0034	0.0088	0.0054	0.5346
110	650	950	0.0049	0.0113	0.0064	0.7040
114	750	1050	0.0067	0.0141	0.0074	0.8436
119	850	1150	0.0088	0.0172	0.0084	0.9996
127	950	1250	0.0113	0.0207	0.0094	1.1938
123	1050	1350	0.0141	0.0245	0.0104	1.2792
93	1150	1450	0.0172	0.0286	0.0114	1.0602
47	1250	1550	0.0207	0.0331	0.0124	0.5828
41	1350	1650	0.0245	0.0379	0.0134	0.5494
27	1450	1750	0.0286	0.0431	0.0145	0.3915
11	1550	1850	0.0331	0.0486	0.0155	0.1705
6	1650	1950	0.0379	0.0545	0.0166	0.0996
0	1750	2050	0.0431	0.0607	0.0176	0.0000
1	1850	2150	0.0486	0.0673	0.0187	0.0187
0	1950	2250	0.0545	0.0741	0.0196	0.0000
2	2050	2350	0.0607	0.0813	0.0206	0.0412
			12 MONTH FAILURE FORECAST S = 10.5			

Prediction intervals around the failure forecast may be added but the author does not recommend them as they may be confusing. See Chapter 7.

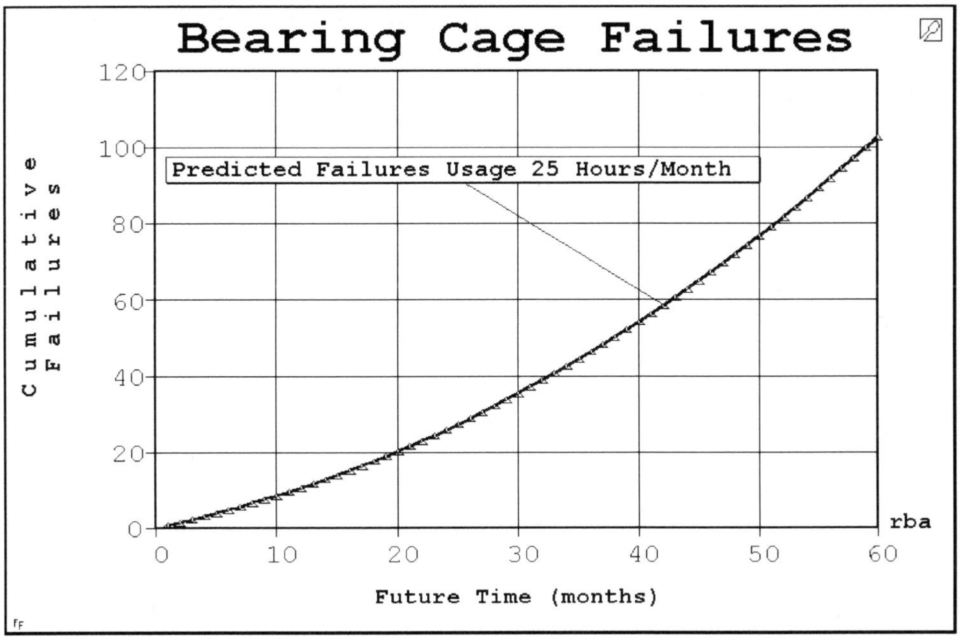

Figure 4-4. Bearing Cage Failures

4.5.2 Case Study 2: Bleed System Failures

Nineteen bleed system failures have been noted and the times and geographical locations of these failures are listed in Table 4-4. Suspensions for Base D are listed in Table 4-5. The high incidence at air base D prompted a 12 month failure forecast.

Table 4-4. Bleed System Failures by Air Base

Air Bases	*Failure (Hours)*	*Air Base D*	*Failure (Hours)*
A	153	D	1198
B	872	D	884
C	1568	D	1251
A	1428	D	1249
A	212	D	708
C	808	D	1082
C	1405	D	884
E	64	D	1105
F	32	D	828
		D	1013

Table 4-5. Base D Units at Risk (Suspensions), Grouped in 100 Hour Intervals

Data Format: (Midpoint Interval x Number of Units)

250 x 2	*550 x 2*	*650 x 2*	*750 x 9*	*850 x 23*
950 x 27	1050 x 20	1150 x 22	1250 x 22	1350 x 11
1450 x 11	1550 x 20	1650 x 8	1750 x 4	1850 x 2
1950 x 3	2050 x 3	2150 x 1		

A Weibull analysis of the fleet failures excluding air base D (Figure 4-5), shows a decreasing failure rate phenomenon, that is, $\beta<1.0$. Weibull analysis of the failures at air base D (Figure 4-6) shows a rapid wear out characteristic. Comparing the β's, it is obvious that the bases are significantly different.

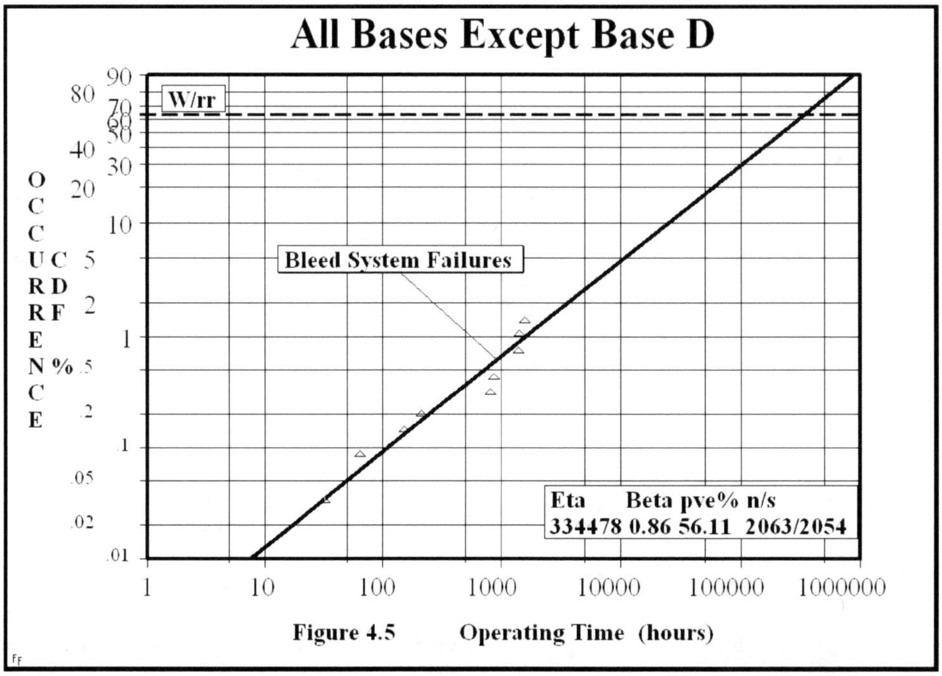

Figure 4-5. All Bases Except Base D

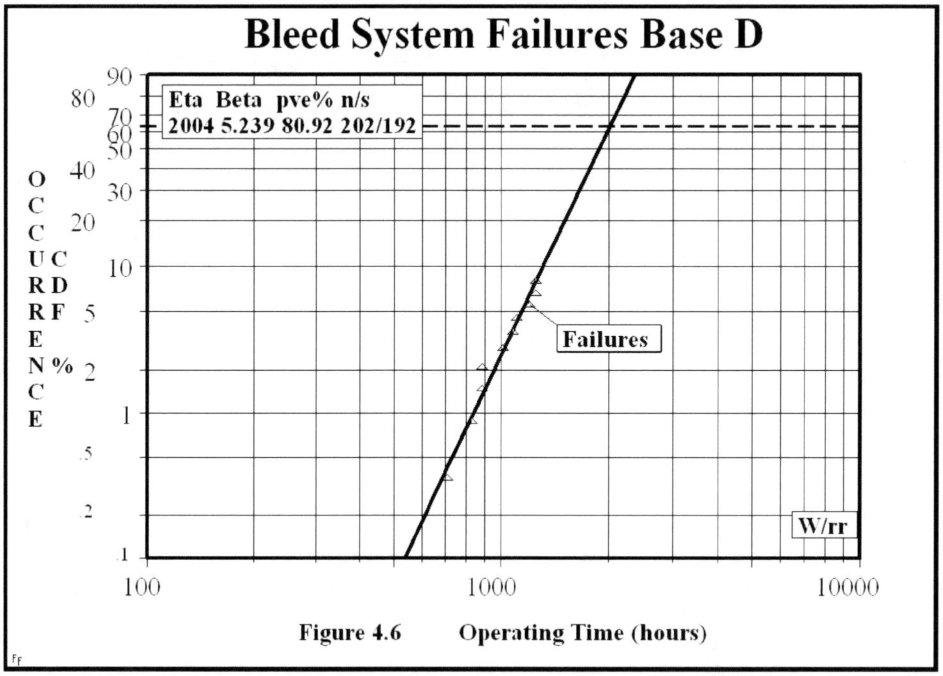

Figure 4-6. Bleed System Failures Base D

Chapter 7 presents methods to statistically show that they are significantly different. Root cause analysis showed the Base D failures were from saltwater corrosion, as the runways were next to the ocean. Since the probability of failure excluding air base D, was acceptable up to 4000 hours (the life limit of the part), a failure forecast for air base D was requested. The failure forecast questions are:

1. What is the expected number of incidents for the next 18 months with usage of 25 hours per month?
Using the histogram of the times on each unfailed bleed system at air base D (Table 4-5) SSW produced Table 4-6, showing the failure forecast for the next 60 months. SSV plotted the result shown in Figure 4-7. The hand calculation is shown (Table 4-7). *Over the next 18 months, 54.7 failures can be expected using a 25 hour per month usage. There is an indication of a batch effect. See Appendix F.*

2. If the usage increases to 50 hours per month immediately, how many more failures can be expected? If the failed units are replaced as they fail what effect would the "renewals" have on the failure forecast?
SSW and SSV were used to plot these figures. Renewal is accomplished by adding a new production unit to the suspension fleet whenever the "without renewal" failure forecast exceeds a unit amount to obtain the "with renewal" failure forecast. The fleet size with renewal is a constant; without renewal the fleet size decreases as units fail. Therefore, the failure forecast with renewal is equal to or higher than without renewal because there are more units at risk. With the lower utilization of 25 hours per month, renewals have no effect as the replacement parts do not age enough within the 60 months to fail.

Table 4-6. Failure Forecast

Weibull Quantity Expected – Risk

Present Risk = 22.02433 Date: M06-D02-YR1998
Bleed System Failures Base D {No Renewal}
Eta = 2004.45 Beta = 5.239307 Total/Suspension = 202/192
Set: #1 Failures Next Expected Occurrence (Day) = +13
Usage Rate [Additional X-Value Age (hours) Each Month / Item] = 25
Expected Additional Occurrence (Cumulative):

Month	Risk	Month	Risk	Month	Risk	Month	Risk
1	2.3	16	47.4	31	105.8	46	157.3
2	4.7	17	51	32	109.7	47	160
3	7.1	18	54.7	33	113.6	48	162.5
4	9.7	19	58.5	34	117.4	49	164.9
5	12.4	20	62.3	35	121.2	50	167.2
6	15.1	21	66.2	36	124.9	51	169.3
7	18	22	70	37	128.6	52	171.3
8	20.9	23	74	38	132.1	53	173.2
9	23.9	24	77.9	39	135.6	54	175
10	27.1	25	81.9	40	139	55	176.7
11	30.2	26	85.9	41	142.3	56	178.2
12	33.5	27	89.9	42	145.6	57	179.6
13	36.9	28	93.9	43	148.7	58	180.9
14	40.3	29	97.9	44	151.7	59	182.1
15	43.8	30	101.8	45	154.6	60	183.2

SSW & SSV will forecast and plot both failures per month and cumulative failures per month as options. With betas greater than one there may be a wave action in the failure forecast plot. Renewals and betas greater than one cause the wave action. Plotting failures per month using SSV shows the wave action, Figure 4-9. The waves are due to the peak failure periods of the original and replacement parts. Maintainability engineers refer to these waves as the "rhythm" of the system. The wave damps out as the system ages beyond several multiples of η. *The steeper the betas, the more multiples it takes to reach the asymptote [Denson]. Figure 4-10 shows the histogram of suspensions from Table 4-5 with the Weibull PDF (probability density function). The intersection of the suspension histogram with the PDF produces failures. This plot is useful to show the proximity of the fleet to the Weibull. Here, the fleet is well into the Weibull producing 20 failures.*

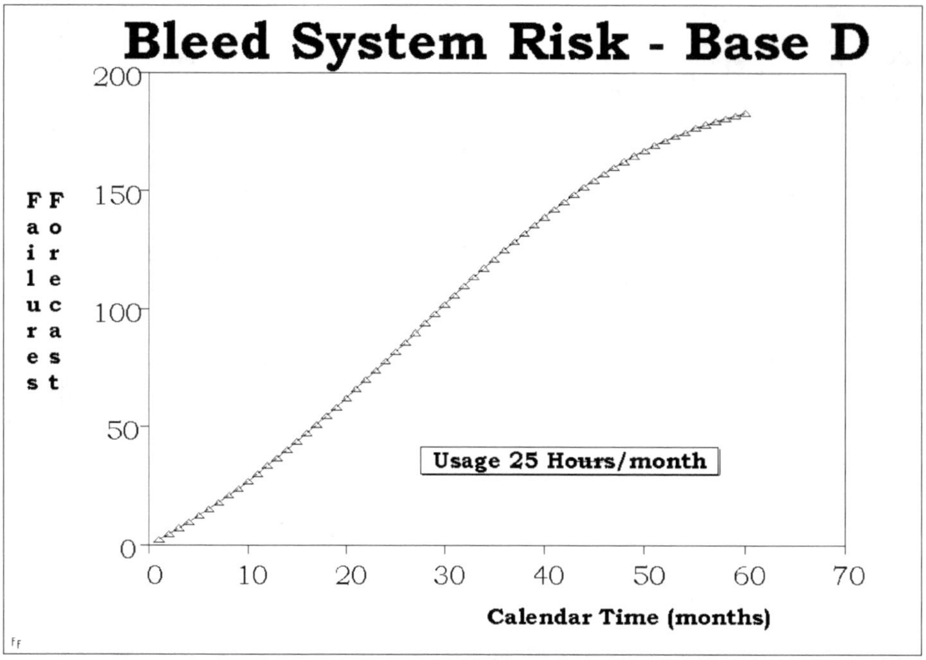

Figure 4-7. Bleed System Risk - Base D

Table 4-7. Bleed System Failure Forecast After 18 Months
(Utilization Rate = 25 Hours Per Month)

Number of Units (N)	*Current Time on Each Unit (t)*	*Time After 18 Months (t + u)*	*(a) F(t)*	*(b) F(t + u)*	*Each Unit's Risk (b - a)/(1 - a)*	*Total Risk (c*N)*
0	50	500	0.0000	0.0007	0.0007	0.0000
0	150	600	0.0000	0.0018	0.0018	0.0000
2	250	700	0.0000	0.0041	0.0041	0.0081
0	350	800	0.0001	0.0082	0.0081	0.0000
0	450	900	0.0004	0.0151	0.0147	0.0000
2	550	1000	0.0012	0.0260	0.0249	0.0497
2	650	1100	0.0028	0.0424	0.0397	0.0794
9	750	1200	0.0058	0.0659	0.0605	0.5445
23	850	1300	0.0112	0.0984	0.0882	2.0286
27	950	1400	0.0199	0.1415	0.1241	3.3500
20	1050	1500	0.0334	0.1965	0.1688	3.3760
22	1150	1600	0.0532	0.2640	0.2227	4.8994
22	1250	1700	0.0810	0.3434	0.2856	6.2832
11	1350	1800	0.1186	0.4328	0.3565	3.9218
11	1450	1900	0.1675	0.5286	0.4338	4.7719
20	1550	2000	0.2287	0.6259	0.5149	10.299
8	1650	2100	0.3023	0.7188	0.5969	4.7752
4	1750	2200	0.3871	0.8016	0.6763	2.7052
2	1850	2300	0.4802	0.8700	0.7499	1.4998
3	1950	2400	0.5774	0.9218	0.8149	2.4446
3	2050	2500	0.6732	0.9573	0.8693	2.6080
1	2150	2600	0.7618	0.9792	0.9125	0.9125
						Sum = 54.7

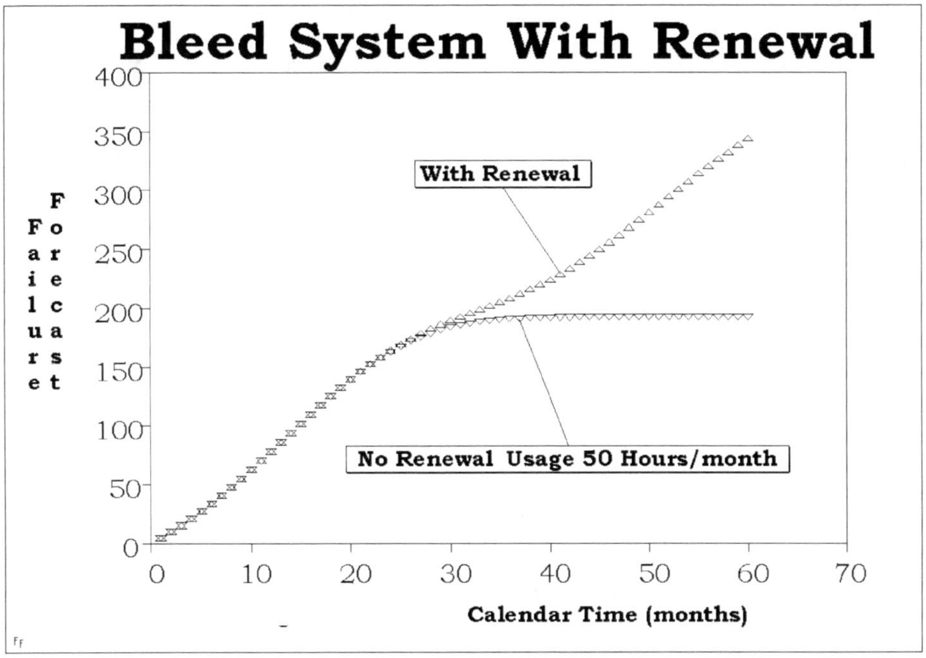

Figure 4-8. Bleed System With Renewal

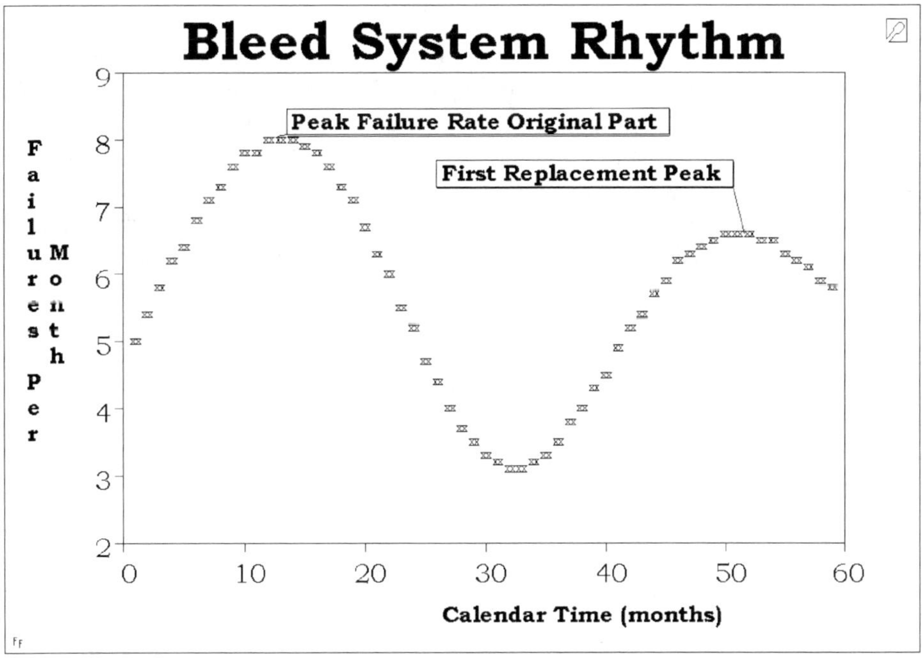

Figure 4-9. Bleed System Rhythm

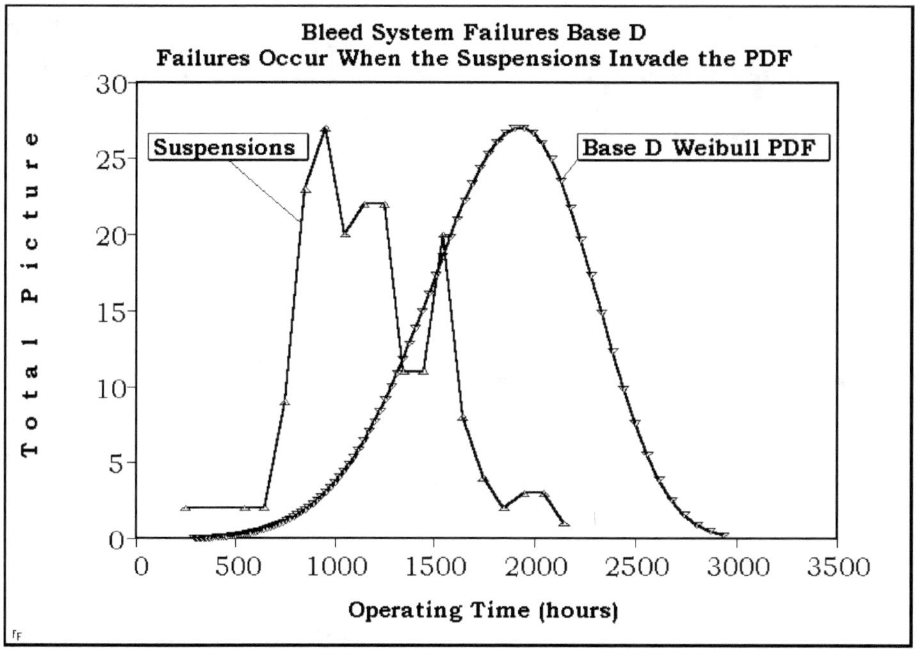

Figure 4-10. Bleed System Failures Occur When the Suspensions Invade the PDF

4.6 System Failure Forecast Without Simulation*

Although the emphasis in this Handbook has been on one failure mode at-a-time, complete system failure forecasts are often required by management. In the past, Monte Carlo methods were required for all system models, but the new SuperSMITH software (SSW + SSV) provides quick, easy solutions under certain conditions that frequently apply. If the system failure modes are (1) renewable, (2) non-catastrophic, and (3) independent, the individual failure forecast predictions may be summed to obtain and plot the system failure forecast with SSV. However, if any of the three conditions are not met, the more complex Monte Carlo simulation may be required. It will be presented in the next section. Examples that may use the simple method include most repairable systems like aircraft, trucks, cars, process plants, refineries, and electric utilities for all failure modes that are not catastrophic. Major and minor subsystems such as air conditioners, pumps, engines, oil systems, bearings, turbines, and compressors may also be modeled in this manner without requiring Monte Carlo simulation. The method is to use SSW to make failure forecasts with renewal for the significant failure modes. SSV sums the individual failure forecast predictions to obtain the system failure forecast.

4.6.1 Case Study 3: Aircraft In-Flight Engine Shutdowns*

Situation: The new top secret stealth fighter, the YF26, is a product of the author's flight test experience and his imagination. The flight test program is just starting. You are the Chief Engineer of the Air Force YF26 Program Office and you are concerned. There are problems with the new F233 engine. Although this is a twin engine airplane, in-flight shutdowns (IFSD) are still hazardous. There is some probability of a dual IFSD, and further, a single IFSD may be non-contained within the nacelle, (engine housing), taking out the other engine, the aircraft, and/or the crew. Crew safety is the top priority. Too many IFSD's would put program approval in jeopardy. There are four known IFSD failure modes:

1. The digital electronic fuel control has been showing quality problems in production acceptance and early service. The control manufacturer has initiated a new "6 Sigma Quality Program" and assures you that the QC problems are fixed. It is true that on your recent visit you saw the assembly people all decked out in new 6 Sigma t-shirts and caps, but you tend to be skeptical. The Weibull parameters are $\beta = 0.5$ and $\eta = 30000$ Engine Flight Hours (EFH).

2. The second stage titanium compressor blade has failed three times, "corncobbing" the rotor. The failure analysis indicates foreign object damage (FOD) from the test stand duct work produced the problem. You are worried that the runway FOD may induce more failures, but the manufacturer is not concerned. You argue that there are other FOD worries such as bird strikes and desert sand storms. You want to start a redesign of a stronger blade. The Weibull parameters are $\beta = 1.0$ and $\eta = 1200$ EFH.
3. The main shaft #2 bearing has failed once. Ball bearings of this type usually have a β of 2.0, based on your Weibull library. Although this failure was contained within the engine cases, there is some probability that this failure could be non-contained. A one point Weibayes provides $\eta = 700$ EFH. The bearing manufacturer says this was just a bad batch of nickel steel and has taken corrective action to prevent any more bad batches. Your bearing expert remarked that he's heard that song before, many times. Still, you hope this problem is under control.
4. The second turbine blade had two failures in R&D testing, both were contained within the engine cases. The cause is stress corrosion cracking. The materials laboratory analysis says a coating will fix the problem but the airframe manufacturer argues that sand ingestion will remove the coating. The Weibull analysis shows $\beta = 6.0$ and $\eta = 600$ EFH. The manufacturer predicts a benign period of two years before the onset of this problem. By then he will have new redesigned blades to retrofit the fleet of 20 engines if you authorize additional funding.

You are worried and all this is keeping you up at night. If the YF26 crashes outside the secure area, there will be a media explosion over this black project and the apparent unreliability. There might be crew fatalities. All this could inspire Congress to cancel the project. Of course, with a two engine plane and fixes on the way, you shouldn't worry so much. The alternative is to delay flight test until the fixes are substantiated by ground test. You will need courage to face the "three star" general with that recommendation. Whatever you decide, you need accurate analysis as the basis for your decision:

Information available for your analysis includes the four Weibulls plus some assumptions:

1. Worst case: all the fixes are "NG," no good.
2. The 20 engines are almost new: 10 have 10 EFH and 10 have 20 EFH.
3. The average usage will be 25 EFH per month. Average flight = 1 hour = 2 EFH.
4. The engine is modular design. Unscheduled failures result in the module being overhauled or replaced, in any event, good-as-new, zero timed.
5. 1% of the failures induce catastrophic crashes. 79% of dual IFSD aircraft land, "deadstick-in," safely...somewhere. Failure events are independent.
6. Scheduled replacement cost of either the bearing or the 2nd turbine blades is $40,000. Unscheduled failures of either component cost $1,000,000.

Weibull Failure Modes

Failure Mode	*Control*	*FOD*	*Bearing*	*Turbine*
η	30000	1200	700	600
β	0.5	1.0	2.0	6.0

Questions and Solutions: The most important consideration in flight test is crew safety.

1. Which failure mode is the driver? Where should we concentrate the effort?

Use SSW to do a failure forecast for each of the four failure modes. Plot the results with SSV. See Figure 4-11. The turbine blade and the bearing failure modes are the worst, almost equally bad.

Note the turbine blade does have an initial benign period. Failures start at about 18 months. These two problems should have priority.

2. How many failures will there be, by failure mode and total, during the first three years? Will things get better or worse?
Things will get worse and worse through the second year. Failures by month are shown in Figure 4-12.

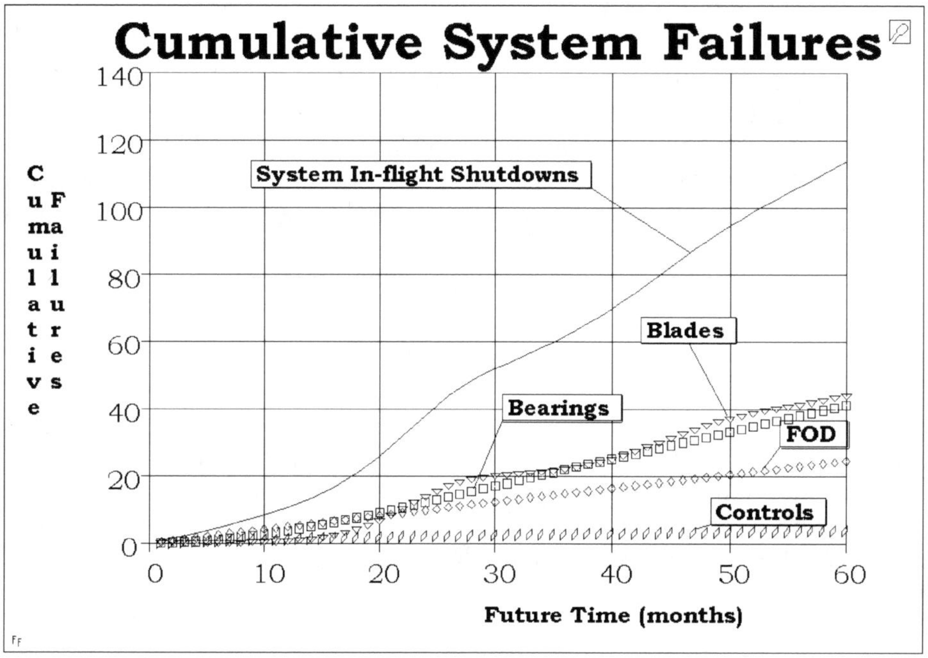

Figure 4-11. Cumulative System Failures

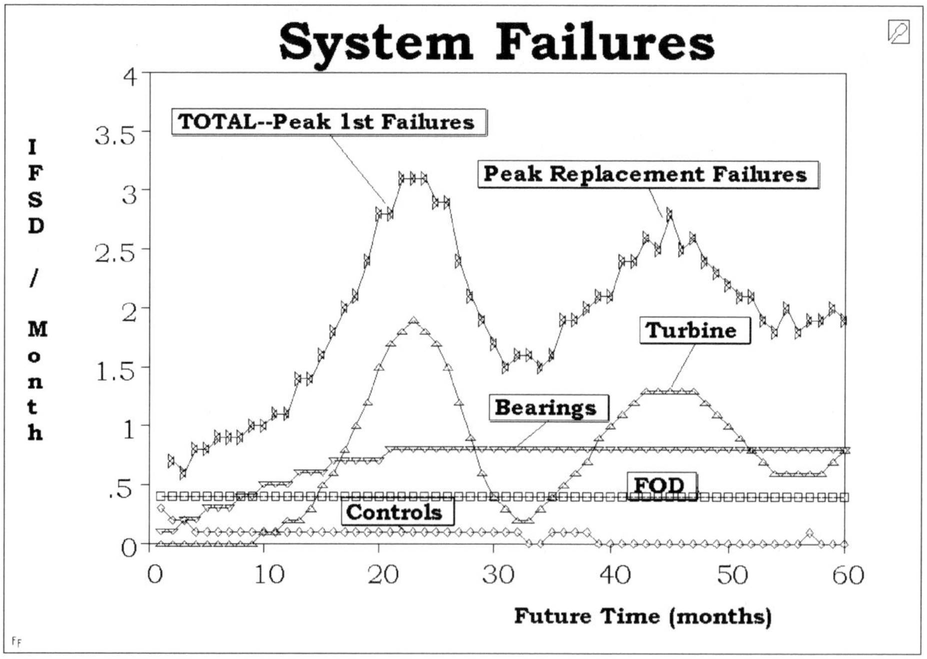

Figure 4-12. System Failures per Month

3. What is probability of dual IFSD, early in the program, say the first year? Does it get better in later years? If the probability of a successful dead stick landing is 0.79, how many dual IFSD crashes do you expect in the first year?
There are 10.6 IFSDs predicted the first year. See Figure 4-11. With 6000 EFH, and one hour flights, there will be 3000 flights. The probability of an IFSD per flight is 10.6/3000 = 0.00353 the first year. This value squared is the probability of a dual IFSD, on a single flight, 0.0000125, which is very small. In 3000 flights, the probability is 0.0368, or about 4 in 100 odds.
$[1-[1 - 0.0000125]^{3000}] = 0.0368$
The probability of a crash from a dual IFSD the first year is 0.0368(1 - 0.79) = 0.008.

4. If 1% of the failures cause catastrophic damage, how many crashes will there be in the first year?
The odds of a crash from a single IFSD inducing a catastrophic failure are higher than the odds of a dual IFSD, 10.6 x 0.01 = 0.1 or about one chance in ten.

5. What test substantiation will you require to be convinced that the fixes work?
Does the control vendor have a Pareto plot by failure cause? The beta hints at inadequate environmental stress screening (ESS). Hats and t-shirts alone will not fix the problem if the ESS is inadequate. Perhaps a production pilot lot of ten controls with more ESS would be in order. Close examination of the production acceptance failures, with and without corrective action is required. FOD may be reduced in flight test by more vacuuming of the runways and eliminating the local bird population. Flight test should be restricted when sand storms or icing are forecast. The compressor blade should be beefed up in the redesign. The bearing and turbine blade are the highest priority as shown in Figure 4-11. Several accelerated mission tests (AMT) should be scheduled for both the new bearing and the new blade. The AMT's duration should exceed 900 EFH, the engine exposure in flight test, by a significant amount, to find unknown failure modes that might occur before 900 hours. 1800 EFH might provide adequate assurance that there are no unknown wear out modes with a tail below 900 hours.

6. Specifically, what do you recommend? Should you delay the flight test program?
There are other considerations besides Weibull analysis. How urgent is the need for the new weapon system? How serious is the threat? Are the required electronics and weapons systems on schedule because if they are late there is less reason to start flight test? Any delay will weaken the USAF public image and encourage Congress to cancel the project. Your position and the YF26 project are on the line. The Air Force Chief of Staff and Congress should understand your analysis and hopefully support your recommendation whatever it is. Is it "go?"

4.7 System Failure Forecasts With Simulation*

Monte Carlo simulation is used to model systems that have failure modes that do not conform to one or more of the characteristics: (1) renewable, (2) non-catastrophic, and (3) independent. Such failure modes may involve dependent failure modes, maintenance actions, or inspections. The analytical solutions are difficult or gross approximations at best. Examples:

- Failure modes that induce secondary damage.
- Catastrophic failures such as missile or rocket engine failures.
- Replacement inspections although this may be handled with complex calculations.
- Inspections looking for cracks or defects which are not always found.
- Complex retrofit or recall programs.
- Opportunistic maintenance for parts replacement, i.e., when you are down for any other problem do the retrofit, replace the parts.
- Variable customer usage, mission mix, abusive maintenance.

Section 4.7.1 provides an example using hand calculations. Fortunately, there is excellent software available that will model these effects easily and quickly although the program availability may change and maintenance/updating of these programs should be double-checked before use.

The SSW Monte Carlo Sampling option provides quick look capability for small system modeling for several different types of systems. This idea was presented by Julius Wang of Chrysler at the 1994 SAE Weibull Users Conference. For example, assume a system has the four Weibull failure modes described in Section 4.7.1: (1) Over temperature, (2) Vane and case cracking, (3) Oil Tube cracking, and (4) Combustion chamber cracking. Table 4-8 shows the results of using the SSW sampling option to generate 5 random failures from each mode with Monte Carlo methods.

Table 4-8. Five Systems Failures From Four Modes
Italics indicate the initial failure in each system

	β	η	*System A*	*System B*	*System C*	*System D*	*System E*
1-Over Temperature	2.09	10193	12723	14802	10346	11476	3949
2-Vane & Case	4.57	2336	*2859*	*2440*	*1884*	*1837*	*2880*
3-Oil Tube	1.885	12050	10220	11778	8158	17279	6783
4-Combustion	4.03	3149	3350	4235	3201	2926	2904

Scenario 1: Assume the data in Table 4-8 are times-to-failure for missile engines. Every failure is catastrophic. [Time is in seconds]. Question: Which mode needs to be fixed most urgently?

Answer: The first failure for each system destroys the system. Failure mode (2), vane and case cracking is responsible in every case and needs urgent attention.

Scenario 2: Concentrating on failure mode #3, these are times-to-failure for oil tube cracking in automotive oil pumps. The responsible engineer proposes changing the oil tubes on the pumps every 5000 hours as it has the shallowest slope and therefore must be the troublemaker. Question: Is this a good proposal?

Answer: The probability of failure for the oil tube at 5000 hours is about 18%. See Figure 4-13. This is too high. The tube should be replaced much earlier than 5000 hours. If the cost of a scheduled replacement is $100 and the cost to replace a burned-out engine is $2000, the optimal replacement is at 2800 hours. (Using SSW Optimal Replacement option, see section 4.8).

Scenario 3. This is the life history of <u>one</u> repairable system, an automobile. Time is in miles. "System A" is the life of the original four parts, "System B" provides the life of the first replacement parts, etc. For example, for the second failure mode, the first replacement part is expected to fail at 2859 + 2440 = 5299 miles. The warranty extends to 10000 miles. Question: How many warranty claims will there be?

Answer: Mode (1) will produce zero claims. Mode (2) four claims. Mode (3) zero claims. Mode (4) two claims. Total = 6 claims.

Sometimes a part's service life will depend on decisions to be made in the future. Monte Carlo simulation is a convenient method for simulating the decision process. It enables an analyst to build a computer model of the decision plan effect on a part's service life. For example, one complication is the <u>repaired</u> unit. The repaired part may have a degraded Weibull compared to a new replacement part. There may be repair or replace decisions depending on the degree of damage and some decisions will be wrong. Scheduled inspections to find cracks, wear or erosion, adding new units, repairs of parts, even the reliability of the inspection process may be simulated. For example, what is the reliability of an inspector detecting a 5 millimeter crack with eddy current, given the crack exists?

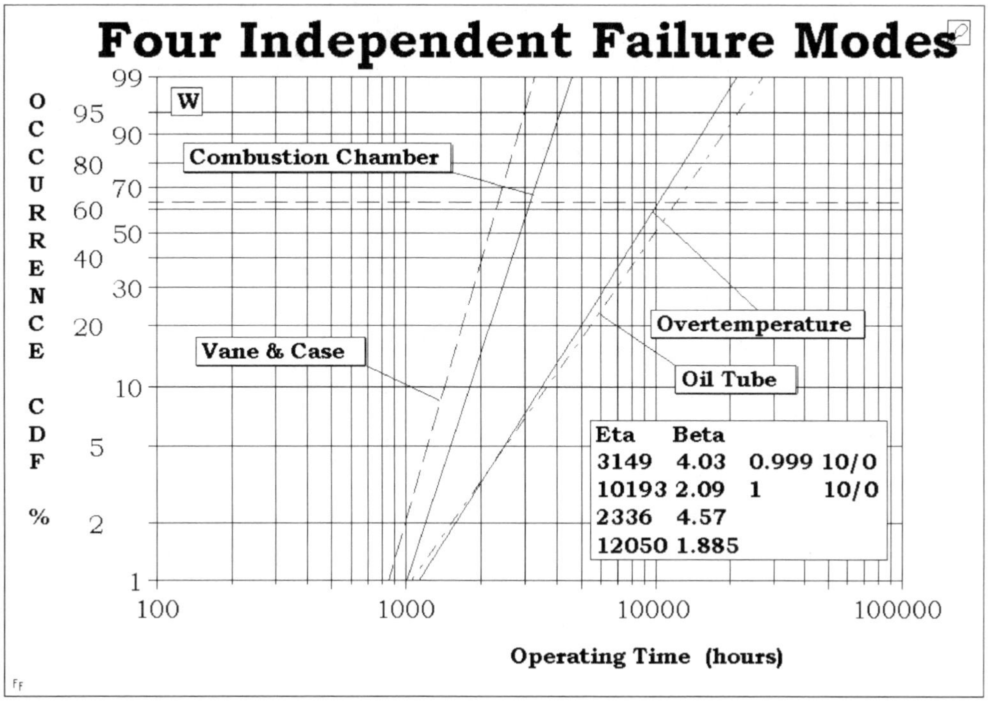

Figure 4-13. Four Independent Failure Modes

More complex inspections may involve several probabilities. Does the part have a flaw? Does the inspection find the flaw? Is there a repair or replace decision? Is the new part identical or a redesign? What is the effectiveness of the redesign? For example, consider Figure 4-8. If the bleed system was redesigned and 100% effective in eliminating the problem, the failure forecast would follow the "no renewal plot." If it is 0% effective, no better than the old design, the forecast will follow the "renewal" forecast. 50% effective will be half way in between.

4.7.1 Case Study 4: System Failure Forecast With Simulation*

Assume a jet engine has four independent failure modes:

- Over temperature: $\beta = 2.09$, $\eta = 10193$ (hours)
- Vane and Case cracking: $\beta = 4.57$, $\eta = 2336$
- Oil Tube cracking: $\beta = 1.885$, $\eta = 12050$
- Combustion chamber cracking: $\beta = 4.03$, $\eta = 3149$.

Estimate the number of failures occurring for each mode before the scheduled inspection. The occurrence of any single failure mode will not affect any other mode. For each unit in the population, use a random number for each of the four Weibull failure distributions to generate a time-to-failure for each failure mode. Use Equation 4-4. Standard random numbers range from zero to one and may be found in Table 4-9. All numbers within the range are equally likely to occur.

Rearranging the Weibull CDF, Equation (2-1), to solve for time-to-failure, you may substitute Random Numbers for F(t):

$$t = \eta * (\ln(\frac{1}{(1-F(t))})^{(\frac{1}{\beta})} \equiv \eta * (-\ln(1-F(t))^{(\frac{1}{\beta})} \qquad \text{Equation (4-4)}$$

Table 4-9. Standard Random Numbers From 0.0 to 1.0

0.007	*0.028*	*0.517*	*0.603*	*0.442*	*0.845*	*0.494*	*0.624*	*0.085*	*0.194*
0.612	0.337	0.393	0.163	0.769	0.620	0.596	0.503	0.857	0.794
0.545	0.945	0.357	0.429	0.124	0.675	0.689	0.203	0.643	0.577
0.232	0.511	0.311	0.213	0.390	0.827	0.354	0.556	0.811	0.811
0.221	0.480	0.345	0.167	0.837	0.987	0.428	0.257	0.298	0.198
0.210	0.457	0.010	0.083	0.133	0.265	0.638	0.948	0.747	0.164
0.519	0.668	0.717	0.230	0.815	0.672	0.658	0.491	0.772	0.676
0.166	0.037	0.971	0.169	0.655	0.876	0.668	0.649	0.205	0.551
0.138	0.601	0.761	0.490	0.748	0.238	0.277	0.123	0.918	0.984
0.214	0.738	0.224	0.706	0.928	0.090	0.389	0.699	0.562	0.761
0.418	0.422	0.402	0.270	0.854	0.982	0.365	0.933	0.323	0.367
0.950	0.469	0.709	0.431	0.364	0.363	0.574	0.630	0.521	0.974
0.202	0.503	0.434	0.394	0.851	0.909	0.168	0.058	0.673	0.012

There is an alternate graphical procedure to determine a random time-to-failure. Enter the Weibull plot horizontal at a B life corresponding to the random number in %, cross over to the Weibull line, and drop vertically to read the time-to-failure. This is a graphical analog of Equation 4-4. After many repetitions of this process, each using different random numbers, the distribution of the expected failures, the failure forecast, is defined.

Figure 4-13 illustrates the failure distribution of each of these modes. Note that the aging parameter is total operating time (TOT). There is a scheduled parts replacement at 1000 hours, TOT. At failure, the failed part is replaced with a zero time part. At scheduled replacement inspections, all parts are replaced with new, zero time parts.

1. How many failures can be expected in each mode over the next 2 years? (Assuming a usage rate of 25 hours TOT/month)
2. How will the risk change if the replacement interval is increased to 1200 hours?

To provide more details on a Monte Carlo simulation based on these ground rules, one engine starting with 0 hours, will be followed in detail to the first scheduled replacement at 1000 hours.

<u>**Step 1**- Draw random numbers and transform them into times-to-failure</u>

Generate random times to failure for each failure mode. First, from the Table 4-9, take four standard random numbers in the 0 to 1 range: 0.007, 0.028, 0.517, and 0.603. Using Equation 4-4:

$$T_1 = \text{Over temperature} = 10193\left[\ln\left(\frac{1}{1-0.007}\right)\right]^{1/2.09} = 951 \text{ hours}$$

$$T_2 = \text{Vane and case cracking} = 2336\left[\ln\left(\frac{1}{1-0.028}\right)\right]^{1/4.57} = 1072 \text{ hours}$$

$$T_3 = \text{Oil tube cracking} = 12050\left[\ln\left(\frac{1}{1-0.517}\right)\right]^{1/1.885} = 10180 \text{ hours}$$

$$T_4 = \text{Combustion chamber cracking} = 3149\left[\ln\left(\frac{1}{1-0.603}\right)\right]^{1/4.03} = 3088 \text{ hours}$$

Steps 2 & 3 - Compare the times-to-failure with the replacement time

The minimum of the times-to-failure and replacement time is 951 hours for an Over temperature failure; therefore, failure occurred before the scheduled replacement at 1000 hours. The other three parts did not fail before they were replaced at 1000 hours.

Step 4 - Draw more random numbers if necessary

This first failure was an Over temperature (F1), recorded as occurring at 951/(25 hours usage) = 38 months in the future. Generate a second time to failure for T_1, using the next random number, 0.442.

$$\text{New } T_1 = 10193\left[\ln\left(\frac{1}{1-0.442}\right)\right]^{1/2.09} = 7876 \text{ hours elapsed time}$$

The clock-time to the second failure is 951 + 7876 = 8827 total hours or 8827/25 = 353 months from zero time. Now, the minimum of (T_1, T_2, T_3, T_4) is greater than 1000 hours, the scheduled replacement. This process can be continued for as many replacement intervals as desired.

For engines with greater than zero hours initially, the Monte Carlo process must be modified. First, the time since the last 1000 hour replacement is the engine's initial age (since engines are "good as new" at each 1000 hour replacement). Note that the first set of four random failure times must be greater than the engine's initial age.

If any are less, it means that the part replaced at inspection has failed and was replaced. To find when the new part will fail, take a second random number using the first failure time as the zero time. Follow the above procedure for each engine in the population. Repeat the entire calculation several times so that an average risk can be calculated.

Table 4-10 shows the detailed results of running the simulation for the 1000 hour replacement. The failure forecasts for the first 24 months are presented. Table 4-11 shows the population at risk, the histogram of 1779 suspensions. Engine availability is 65.7% so engines average 16.42 hours per month TOT, (25 hours times .6571). Summary results for 1200 hour replacement are also shown. Compare the totals on the bottom of each table. As all four parts have wear out failure modes, $\beta > 1$, a longer inspection period should increase the number of failures. Increasing the replacement interval to 1200 hours increases the expected number of failures from 25 to 34, a delta of 9, by the end of the second year. Parts that have infant mortality or random failure modes should not be replaced as these parts do not degrade as they age. See Case Studies in Chapter 11 for more failure forecast analysis examples.

4.8 Optimal (Lowest Cost) and Block Replacement Intervals[1]*

Optimal Replacement Intervals:

If a part has a Weibull wear out distribution and the cost of unplanned failures is much greater than the cost of planned replacements, there is an optimal replacement interval. Replacing the part at the optimal replacement interval results in minimum cost. If the interval is too short, the replacement costs are too high; if it is too long, the unplanned failures drive the total cost too high. Weibull replacement strategy finds the most cost effective interval for replacing the hardware. When a breakdown occurs before or at the optimum replacement interval, replace the part.

[1] Advanced material may be omitted on first reading.

Table 4-10. Cumulative Failure Forecast for 1000 Hours Replacement

Month	Hours/Month*	Oil Tube β = 1.885, η = 12050	Vane Case β = 4.57, η = 2336	Overtemp β = 2.09, η = 10193	Combustor β = 4.03, η = 3149
Jan	29,225	0.00	0.00	0.00	0.00
Feb	29,225	0.17	0.33	0.21	0.17
Mar	29,225	0.38	0.67	0.47	0.34
Apr	29,225	0.60	1.15	0.74	0.58
May	29,225	0.78	1.47	0.95	0.74
Jun	29,225	0.92	1.71	1.13	0.87
Jul	29,225	1.22	2.27	1.49	1.15
Aug	29,225	1.46	2.81	1.77	1.41
Sep	29,225	1.66	3.16	2.02	1.60
Oct	29,225	1.95	3.90	2.36	1.95
Nov	29,225	2.07	4.03	2.51	2.03
Dec	29,225	2.38	4.90	2.87	2.45
Jan	29,225	2.66	5.55	3.19	2.77
Feb	29,225	2.77	5.83	3.32	2.91
Mar	29,225	2.87	6.13	3.44	3.05
Apr	29,225	3.07	6.68	3.67	3.31
May	29,225	3.28	7.33	3.91	3.62
Jun	29,225	3.37	7.48	4.02	3.69
Jul	29,225	3.64	8.26	4.33	4.06
Aug	29,225	3.70	8.45	4.40	4.15
Sep	29,225	3.76	8.59	4.47	4.21
Oct	29,225	3.80	8.59	4.47	4.21
Nov	29,225	4.16	9.40	4.95	4.62
Dec	29,225	4.44	9.96	5.29	4.90

Failure Forecast for 1000 Hour Replacement From Table 4-10

1000 Hour 1 Year Total = 2.38 + 4.90 + 2.87 + 2.45 = 12.6 Failures

1000 Hour 2 Year Total = 4.44 + 9.96 + 5.29 + 4.90 = 24.59 Failures

***Hours/Month = 1779 Engines x 25 Hours/Month x 0.657 Availability = 29,225**

Failure Forecast for 1200 Hour Replacement

1200 Hour 1 Year Total = 2.35 + 8.25 + 3.08 + 3.69 = 17.37 Failures

1200 Hour 2 Year Total = 4.46 + 16.47 + 5.32 + 6.94 = 33.19 Failures

Table 4-11. 1779 Suspensions in 50 Hour Intervals

Data Format: (Midpoint in Hours x Number of Units)

25 x 118	75 x 14	125 x 39	175 x 44	225 x 34
275 x 34	325 x 47	375 x 53	425 x 45	475 x 50
525 x 43	575 x 40	625 x 29	675 x 44	725 x 43
775 x 39	825 x 44	875 x 40	925 x 45	975 x 44
1025 x 47	1075 x 55	1125 x 57	1175 x 66	1225 x 58
1275 x 59	1325 x 51	1375 x 67	1425 x 59	1475 x 50
1525 x 72	1575 x 56	1625 x 32	1675 x 31	1725 x 35
1775 x 24	1825 x 15	1875 x 15	1925 x 13	1975 x 7
2025 x 4	2075 x 5	2125 x 5	2175 x 3	2225 x 3
2275 x 0	2325 x 0	2375 x 0	2425 x 0	2475 x 1

The optimal interval is the age with the minimum ratio of the mean cost to the mean time to failure, C(t). The following is an approximate method as it depends on the size of the intervals that are numerically integrated. SSW uses 60 intervals. The input is:

C(t) = Cost per unit time
U = Cost of an unplanned on-line replacement,
P = Cost of a planned off-line replacement before failure, where P<U
Weibull parameters β, and η, provide the reliability parameter:

$$R(t) = e^{-(t/\eta)^{\beta}}$$

The cost per unit time is the ratio of the mean cost to MTTF. The function can be expressed as follows:

$$C(t) = \frac{P*R(t)+U*(1-R(t))}{\int_0^t R(x)dx} = \frac{P*R(t)+U*(1-R(t))}{\int_0^t e^{-(x/\eta)^{\beta}}dx} \qquad (4\text{-}5)$$

where t = the replacement period and x is integrated from 0 to t. The first term in the numerator is the planned replacement cost multiplied by the fraction surviving the period t. This term decreases with time. The second term in the numerator is the unplanned replacement cost multiplied by the fraction failing during the interval. The term increases in value with time. The denominator is the MTTF within the replacement interval, (the area under the reliability curve from 0 to the replacement interval, t). If β is greater than one and the cost ratio is greater than one, the cost per unit time will have a practical minimum [Glasser 1969], [Stanton & Sharp 1991]. The first equation in (4-5) applies to ANY distribution, while the second equation is specific to Weibull, substituting in **R(t)** from above.

This method is not practical for replacement intervals beyond the characteristic life as the cost savings will be negligible.

For example, let β = 6, η = 100, U = $10, P = $1, and let t go from 1 to 90 time units. Table 4-12 shows the SSW output. SSV plotted Figure 4-14, which shows an optimal replacement interval of about 53 time units. Note there is a range of time with little change in the cost parameter, allowing some flexibility in the maintainability plan. The cost ratio of 10:1 is small in the author's experience. Twenty to one is more common and for expensive systems like large trucks and aircraft the ratio may exceed 100:1. Higher cost ratios and steeper betas make the optimum better defined. See also Chapter 11 Case Studies. The more complex equations for the three parameter model have not been developed.

If a system or subsystem has more than one independent wear out failure mode, there may be an optimal system replacement interval. In other words it may be cost effective to consider shutting down the system once for all the replacements instead of several replacement shutdowns. Again the individual cost per unit time plots may be summed. Here the cost of the planned replacement for each mode should be reduced to reflect the lower cost of a single system replacement. SSV will add the costs (C(t)) from each mode to produce the system optimum.

Table 4-12. Weibull Optimum Replacement Interval

RESULTS Date: M08-D07-YR1998
Eta = 100 Beta = 6
Cost / Item [$] Planned = 1 Cost / Item [$] Not Planned = 10
At Optimum (52.5) Steady-State: Replacement Rate [/Unit-Units] = 1.865293E-02
Failure-Rate (FR) [/Unit-Units] = 3.946914E-04 MTBF (1/FR) = 2533.625
Cost/Time [$/Units]:

Units........	Cost	Units........	Cost	Units........	Cost	Units........	Cost
1.5	0.6667	24	0.04174	46.5	0.02349	69	0.02829
3	0.3333	25.5	0.03931	48	0.02316	70.5	0.02947
4.5	0.2222	27	0.03717	49.5	0.02291	72	0.03078
6	0.1667	28.5	0.03526	51	0.02275	73.5	0.03221
7.5	0.1333	30	0.03356	52.5*	0.02267	75	0.03377
9	0.1111	31.5	0.03203	54	0.02269	76.5	0.03547
10.5	0.09524	33	0.03066	55.5	0.02279	78	0.03729
12	0.08334	34.5	0.02943	57	0.02299	79.5	0.03925
13.5	0.07408	36	0.02833	58.5	0.02328	81	0.04133
15	0.06667	37.5	0.02735	60	0.02367	82.5	0.04354
16.5	0.06062	39	0.02647	61.5	0.02417	84	0.04587
18	0.05557	40.5	0.02569	63	0.02477	85.5	0.04831
19.5	0.05131	42	0.025	64.5	0.02548	87	0.05087
21	0.04766	43.5	0.02441	66	0.0263	88.5	0.05352
22.5	0.0445	45	0.02391	67.5	0.02723	90	0.05626

*indicates the optimal replacement interval

The effect of optimal parts replacement on the failure forecast is usually significant. SSW provides failure forecasts with and without optimal replacement. See Figure 4-15 for a comparison using the optimal replacement calculated above.

SSW will also estimate the improvement in MTBF for planned replacements. This technology [Weiss] was brought to our attention by **Randy Pirtle of Honeywell.**

$$\boldsymbol{MTBF = \left[\int_0^T R(t)dt\right] / \big(1 - R(t)\big)}$$

where T is the planned replacement interval.

This equation gives the Mean Time to Failure for product undergoing renewal at intervals of T. The assumption is that the renewal returns a product good as new. With this capability a comparison can be made between MTTF with and without replacement. [Note: The use of Mean Time To Failure in this paragraph is not consistent otherwise with its use herein. We would call this MTBF as it is a repairable system. However, Honeywell insists on calling it Mean Time to Failure instead of MTBF. In Weiss's paper he refers to it as "the expected time to failure" if the part is replaced at time T.]

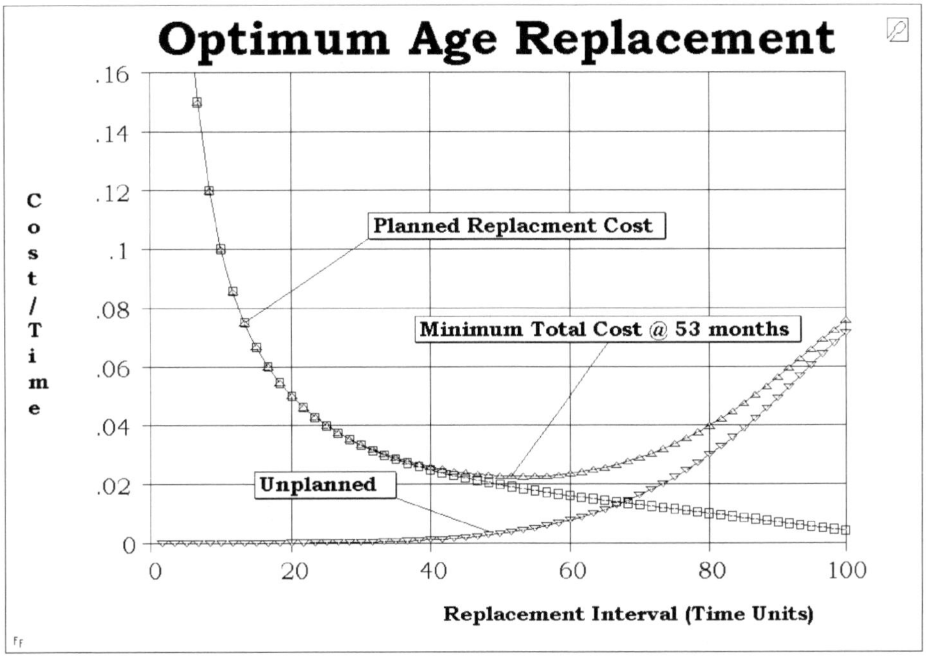

Figure 4-14. Optimum Age Replacement

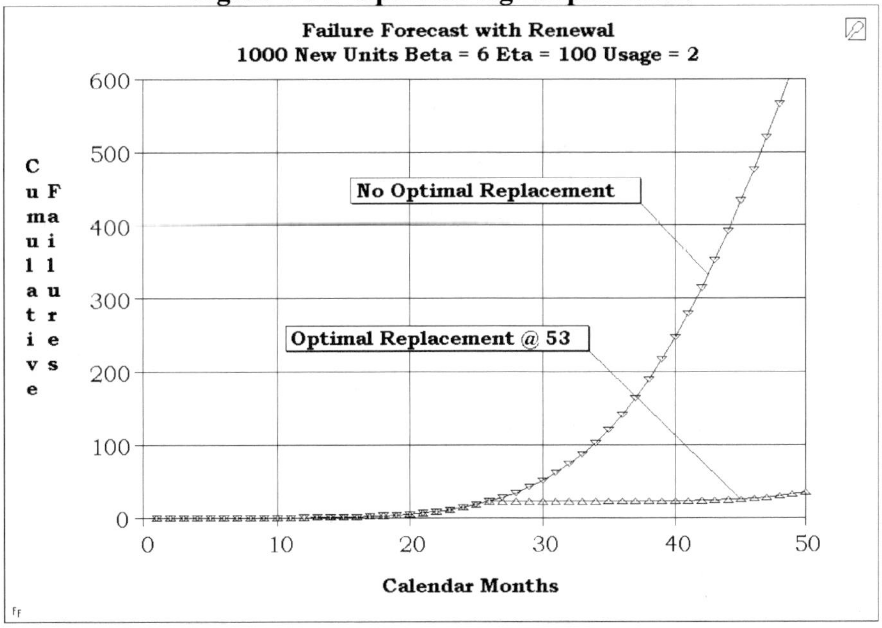

Figure 4-15. Failure Forecast with Renewal 1000 New Units Beta = 6 Eta = 100 Usage = 2

For example, consider the aircraft in-flight engine shutdown case study 3, Section 4.6.1. The manufacturer recommends replacing the bearings and turbine blades every 300 hours until they get the new designs. Would you approve this recommendation? *Do an optimal replacement table using SSW for the two wear out failure modes. Use the SSV transform function to sum the Y values to obtain the system optimal interval. See Figure 4-16. The analysis does not confirm the manufacturer's 300 hour recommendation. It would appear that 200 hours would have a slightly lower cost. Removing the engine once to replace both parts is cheaper than two removals. Further analysis should estimate the number of IFSD's, and catastrophic events at 200 hour versus 300 hour replacement. Spare parts availability and maintenance labor may also be constraints. More spare parts and labor are needed for the 200 hour replacement than a 300 hour replacement.*

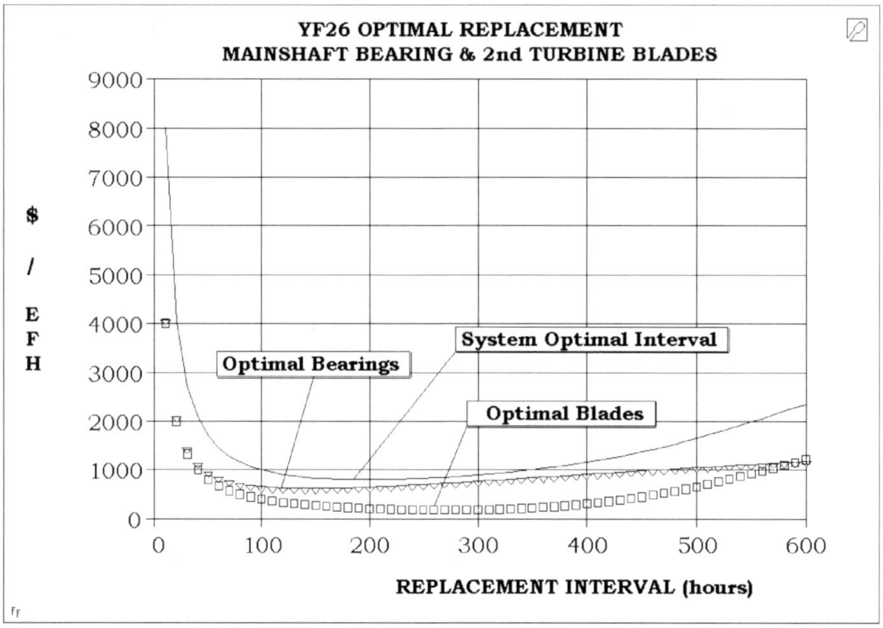

Figure 4-16. YF26 Optimal Replacement Mainshaft Bearing and 2nd Turbine Blades

Table 4-13. Block Replacement

Weibull Optimum Replacement Interval (Block)03-11-1996							
Eta = 100 Beta = 6 Cost Each ($) Planned = 1 Cost Each ($) Not Planned = 10 Cost/Time [$/(Units)]:							
(Units)........Cost		(Units)........Cost		(Units)........Cost		(Units)........Cost	
1.5	0.6667	24	0.04175	46.5	0.02367	69	0.02932
3	0.3333	25.5	0.03932	48	0.02337	70.5	0.03057
4.5	0.2222	27	0.03718	49.5	0.02315	72	0.03195
6	0.1667	28.5	0.03528	51	0.02303	73.5	0.03345
7.5	0.1333	30	0.03358	52.5	0.02299	75	0.03507
9	0.1111	31.5	0.03206	54	0.02305	76.5	0.03681
10.5	0.09524	33	0.03069	55.5	0.02321	78	0.03867
12	0.08334	34.5	0.02947	57	0.02346	79.5	0.04064
13.5	0.07408	36	0.02838	58.5	0.02381	81	0.04272
15	0.06667	37.5	0.02741	60	0.02426	82.5	0.0449
16.5	0.06062	39	0.02654	61.5	0.02482	84	0.04717
18	0.05557	40.5	0.02578	63	0.02549	85.5	0.04952
19.5	0.05131	42	0.02511	64.5	0.02627	87	0.05194
21	0.04766	43.5	0.02454	66	0.02717	88.5	0.05441
22.5	0.0445	45	0.02406	67.5	0.02819	90	0.05692

Block Replacement:

The block replacement time for a part may be useful when there are no records on parts life or if maintenance is only allowed at fixed intervals. Every t units of time, replace the part, despite its age. An automobile oil filter is an example. It is usually replaced at every oil change at some block of mileage. Nuclear reactors are refueled every 18 months so components within the containment structure use 18 month block replacement intervals. Again the method is not valid beyond eta. For intervals less than η, the following input is needed: U = Cost of an unplanned on-line failure, P = Cost of a planned replacement at block time t, and Weibull parameters β, and η. The cost per unit time is the ratio of mean cost to mean life. SSW provides block replacement analysis. See Table 4-13.

$$\textbf{Cost per unit Time} = \frac{P + U*(1-R(t))}{t} \qquad (4\text{-}6)$$

4.9 Problems

Problem 4-1. A fleet of 100 engines is subject to a Weibull failure mode, $\beta = 3.0$ & $\eta = 1000$ hours. The current engine times are:

Number of Engines	*Engine Time*
20	150 hours
20	200 hours
20	250 hours
20	300 hours
20	350 hours

a. What is the expected number of failures now, the Present Risk?
b. How many additional engines will fail in the next 6 months if the utilization is 25 hours/month and there is no replacement of failed parts?

Use the Weibull equation to get F(t) or read F(t) on Figure 4-17.

$$F(t) = 1 - e^{-(t/1000)^3}$$

Engine Time (t)	**F(t)**	**# Engines**	**Product = 20 x F(t)**
150	0.0033	20	
200	0.008	20	

Expected Failures Now, Present Risk = Sum = ________________

u = 6 months * 25 hours/month = 150, FF = {F(t + u) - F(t)}/{1 - F(t)}

T + u	**F(t + u)**	**FF = {F(t + u) - F(t)}/ {1 - F(t) }**	**20 x FF**
300	0.0266	0.0234	0.468
350	0.0420		

Expected Failures Next 6 Months = Sum = ______________

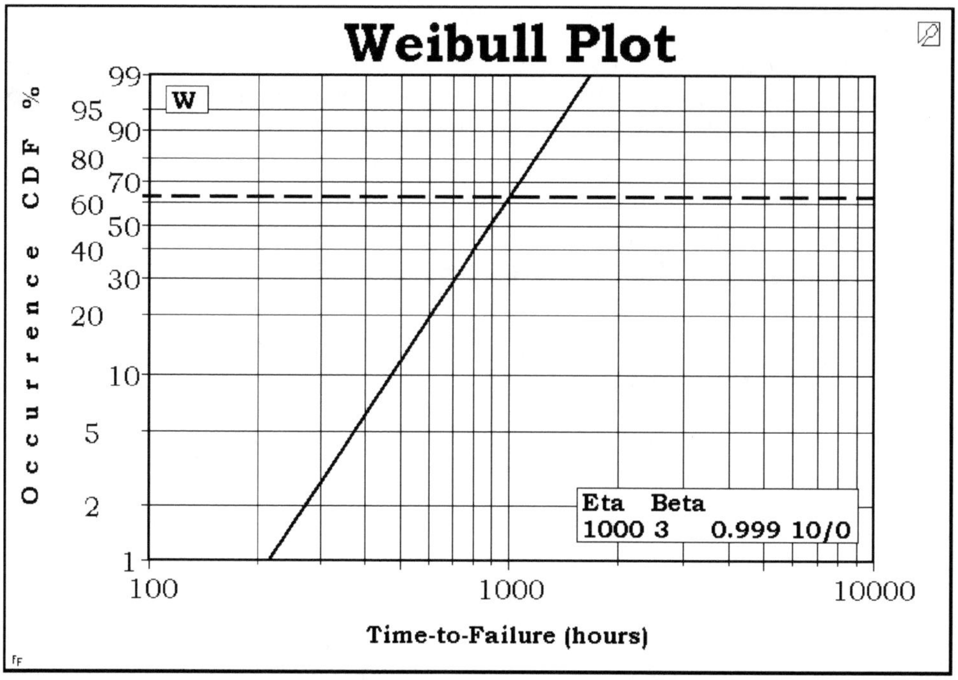

Figure 4-17. Weibull Plot

Problem 4-2. A turbine airfoil has caused unscheduled engine removals at the following times and locations.

Timeat Failure/Location				
684 (hours)/A	821/A	812/A	701/A	770/A
845/A	855/B	850/C	806/E	756/G
755/H	741/G	681/E	667/C	649/B
603/B	600/C	596/G	576/D	504/E
476/H				

a. Generate Weibull plots using the suspension populations in the following table, 158 overall less Location A and 25 at Location A. Put the suspensions at mid-interval, i.e., 50, 150. How do these Weibulls compare?

Interval	*Overall-Less A*	*Location A*	*Interval*	*Overall-Less A*	*Location A*
1-100	9	0	1201-1300	5	2
101-200	7	0	1301-1400	5	1
201-300	14	0	1401-1500	6	1
301-400	8	1	1501-1600	6	2
401-500	7	0	1601-1700	6	1
501-600	9	0	1701-1800	7	1
601-700	8	1	1801-1900	6	2
701-800	7	1	1901-2000	2	2
801-900	5	2	2001-2100	1	2
901-1000	4	2	2101-2200	2	0
1001-1100	4	2	2201-2300	1	0
1101-1200	4	2	Total	133	25

b. How many failures can be expected in the next 12 months?, the next 24 months?, from each population? (Use 30 hours/mo.)

Problem 4-3. Given a control failure mode with $\beta = 1.26$ and $\eta = 19735$ with a population of 1308 nonfailed units:

a. How many failures can be expected when each unit has reached 1000 hours?

b. 2000 hours?

c. If the life of a control is 4000 hours, what is the projected total number of failures in the life of the control if no further controls are added to the population?

d. If inspections "zero-time," or make the control units "good-as-new," how many failures are projected to occur in this population by 4000 hours with a 1000 hour inspection?

e. With a 2000 hour inspection?

Problem 4-4. Using Table 4-9, 0-1 random numbers, and the three Weibull failure modes:

a.	$\beta = 0.76$	$\eta = 96587$ hours
b.	$\beta = 2.638$	$\eta = 4996$ hours
c.	$\beta = 7.4$	$\eta = 1126$ hours

Assume two scheduled inspections, at 1000 hours and 2000 hours, that make modes a and c "good-as-new," while not helping mode b. A usage rate of 25 hours per month is assumed.

The following population of 5 engines is at risk: 1 engine at 100 hours, 1 engine at 200 hours, 1 engine at 500 hours, 1 engine at 700 hours, and 1 engine at 900 hours.

a. How many failures will occur over the next 48 months? Use the Monte Carlo simulation technique to solve this problem.

b. Would it be advisablc to drop thc 1000 hour inspection? Solutions to these problems are in Appendix K. (No peeking!!)

"Statistics show that of those who contract the habit of eating, very few survive." -- George Bernard Shaw

CHAPTER 5: MAXIMUM LIKELIHOOD ESTIMATES & OTHER ALTERNATIVE METHODS

"For real problems I prefer to have several tools available for my analysis of which the likelihood function is the Swiss Army knife and common sense the vice to hold the problem."

[Professor Dennis Farlie in a recent note to the author. Professor Farlie was one of the author's teachers at the University of London in 1963-1965 along with Professor George Barnard.]

5.1 Introduction

The standard method and **best practice** for Weibull analysis is median rank regression, regressing X on Y, as described in the preceding chapters. It is the **best practice** for the vast majority of data sets. However, there are other methods that under some conditions will provide more accurate results. Some of these alternatives like maximum likelihood are complex and therefore, they are not introduced until this point in the Handbook and the second day of the Weibull Workshop.

5.2 Maximum Likelihood Estimation (MLE)

Statisticians prefer maximum likelihood estimates (MLE) over other estimates because MLE have excellent statistical characteristics *in general, particularly for large data sets.* With the Weibull distribution there are some difficulties with MLE. The method consists of finding the values β and η which maximize the "likelihood" of obtaining the observed data. The likelihood function is a function of the data. It consists of the product of the probability density functions for each failure data point times the product of the cumulative density function for each suspension with the distribution parameters unknown. The function is complicated with many terms and is always evaluated with logarithms. See Appendix C for details on the log likelihood function. With two parameters, the log likelihood is a three-dimensional surface shaped like a mountain. The top of the mountain locates the maximum likelihood values. According to the theory the MLE values are the most likely to be the true values.

For the Weibull parameters, maximum likelihood estimation, (MLE), is complex. A computer is required and the solution for beta is iterative. There is no known correct method for plotting the data on an MLE Weibull plot. If median rank regression (MRR) plotting positions are used the "fit" to the MLE line will appear to be poor, particularly for small data sets. The lack of a good graphic display of the data is a major deficiency of maximum likelihood estimation. Engineers like to see the data plotted. Also, as the calculation is complex and iterative, convergence may not always occur. MLE solutions are possible with a single failure if there are some right suspensions. (This may seem illogical.) Median rank regression cannot do this. See Appendix C for more detailed comments on MLE versus MRR.

In addition to the MLE estimates, the likelihood function provides *likelihood ratio confidence intervals, likelihood contours, and the likelihood ratio test.* These are important useful methods that are described in Chapter 7. The likelihood function is a sufficient statistic, which means that all of the information in the data is in the function. The most desirable likelihood functions are tall thin mountains with the MLE located near the true values; the inference would be good tight precision with small bias errors, in other words, accurate estimates. The SSW software provides the ability to look at the cross sections of the likelihood surface for different confidence levels in a 2D plot and to look at the surface with a 3D plot.

Figure 5-1 is an attempt to portray the Weibull likelihood function. The horizontal slices through the surface represent the joint confidence regions for eta and beta. Imagine the three-dimensional surface covering these ellipses. For example, the bottom slice might be 99% confidence, the next above 95% confidence and so on. For small samples the Weibull likelihood slices are triangular, but they are elliptical as shown here for moderate and large samples. The higher up the mountain the "more likely" the eta and beta coordinates are the true unknown values. The MLE, (maximum likelihood estimates for beta and eta), are found by dropping a vertical line from the top of the mountain to the bottom to determine the most likely estimates (MLE) values of eta and beta.

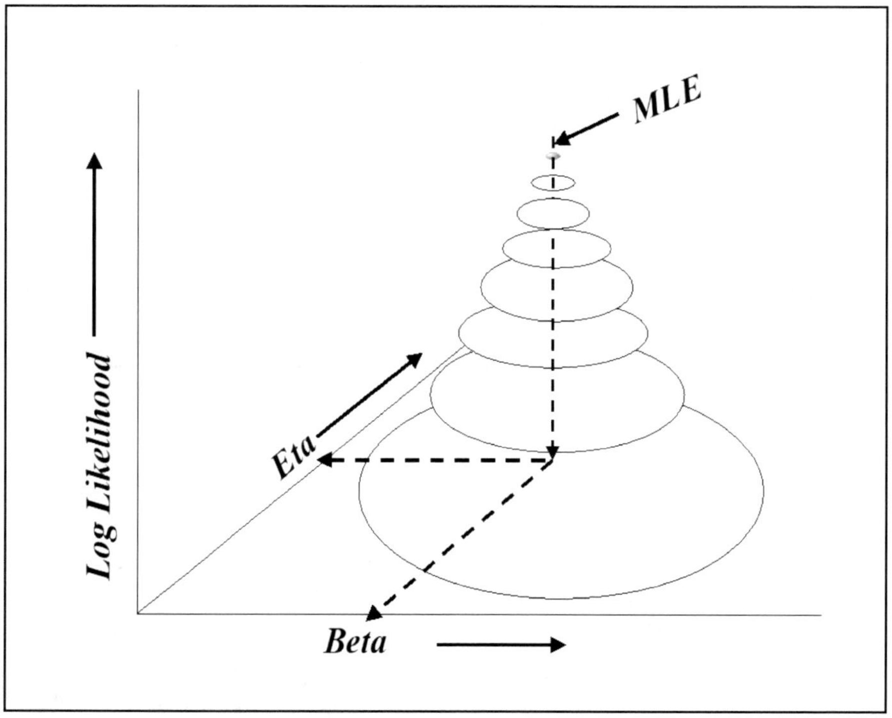

Figure 5-1. The Weibull Likelihood Function

The pros and cons of the median rank regression versus maximum likelihood methods will be indicated herein and in Appendix C. For the three parameter Weibull, all three parameters can be found by maximizing the value of the likelihood function in SuperSMITH Weibull (SSW).

For small and moderate size samples, less than 100 failures, the maximum likelihood estimates tend to be biased optimistically, less accurate, as indicated in Table 5-1. The table is based on 3000 Monte Carlo trials for 4 to 100 failures and 1000 trials for 500 and 1000 failures. Suspensions are not considered in this table; other research [Chi Chua Liu, Wenham, Lam & Spelt] reach the same conclusions with suspensions. For very large samples, MLE will be slightly more accurate than rank regression. For small and intermediate size samples *MLE is not recommended* as the results will be optimistically biased. Table 5-2 contains a root mean square error comparison. Although regression shows a slight advantage, it is essentially a standoff for large samples. The MLE bias error also exists for samples *with suspensions* for small and moderate sample sizes as will be shown later.

When the Weibull failure mode does not apply to the entire fleet, but only to a subset of the fleet, this is called a *batch problem*. Clues indicating batch problems are summarized in Chapter 8. In Section 4.4.1, the "Now Risk" method for detecting batch effects using median rank regression was described. Appendix F presents the Aggregate Cumulative Hazard (ACH) method for detecting batch effects. This is the newest and most sensitive method for analysis of batch data and works with both MLE and MRR. Two comments on MLE with batch data: (1) The "Now Risk" method does not work with MLE; even with a batch problem the expected number of failures "now" will be close to the observed number. (2) If there is a batch problem, the MLE beta will tend to be less than the MRR beta, reversing the bias errors described above. See Section F.2. Therefore, if the MLE beta is less than the MRR beta this may be another clue that there is a batch problem with the data.

In Section 5.3 the new reduced bias adjustment (RBA) will be described. This method (MLE-RBA) produces unbiased estimates of beta from the MLE estimate and is much more accurate than MLE for small and intermediate size samples.

Note the author uses median bias estimates to gage bias and mean square error as these small sample distributions are highly skewed and therefore, the mean bias is not representative. If mean bias were used most of these errors would be much larger.

Table 5-1. Bias Comparison - X on Y Regression versus Maximum Likelihood Estimates
Median Estimates- *Most accurate are in bold Italics*

True	η = 1000		β = 3.0		B1 = 216	
Sample Size	X on Y Eta	MLE Eta	X on Y Beta	MLE Beta	X on Y B1	MLE B1
N=4	***997***	981	***3.03***	3.99	***216.4***	310.2
N=10	***996***	992	***2.95***	3.32	***209.9***	250.9
N=30	***998***	***998***	***2.985***	3.338	***208.63***	225.5
N=100	***1000.0***	999.9	2.970	***3.024***	211.8	***218.4***
N=500	***999.9***	1000.4	2.983	***3.004***	213.1	***216.5***
N=1000	1000.8	***999.8***	***3.001***	3.004	***215.85***	***216.5***

Table 5-2. Root Mean Square Error/True Value
Most accurate are in bold Italics

True	η = 1000		β = 3.0		B1 = 216	
Sample Size	X on Y Eta	MLE Eta	X on Y Beta	MLE Beta	X on Y B1	MLE B1
N=4	0.1763	***0.1762***	***0.8750***	0.9467	0.8170	***0.8165***
N=10	***0.1111***	0.1129	***0.3407***	0.3681	***0.5147***	0.5232
N=30	0.06496	***0.06483***	0.1809	***0.1599***	0.30103	***0.30042***
N=100	***0.03526***	0.03596	***0.1010***	0.1025	***0.16341***	0.16662
N=500	0.01614	***0.01556***	0.048	***0.0350***	0.07772	***0.05966***
N=1000	***0.01177***	0.01556	***0.00009***	0.00010	***0.01151***	0.01287

5.3 MLE With Reduced Bias Adjustment (RBA) for Accurate Results

When the author toured Ireland in 1998, he visited the Guinness brewery in Dublin, where W.S. Gosset did his famous work on Student's t and later the F-test. A century ago, this brew master was trying to improve the accuracy of small sample statistical methods to improve the quality and taste of Guinness, an admirable objective. Like most of us he was forced to use small samples. He focused on maximum likelihood estimate (MLE) of the standard deviation (sigma) from the normal distribution and he knew about the large errors. His derivation of the Student's t statistic indicated not to use the MLE of sigma, but rather the square root of an unbiased estimate of variance or SQR(Var-U). This square root is still biased, but much less so than the MLE sigma. When Gosset showed his derivation to Sir Ronald Fisher, there was disagreement. Fisher said use the MLE sigma. Four years later in 1912 Fisher changed his mind and agreed that Student's t should be used with SQR(Var-U). To convert an MLE variance to SQR(Var-U), we multiply by N/(N-1). Today engineers always use this N/(N-1) correction because the sigma estimate is more accurate, even though it does not completely unbias the MLE sigma.

With small sample life data analysis, the Weibull, the normal and the lognormal MLE estimates are all biased by the probability plot line being too steep and therefore, too optimistic, predicting longer life. This is the same problem Gosset worried about long, long ago. For small samples most of the uncertainty is in the Weibull beta and the normal sigma; Weibull eta and normal mu have much less uncertainty. If the three distributions have the same disease, would Gosset's solution help our problem with the bias in the plot slope? His work inspired me to look for reduced bias adjustments for beta and sigma. His work also inspired Ronald Fisher to improve the accuracy of small sample statistics. See [Joan Fisher Box] and [Pearson 1990]. See the last page in this chapter for more on Gosset.

5.3.1 The RBA Factor for Normal and Lognormal Distributions: The author's research showed the N/(N-1) factor reduced but did not eliminate the bias in either sigma. In statistical quality control, there is a C_4 factor, which eliminates the bias in SQR(Var-U). It is tabled in Appendix L and in several texts. For example, see [ASTM] and [Duncan]. Even earlier than these publications, in October 1923, William S. Gosset was calculating a table of C_4 for Ronald Fisher and Karl Pearson. [See "Student," by E.S. Pearson].

$$C_4 = \sqrt{\frac{2}{n-1}}\frac{\left(\frac{n-2}{2}\right)!}{\left(\frac{n-3}{2}\right)!} = \sqrt{\frac{2}{n-1}}\frac{\left(\Gamma\left(\frac{n}{2}\right)\right)}{\left(\Gamma\left(\frac{n-1}{2}\right)\right)}$$

n = number of failures not including suspensions

Therefore, the Reduced Bias Adjustment (RBA) factor for the normal and lognormal standard deviation is the square root of N/(N-1) divided by C_4. The product of the MLE sigma and the RBA_σ factor is an unbiased estimate of sigma (S_U) for the normal and the lognormal distributions. (Note: ^ is the symbol for MLE)

$$S_U = \hat{S} * RBA_\sigma$$

$$RBA_\sigma = \left(\sqrt{N/(N-1)}\right)/C_4$$

This is well-established theory for statistical quality control. See for example: [J.M. Juran 1951] which includes a table of RBA_σ called C_2 with references to the 1945 edition of the ASTM report listed above. Herein it is applied to life data analysis *for the first time*. C_4 is tabled in Appendix L. Table 5-3 shows that S_U is unbiased within simulation precision while the MLE sigma is highly biased. The S_U with the MLE mean (mu) provides accurate life data analysis.

Mean Bias versus Median Bias: Gosset eliminated the mean bias with the above corrections. The mean bias is the difference between the true value and the mean value of many estimates. For symmetrical distributions like the normal and lognormal the mean bias and median bias are the same. In this Handbook we are primarily concerned with the Weibull and the bias in beta and eta. As beta is highly skewed, it is more logical to use the median bias for Weibull, the difference between the true value and the median value of many estimates. This is consistent with all the studies in this Handbook except for Gosset's symmetrical distribution work above. We will use median bias to address the problem with beta. However, the study that follows is repeated in terms of mean bias in Appendix C.

Table 5-3. Comparison of Methods for the Normal (and Lognormal) Distributions
Mean Values from 20,000 Monte Carlo Trials, True Values mu = 1000, sigma = 3.0, B1 = 993.03

Sample	Median Ranks			MLE			Reduced	Bias Adjustment	
Size	Mu	Sigma	B1	Mu	Sigma	B1	mu	Sigma	B1
4	999.85	3.072	992.85	***1000.00***	2.383	994.61	***1000.00***	***2.996***	***993.05***
6	***1000.00***	3.067	992.88	***1000.00***	2.613	993.93	***1000.00***	***2.991***	***993.04***
10	***1000.00***	3.065	992.88	1000.01	2.773	993.57	1000.01	***3.004***	***993.02***
30	***1000.00***	3.041	992.94	1000.01	2.921	993.22	1000.01	***3.003***	***993.02***
100	***1000.00***	3.02	992.98	***1000.00***	2.977	993.08	***1000.00***	***2.998***	***993.04***
	....(Root Mean Square Error)...								
Sample	**Median Ranks**			**MLE**			**Reduced**	**Bias Adjustment**	
Size	**Mu**	**Sigma**	**B1**	**Mu**	**Sigma**	**B1**	**MLE mu**	**Sigma**	**B1***
4	***1.500***	1.306	3.401	***1.500***	***1.180***	***3.123***	***1.500***	1.262	3.285
6	***1.232***	1.005	2.630	1.236	***0.921***	***2.469***	1.236	0.963	2.558
10	0.950	0.740	1.968	***0.947***	***0.697***	***1.881***	***0.947***	0.719	1.926
30	0.548	0.406	1.090	***0.544***	***0.395***	***1.066***	***0.544***	0.396	1.072
100	0.299	0.218	0.589	***0.302***	***0.212***	***0.575***	***0.302***	0.213	0.579

5.3.2 The RBA factor for the Weibull distribution: For life data the more important problem is the Weibull beta; eta's problem is smaller. However, the Weibull distribution was unknown in Gosset's time. Would Gosset's correction fix beta as well as sigma? Trying RBA_σ from above did not work. New research by Paul Barringer shows that empirically the *RBA factor for the Weibull beta* is C_4 raised to the 3.520 power. If we multiply the MLE beta by $c_4^{3.520}$ it eliminates the median bias in the MLE beta. (To eliminate the mean bias a larger correction is needed, multiply the MLE beta by c_4^{6}).

$$\beta_{Unbiased} = \hat{\beta}\left(RBA_{\beta-factor}\right) = \hat{\beta}\left(C_4^{\,3.520}\right)$$

$$\eta_{Unbiased} = \hat{\eta}\left(RBA_{\eta-factor}\right) = \hat{\eta}\left(C_4^{\,-0.217}\right)$$

Table 5-4A shows the results of applying the RBA factor for Weibull data sets. Note the large improvement in the RBA beta bias compared to the MLE beta. Further compare the resulting accuracy of B1 life. The MLE η_U and β_U provide a *significant improvement in accuracy for Weibull analysis* of life data. Table C-1 in Appendix C shows a similar comparison for mean bias corrections.

There were two earlier attempts to unbias the MLE beta: [McCool 1969] and [Thoman, Bain & Antle, 1969]. [Barmoav 2009] presents a distribution that provides almost identical corrections compared with RBA. Their tabled corrections are almost identical to the mean bias correction, c_4^{6} and it is shown that "the percent bias in the MLE beta is independent of the true value of eta and beta". The latter paper is also the basis for "***ASTM C1239-95 Standard Practice for Reporting Uniaxial Strength Data and Estimating Weibull Distribution Parameters for Advanced Ceramics***." This ASTM standard recommends using unbiased estimates of beta. [Wang 2001] is the latest contribution to this technology. [Lam and Spelt] provide Monte Carlo results that are similar to Table 5-4A.

Table 5-4A. Comparison of Methods for the Weibull Distribution
Median Values from 25,000 Monte Carlo Trials; N = 4 to 100; Complete Sample
True Values: Eta = 1000, Beta = 3.0, B1 = 215.8 ... Bias = Median - True value ...

Sample	MRR			MLE			MLE With RBA		
Size (N)	**Eta**	**Beta**	**B1**	**Eta**	**Beta**	**B1**	**Eta**	**Beta**	**B1**
4	993.16	3.05	218.59	979.71	3.99	309.04	***997.29***	***2.99***	***213.91***
6	994.62	2.99	211.55	986.64	3.57	272.34	***997.34***	***3.00***	***215.19***
10	997.79	2.96	210.01	993.40	3.31	248.27	***999.39***	***3.01***	***216.30***
30	***999.64***	2.95	210.17	997.85	3.10	225.77	1000.43	***3.00***	***216.30***
100	***1000.3***	2.97	212.57	999.34	3.03	218.98	1001.35	***3.00***	***216.46***

5.3.3. Best Practice: Which is the **best practice**, MLE with RBA factors or Median Rank Regression, X on Y? For engineers, the author recommends MRR as **best practice** as it provides a good graphic plot and is the simplest method. Engineers want a graphic plot. For the mathematical analysts that do not use graphic plots, the new MLE with RBA factors will be their choice. Why not use both methods, MRR and MLE-RBA? If the results are similar this is additional evidence that the fit is good. If the MLE-RBA beta is much less than the MRR beta, there may be a batch problem. MLE is not recommended. (The author wants to thank Paul Barringer, Wes Fulton and Richard J. Petshauer for their help with this research.)

CASE STUDY 5.3.1 FORECASTING WITH MRR, MLE AND MLE-RBA
Contributed by C. Richard Unkle, Senior Reliability Engineer, GE Transportation Systems

Locomotive component defects and suspension data from 1999 were used to forecast defects in the next year, 2000. Based on the four defects in 1999 in 321 components the forecast with replacement were made using the three different methods, MRR, MLE and MLE-RBA. The actual number of defects in the year 2000 was seven. As predicted the MLE forecast is optimistic. **The MRR and MLE-RBA forecasts of 7.3 and 7.4 respectively are much more accurate than the MLE forecast of 3.5**. As usual MLE provides optimistic estimates. The results are shown on Figure 5-2. There is mild evidence of a batch problem in the 1999 data, but it is not significant at the 90% level.

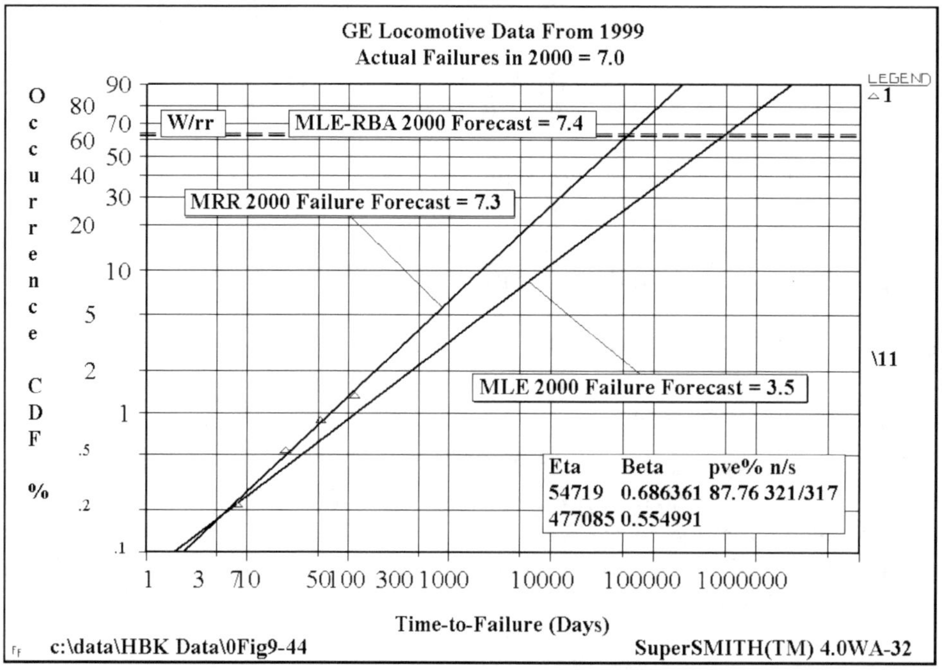

Figure 5-2 Failure Forecast Comparison MLE with Batch Data

[Meeker] provides another MLE-MRR comparison with Example 12.10 in his text using the bearing data shown herein in Chapter 4. Using median rank regression our 12-month forecast shown in Section 4.5.1. is **10.5 failures**. [Meeker] using MLE predicts **5.057 failures** for the same period with an **upper 95% bound of 9.0**. This was an aircraft engine in-flight shutdown failure mode. As human safety was involved, optimistic MLE estimates like these are unacceptable. MLE-RBA predicts 9.94 failures.

5. 4 Median Rank Regression: X on Y Versus Y on X

Median rank regression determines the best-fit straight line by least squares regression curve fitting, instead of by eye. This is the most popular method and is generally the default method for Weibull software. However, from the standpoint of statistical rigor the method may be criticized because the residual variation around the line will not be uniform. (The data is said to be "heteroscedastic" in contrast to data with uniform residuals called "homoscedastic.") The scatter is larger on the low end that tends to overweight the data in this area. Some statisticians sometimes suggest omitting the first few failures for purposes of curve fitting. The author disagrees. For engineers, the lower end of the line is the area of greatest interest and therefore overweighing the first failures is acceptable, and in fact, desirable, even though it lacks rigor.

Another problem with using median rank regression to fit Weibull life data is that there are two regressions, Y on X and X on Y. They give different answers. Y on X minimizes the sum of squares of the residual variation in the Y direction, while X on Y minimizes it in the X direction. Which line is most accurate? For Weibull plots, the X variable, time-to-failure, has much more scatter and error than the Y variable, median ranks. The research by [Berkson] is recognized as the best solution to this problem. [Berkson] recommends selecting the scale with the largest variation as the dependent variable. For Weibull analysis of life data, the times-to-failure, X'S, always have much more statistical scatter (random error) than the median ranks. In addition, there is often real error in the data, particularly for in-service failures. Therefore, the best method for Weibull failure plots is to regress X on Y, not Y on X.

The X on Y line is more accurate than Y on X line for life data. However, many Weibull computer programs do it the wrong way because the theory of errors has not been well publicized. Selecting Y on X instead of X on Y introduces errors in the Weibull analysis. Table 5-4B shows a comparison of X on Y versus Y on X for 1000 sets of various sample sizes without real error. The data was produced with Monte Carlo simulation. If you compare these median estimates with the true values, X on Y is the clear winner providing more accurate estimates of η, β, and B1.

Table 5-4B. Comparison of X on Y versus Y on X Regression Most Accurate are in bold *Italics*

True	$\eta = 1000$		$\beta = 3.0$		B1 = 216.0	
Sample Size	**X on Y Eta**	**Y on X Eta**	**X on Y Beta**	**Y on X Beta**	**X on Y B1**	**Y on X B1**
N=4	***994***	1009	***3.02***	2.75	***213***	190
N=10	***996***	1014	***2.95***	2.79	***210***	193
N=30	***999***	1011	***2.98***	2.86	***212***	204
N=100	***1000***	1006	***2.98***	2.92	***212***	208
N=500	***1000***	1002	***2.99***	2.97	***214***	213
N=1000	***1000***	1001	***2.99***	2.98	***215***	213

Weibull computer programs that differ with SSW results may be incorrectly regressing Y on X. SSW allows the choice of X on Y or Y on X regression to facilitate comparisons. The default is X on Y for improved accuracy.

Weibull $[1961] said, "The parameters may now be determined by fitting a straight line to the data points in such a way that the sum of the squared deviations ($d_i = [x_1 - x_{line}]...[x_n - x_{line}]$) is minimized...If, however, the straight line was fitted by minimizing the squared deviations in the y-direction, the conclusions would not hold, and biased estimates would result." See also [Berkson], [Natrella], and [Mann, Shafer, and

Singpurwalla]. NOTE: The "$" symbol preceding a reference means the document may be downloaded from Paul Barringer's website (see References).

The author strongly favors Median Rank Regression (MRR), X on Y, with and without suspensions, as the engineering standard because it is most accurate, simple to understand and provides a useful graphic plot. MLE provides optimistically biased answers for all but huge samples. MLE is not recommended; MLE-RBA is the recommended alternative to MLE. For 20 failures or less, with and without suspensions, Weibull analysis using **MRR, X on Y, is recommended as best practice**. MRR is recommended for both Weibull and lognormal samples up to 100 failures. This recommendation is based on numerous studies and experience over four decades showing MRR results are more accurate than MLE. However, for mathematical statisticians that persist in recommending MLE, we recommend MLE-RBA.

Waloddi Weibull $[1967] states, "The Maximum Likelihood Method is the most efficient one of the general methods, but it requires complicated equations, and, above all, the estimates from small samples are heavily biased."

Two new comprehensive studies, [Wenham] and $[Liu], have reviewed all the methods described herein plus other methods, with and without suspensions in various proportions, and reached the same conclusion; median rank regression, X on Y, is the **best practice**. Dr. Maggie Wenham, Corporate Statistician GKN Automotive concluded, "The Rank Regression (RR) method calculates Weibull parameters with the best combination of accuracy and ease of interpretation over more of the range of conditions thought to be common within GKN than any other method investigated." Further, under Recommendations: "The RR method should be adopted for Weibull analysis throughout GKN." The study was based on Weibull data, small and large samples with and without late suspensions in various proportions.

Chi-Chao Liu completed the largest and most extensive study of life data analysis known to the author using both Weibull and lognormal data sets, with and without all kinds of suspensions. Liu had four conclusions:

(1) Median rank regression, X on Y, is more accurate than MLE except for very large sample sizes.
(2) The two best measures of goodness of fit are the p-value of r square and the likelihood ratio test.
(3) The Weibull two parameter is always more conservative in the lower tail than the lognormal.
(4) With 20 failures or less there is not enough information to make credible distribution analysis.

Considering the third and fourth conclusions, for engineering analysis, estimating B5 life or less or failure forecasting, the author states that **best practice** for small samples, 20 failures or less, is **always use the Weibull two-parameter**. In recent discussions with Chi-Chao Liu he agrees.

Tables 5-1 & 5-2 show comparisons of MRR and MLE for complete samples, no suspensions, the results heavily favoring MRR as more accurate. Some statisticians accepted these results for complete samples but still recommend MLE for small samples that have suspensions because they believe MLE handles the suspensions "better" than MRR in spite of the two studies above. With Monte Carlo simulation it is easy to show that to be false. Table 5-5 shows the results with 95% Type II (right) suspensions, small sample of 5 failures and moderate sample of 30 failures. The most accurate estimates are in bold italics. The median values from 1000 trials are shown. MRR is the clear choice. Chi-Chao Liu's thesis confirms these results with all types of suspensions.

Table 5-5. Comparison of MRR with MLE for Data Sets With Type II Suspensions Based on 1000 Data Sets - Weibull True Values β = 3.0 η = 1000 B1 = 215.8

Sample Size	Eta	Beta	B1
5F/95S	------------------------	**---------Small Sample-------**	--------------
MRR	***935***	***3.20***	***215***
MLE	730	4.09	240
30F/570S	-----------------------	**----Moderate Size Sample-**	--------------
MRR	***1015***	***2.96***	***214***
MLE	953	3.11	219

5.5 Plotting Positions – Estimates of F(t)

What is the Y coordinate when we plot on probability paper? If (i) is the rank order number, at first i/N would seem correct except that puts the last point at 100% which is off scale. The second solution is often (i-1)/N but here the first point is at 0%, off scale. Hazen is given credit for selecting the midpoint of the first two attempts and many statisticians use his plotting position because they believe it has minimum mean square error. As shown in Table 5-6A this is not true.

Hazen's Plotting Position = (i - 0.5)/N

Median rank plotting positions are by far the most popular and most accurate for engineering analysis as recommended by Leonard Johnson with the later endorsement by Waloddi Weibull and the author. See Table 5-6A. They may be calculated exactly as in Appendix I using the beta-binomial method described in Excel language in the same appendix. Alternatively median ranks may be accurately approximated with Benard's formula. Median ranks are **best practice.**

Benard's Plotting Position = (i - 0.3)/(N + 0.4)

Weibull originally used mean ranks but later switched to median ranks.

Mean Rank Plotting Position = i/(N + 1)

All these plotting positions are available in SSW with median ranks as the default and **best practice**. They may be compared with Monte Carlo simulation. Both bias and mean square error are used as a gage of accuracy and are shown in Table 5-6B. Sample size four appears inconsistent but other studies show it is not. Median rank regression, X on Y, is the **best practice**.

Table 5-6A. Comparison of Plotting Positions
Median Values from 3000 Monte Carlo Trials, True Values Eta = 1000, Beta = 3.0, B1 = 215.8
...Bias = Median - True value...*Italics indicate most accurate*

Sample Size	Median Ranks			Mean Ranks			Hazen's Ranks		
	Eta	Beta	B1	Eta	Beta	B1	Eta	Beta	B1
4	***996.54***	***3.032***	***216.39***	1013.53	2.568	168.73	983.23	3.544	266.67
6	***992.26***	***2.942***	***207.07***	1009.83	2.566	167.57	986.30	3.333	246.67
10	***996.00***	***2.985***	***209.93***	1010.90	2.667	179.03	995.30	3.236	237.53
30	***997.83***	***2.940***	***208.63***	1004.57	2.778	192.47	997.73	3.102	224.93
100	***1,000.02***	***2.970***	211.77	1001.77	2.889	203.40	998.63	3.035	***218.63***

Table 5-6B (Mean Square Error) / (True^2)									
Sample	Median Ranks			Mean Ranks			Hazen's Ranks		
Size	Eta	Beta	B1	Eta	Beta	B1	Eta	Beta	B1
4	0.03108	0.76556	0.66745	***0.03073***	***0.66751***	***0.65980***	0.03231	1.08324	0.69380
6	***0.00897***	0.31546	***0.19256***	0.01992	***0.25766***	0.42778	0.01943	0.41750	0.41731
10	0.01234	0.11608	0.26492	***0.01190***	***0.11479***	***0.25549***	0.01224	0.16162	0.26283
30	0.00422	***0.03272***	0.09062	0.00421	0.03436	0.09036	***0.00419***	0.04028	***0.09002***
100	***0.00124***	***0.01020***	***0.02670***	0.00128	0.01050	0.02754	0.00132	0.01121	0.02828

5.6 Special Methods: Gosset's Student's t

There are special MLE methods available in SSW for the normal distribution. There are two types of data that may employ the normal confidence bounds based on Student's t: (1) complete data, no suspensions and (2) non-life data, the typical application of the normal distribution for analyzing measurements. For this case, a modified MLE solution that minimizes bias is found in textbooks. The MLE mean value is simply the average of the data values. The modified MLE (MMLE) mean is the same. The MLE standard deviation is found by taking the root sum of the squares of deviations from each data point about the mean and dividing by the number of data values (n). The Modified MLE (MMLE) standard deviation is the MLE standard deviation multiplied by the square root of ((n)/(n-1)). Normal regression (nr) confidence bounds are recommended for MMLE fit. These bounds are based on Student's t and are interval estimates for the location of the line as well as B values. Of course, MLE-RBA will be even more accurate for the normal analysis.

5.7 The Dauser Shift - Unknown Suspension Times

What can be done if the times on each failure are known but the times on the population of unfailed units are unknown? This is a very common form of dirty data; for example, warranty data usually falls in this category. Non-serialized parts have unknown suspension times. The following method is **not** best practice, is not recommended, and should be considered only as a last resort. The method has no mathematical rigor but often works well. It is included because it is widely used, and some experts consider it to be valuable. It seems impossible to generate a Weibull under these circumstances. Fred Dauser (Pratt & Whitney Aircraft) conceived a method to "adjust" the failure Weibull to approximate the Weibull that would have been obtained if the suspension times were known. Note that if a reasonable estimate of the suspension times can be estimated, it will probably be more accurate than the Dauser Shift. The steps in the Dauser Shift method are:

1. Plot the failure data on Weibull Probability Paper to estimate β.
2. Estimate the mean-time-to-failure (MTTF) with the simpler mean-time-between-failure (MTBF), the sum of the failure times divided by the number of failures.

$$\textbf{MTBF} = (\Sigma \textbf{ Failure Ages})/(\textbf{Number of Failures})$$

3. Draw a vertical line on the plot at the MTBF.
4. Calculate the proportion failed (%), (number of failures/sum of failures plus suspensions).
5. Draw a horizontal line at the cumulative %.
6. At the intersection of the vertical and horizontal lines draw a line parallel to the failure Weibull. This is an estimate of the "true" Weibull.

Example: Five sports cars failed the differential gear at the following mileages: 3165, 3579, 4098, 5134, and 5402 kilometers. The entire fleet has 5000 cars. Estimate the fleet Weibull. The failure Weibull is shown on Figure 5-3. Draw a vertical line at the MTBF of 4276 km. The mean may be estimated by summing the failure ages and dividing by the number of failures. Five out of 5000 is 0.1%. Draw a horizontal line at 0.1%.

At the intersection construct a second Weibull with the same β, 4.656. SSW can add this Dauser solution to the plot automatically with a Marker Line, if that option is activated.

Another Warning: the Dauser Shift is a last resort, though it is widely employed, there is no theoretical justification for this method. If the reader can make a reasonable estimate of the suspension times, perhaps a histogram, this will be preferred and will be more accurate. The reader should keep his natural skepticism. The shape and proximity of the suspension fleet to the Weibull may invalidate the Dauser method. Still, there have been many cases where the Dauser Shift works well. For example, the bleed system failure case study in the Chapter 3 provides good results with the Dauser Shift.

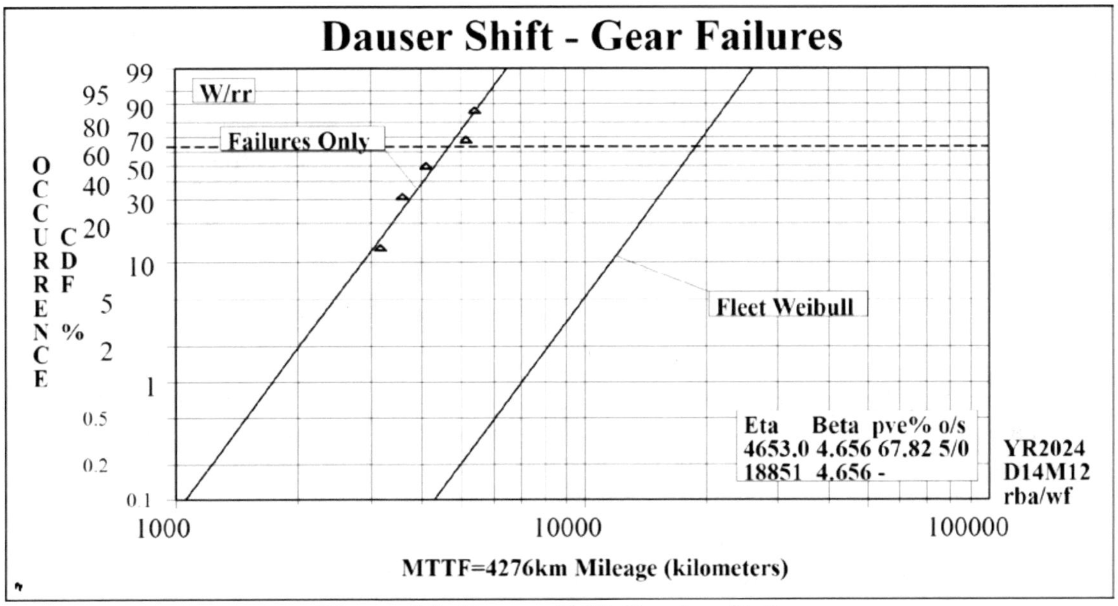

Figure 5-3. Dauser Shift Gearbox Failures

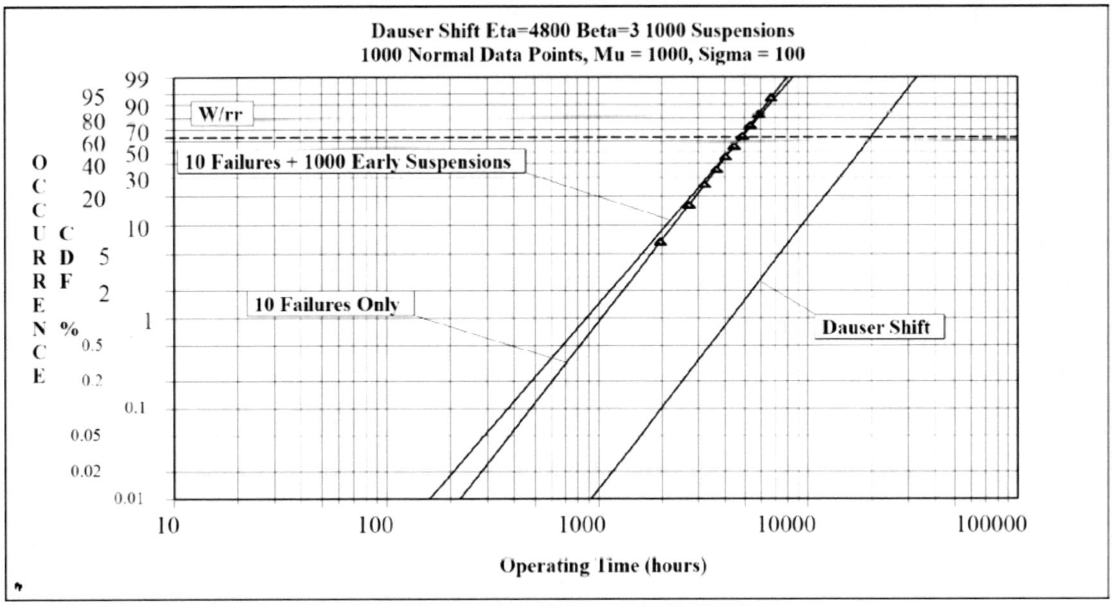

Figure 5-4. Dauser Left Shift

SSW may be used to establish limits for the true Weibull. Consider ten failures from a Weibull. With η = 4800 and β = 3. In Figure 5-4 a fleet of 1000 normally distributed suspensions, mean = 1000, range 500-1500, is added to the ten failures and a Dauser shift is provided for comparison. The suspensions are "early" suspensions, with ages earlier than the first failure. This is a left shift, the maximum possible shift if all the suspensions were early. The effect of early suspensions is negligible and the Dauser shift is optimistic.

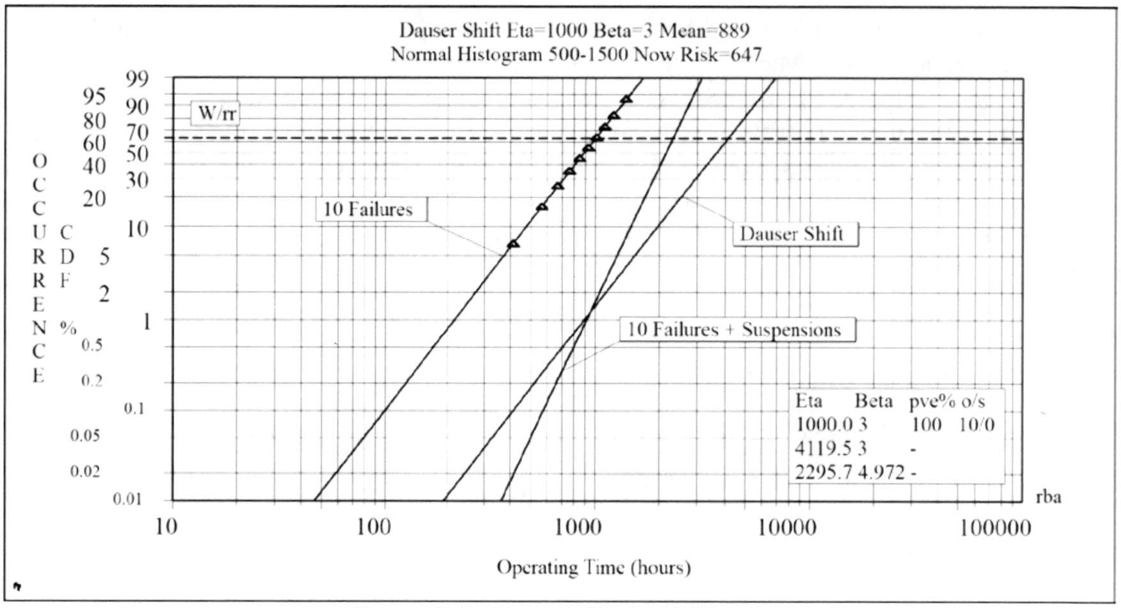

Figure 5-5. Dauser Right Shift

In Figure 5-5, the same example is repeated but here the failures come from a Weibull located in the middle of the suspensions, $\eta = 1000$. Here most of the suspensions are random or late suspensions, older than the last failure. The Dauser shift provides a better answer but the situation may be artificial as many late suspensions may indicate a batch problem

5.8 Special Methods for Inspection, Interval and Grouped Data

Sometimes we do not know the exact times-to-failure but rather the data is grouped into intervals. All interval data has deficiencies. We do not know exactly when the failure occurred. For example, inspection-interval data depicts benign failures that are only found when the system is shut down and inspected. This is interval or grouped data. The failures discovered at an inspection occurred earlier, before the inspection. This increases the uncertainty of fitting a Weibull. There are five special methods for improving the accuracy of inspection data labeled Inspection Options 1-5:

5.8.1. Use the **Inspection Option #1** for obvious failures found at Inspection and coarse data collection Also widely used for predicting warranty claims by the age or mileage of the unit.

5.8.2. & 5.8.3 Probit #2 & 3 are used for Non-Destruction Evaluation (NDE) when the same number or parts are inspected at each inspection interval. It is also used for Destructive Inspection.

5.8.4 Kaplan-Meier #4 is a robust analysis method tolerant to dirty data situations and used for examples to help with missing data and with very large suspension quantities.

5.8.5 Interval MLE #5 is used for types of interval data, even variable inspection intervals.

Options 2-5 have special input formats in SSW. If confused, use **Logic Diagram-Chapter 10**

5.8.1 Inspection Option #1

The standard plot for interval and grouped data will show vertical columns of points. Some examples are inspection data, coarse data, and instrumentation calibration intervals. See Figure 5-6. The vertical columns of data points imply the lower-level points were associated with earlier times. If the exact times were recorded, the data would have fallen to the left of the column and scattered about the trend line. Many experts, like [Sherwin 1980], suggest a better fit may be obtained by ignoring the lower points in each column, as the uppermost point is the most accurately located. (Note: Sherwin did not invent this method as it was in use long before his paper.) The solid line on Figure 5-6 represents this solution.

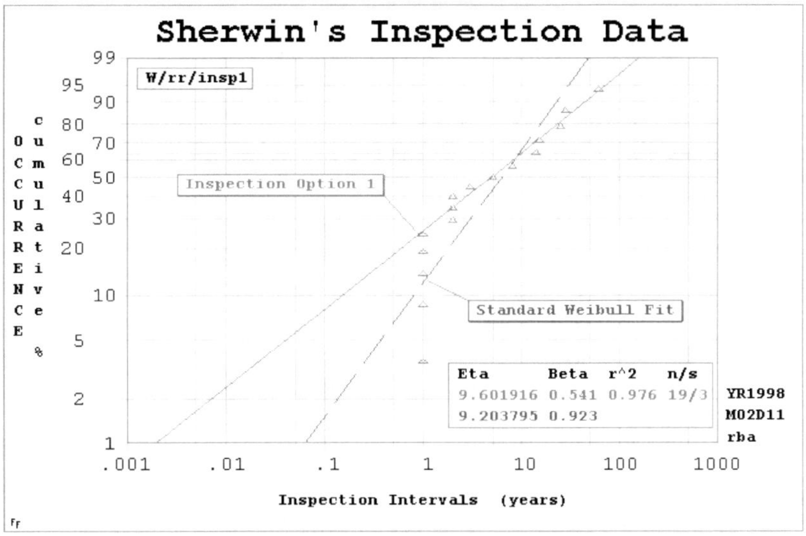

Figure 5-6. Inspection Option 1

The author has had excellent success with this method. It is Inspection Option 1 in SSW. Note that it is an option. The method may or may not improve the fit of the data depending on whether there are duplicate data points. This option should improve the fit any time some of the data values are exactly repeated. The improvement may be judged by eye or by utilizing the "pve%" metric from correlation as explained in the section for distribution analysis below. Another application for the same inspection technique is coarse data. Perhaps failures recorded by month should have been recorded by day or hour. This will appear as interval data and the same technique may be employed to improve the fit. The Inspection Option is used for warranty claims forecasting which will be treated in Chapter 9.

5.8.2 & 5.8.3 Probit Analysis Inspection Options #2 & 3

The author borrowed the term *"Probit"* from [Fisher] who developed the original method with D.J. Finney many decades ago. They assumed normality for the data (probably because the Weibull was unknown.) Herein we do not make that assumption.

[E.S. Pearson] defines probit: "The method of probit analysis is applied in situations where the "individuals in a population require a varying 'stimulus' to produce a characteristic quantal "response" and where our objective is to estimate the parameters of this distribution of critical stimuli. Such problems arise, for example, in dosage-mortality studies, in biological assay and in tests on the detonation of explosives."

The Probit method is recommended when the uncertainty is greater in the Y direction then the X direction and the cumulative percent failure or occurrence can be estimated. Two common types of Probit data are (1) non-destructive evaluation (NDE) data from eddy current, fluorescent penetrant, and ultrasonic inspections and (2) destructive inspections.

In the usual Weibull application to service failures, the uncertainties are greater in the time to failure, (X), than the % Failure, (Y), and the regression is done X on Y. With Probit data, the uncertainties on the Weibull plot are larger in the % Failure, (Y) direction than in the time of the inspection, (X), and the regression is done Y on X. For example, with NDE data the uncertainties are greater in the (Y) direction as the small crack or defect may not be found even if present. See Figure 5-7. Probit data analysis is even easier than the point-to-point analysis we have been using so far, because there is no question where the plotting position is. Median ranks or other plotting positions are not used to plot this data. Instead, the actual cumulative % failure is used as the plotting position. The method for estimating the cumulative % failed depends on the data format. There are at least two different data formats commonly encountered where the probit method is recommended as the **best practice**:

(Format 1) The same units are inspected each time so the quantity of units in the population inspected is a constant, say boiler tubes for example. The cumulative number of failures at each inspection divided by the constant number in the population estimates the cumulative failure rate. For example, **NDE inspections,** with nuclear steam generators, there is a scheduled outage every 18 months to refuel the reactor. The number originally inspected would be 4000, the number of tubes in the population. The cumulative failures would be the number of cracks found in the current inspection plus the number of tubes previously plugged. The ratio of the cumulative failed to 4000, the number of tubes inspected one or more times, is used to estimate the cumulative failure rate. The data for Figure 5-7 is shown in Table 5-8. The data for each point has three elements: inspection time, cumulative number failed, and the constant number inspected in the population for Probit #2. If the cumulative number failed is expressed as a percentage this is Probit #3. SSW now has an option to convert standard data into the Probit format and then into the Interval MLE format. This major advance allows easy comparisons between methods.

(Format 2) The same units are inspected more than once and the sample at each inspection may vary in size. There are several applications that have this format. **Destructive inspection** produces this format. For example, each year the U.S. Navy takes samples of bombs, missiles, and projectiles from the fleet representing each production lot and fires them to estimate shelf life. Duds and misfires are failures. Each sample is a binomial data set, "n" units tested with "f" failures. For destructive inspection, the number failed in each inspection-sample divided by the sample size is used to estimate cumulative % failed.

Production batch failures may also use the Probit destructive data format. For example, assume a batch of 60 units are 15 months old and have experienced 3 failures during the 15 months. Another batch of 40 units is 25 months old and one has failed. For nine batches, the SuperSMITH input might look like 15 x 3 x 60, 25 x 1 x 40, 35 x 6 x 45, 45 x 7 x 40, 55 x 8 x 35, 65 x 10 x 30, 75 x 12 x 25, 85x 13 x 20, 90 x 15 x 25. Here the cumulative % failure may be estimated as 3/60, 1/40, 6/45, etc., and plotted with Probit Option 2. See Figure 5-8. Note that the cumulative failure rate may not monotonically increase; it may sometimes decrease due to statistical scatter. This produces a bad fit in many cases as in Figure 5-8. As each probit inspection represents binomial trials for formats (1) and (2), 95% confidence intervals are calculated from two binomial standard deviations. See Chapter 8. The plot and confidence intervals are obtained from SSW as Inspection Option 2 or 3. Option 3 allows the cumulative failure estimates to be input.

5.8.4 Kaplan-Meier (KM) Inspection Option #4

This method will be described in detail in Chapter 8. *Kaplan-Meier* is widely used in the American medical industry for substantiating new drugs and surgical implants. It is distribution free and is sometimes used for inspection data. Unfortunately, the Actuarial Correction is applied to KM grouped data including inspection data and this modifies the number of suspensions complicating Abernethy Risk predictions. Therefore, it is the least used inspection option.

5.8.5 Interval Maximum Likelihood Estimation (MLE) Inspection Option #5

There is an *MLE Interval* solution for interval or group data available in SSW. This method is quite different from the standard MLE and is not subject to the adverse comments on MLE. On the contrary it is the most accurate of the inspection options. It will handle all kinds of interval data even when the inspection intervals vary from unit to unit, where each unit has its own inspection schedule. There are four main data types, **1) occurrence, 2) suspension, 3) discovery, and 4) interval**.

1. An **occurrence** is a failure age or event recorded precisely at a point on the time scale **(t).** This type of data is only available when the failure can be noticed immediately from sounds, visual cues, or other sensory information.
2. A **suspension** is an unfailed unit where the true time to failure is greater than **(>t)** the age attained so far.

3. A **discovery** is the analytical opposite of a suspension, occurs when a benign (or dormant) failure is discovered upon first inspection. With a discovery, the true time to failure is less than **(<t)** the age recorded at first inspection.
4. Finally, inspection data for benign failures is in the **interval** between the last inspection time (t1) and the next inspection time (t2) where failure was detected. In this case the true failure time is greater than the previous inspection age and less than the detection age **(>t1<t2)**.

Additionally, each data type entry can have a frequency number associated. For example, ">t1<t2x5" indicates that there were five failures found between the ages of t1 and t2. See Appendix C, page C-7 for details on the interval MLE method. Interval MLE equations can be established and solved for problems involving all of these data types intermixed, although the solution is significantly more complex so that convergence is not guaranteed. Many real data situations include a combination of all of these data types. See Appendix C for details on Interval MLE solution for all data types. Todd Marquart has completed a massive Monte Carlo study at the author's request to compare the three methods for interval analysis plus many others. His conclusion is that all three methods are accurate but overall from best to worst the ranking is Interval MLE, Probit, and the Inspection Option, but the differences are small. His complete paper includes results for many different sampling conditions and methods of analysis. Here are typical median results for 1000 trials, three early inspection intervals at 168, 336, and 504, eta of 1000, beta of 2.0, B1 life of 110.3. There were approximately 200 plus failures discovered.

Table 5.7 Simulation Comparison of Interval Data Methods

Parameter Estimation Method	Beta	Eta	B1
1 (MLE)	3.722	736.0	213.5
2 (Interval MLE)	***2.012***	***999.0***	***101.4***
3 (RR, Inspection, Y on X)	2.024	991.0	102.8
3 (RR, Inspection, X on Y)	2.025	989.4	103.0
4 (RR, Probit, Y on X)	2.016	993.4	102.0
4 (RR, Probit, X on Y)	2.016	992.2	102.2

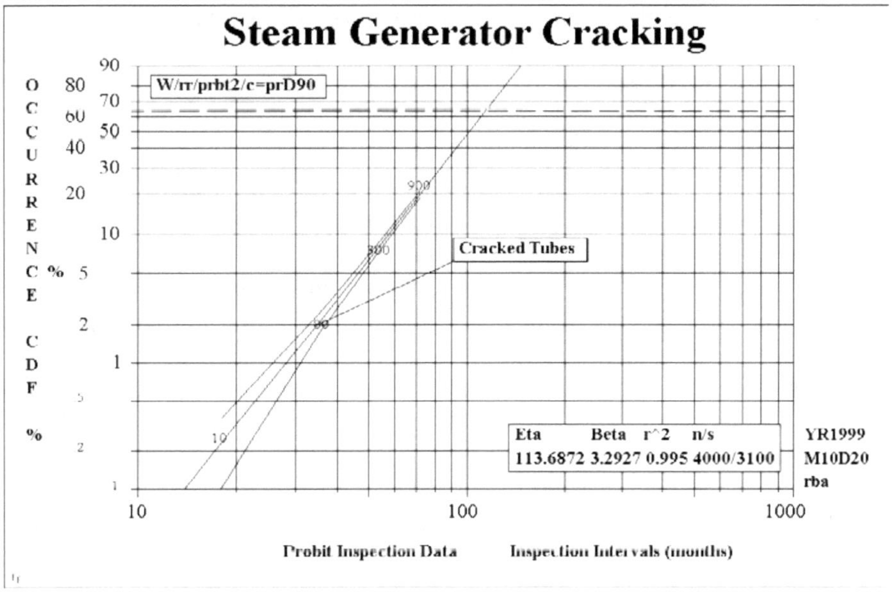

Figure 5-7. Steam Generator Cracking. Note … r^2 of 0.995 gives pve%=73.29

Table 5-8. Nuclear Steam Generator Cracked Tubes

Inspections	18 Months	36 Months	54 Months	72 Months
Cum Failed	10	80	300	900
Cum # Tubes Inspected	4000	4000	4000	4000
SSW Input	18 x 10 x 4000	36 x 80 x 4000	54 x 300 x 4000	72 x 900 x 4000

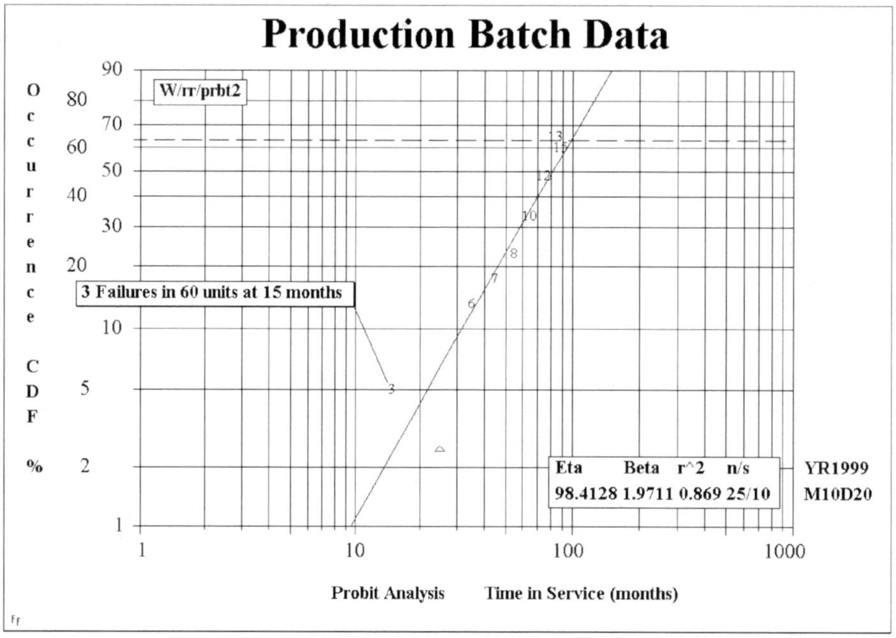

Figure 5-8. Production Batch Data. Note ... r^2 of 0.869 gives pve%=.001 (not good)

5.9 Distribution Analysis

If the Weibull plot is a poor fit it is possible that another distribution may be a better choice. If there is no prior information on what distribution best fits this type of data in the past, a statistical distribution analysis may be helpful. For life data the best alternatives are three parameter Weibull, and the lognormal. The statistical clue will be curvature on the Weibull plot while the engineering clues will come from identifying the physics of failure and prior experience if available. There are complex statistical measures of goodness of fit, but the author prefers the simple correlation coefficient as explained in Chapter 3. The distribution with the highest p-value, "pve%", for r and r^2 (r squared) is the *best statistical* choice. Knowledge about the failure mode, and engineering analysis, will hopefully support the statistical choice and should be weighed at least equally with the statistical choice. SSW will provide the pve%, the percentile of r and r^2 for each distribution. (pve% is smoothed model result from many Monte Carlo simulations and prr% is the similar result but from a single quick simulation.) See Appendix D for new research on goodness of fit.

For small samples (twenty failures or less) distributional analysis may be deceiving and is not recommended. In this case the two parameter Weibull is the **best practice for engineering analysis.** It provides the most conservative B lives and predictions. For intermediate and large size samples, twenty-one or more, there may be enough information to decide if other distributions may be more appropriate.

For MLE-RBA estimates the value of the likelihood function evaluated with the MLE-RBA parameter estimates may be used to gauge the best distribution between the two-parameter versions of the Weibull, normal, and lognormal [Dumonceaux]. The distribution with the highest likelihood value is the best choice. The three-parameter Weibull distribution can be added to the comparison if an adjustment is made for the reduction in degrees of freedom. SSW will provide the distribution analysis using the likelihood function if MLE-RBA or MLE estimation is selected. These two approaches to distribution analysis, p-value for rank regression and the likelihood value for MLE-RBA, are the **best practice**. They employ methods that are entirely independent, so it is worthwhile to compare the results.

A study mentioned earlier in this chapter $[Liu] included a comparison of all of the popular goodness of fit methods and concluded: "...the Maximum Likelihood Test is most powerful. The graphical goodness-of-fit is much more powerful than the Kolmogorov-Smirnov (K-S), and the Chi-Squared tests." The graphical goodness-of-fit is the comparison of the pve% value of r and r-squared from the distributions considered. The distribution with the highest pve% value is the best statistical choice. The Likelihood Test is the test described immediately above. In other words, SSW provides the two best methods for distribution analysis as **best practice**. (Liu) also concludes, "Erroneously classifying data and using one or other model is less damaging if the Weibull model is chosen than if the lognormal model is used especially when infant mortality and useful life periods are concerned. This might be regarded as the most important general conclusion to arise from my study." His discussion indicates that the Weibull is the conservative choice for small samples, twenty or less, even if the data is lognormal. Therefore, distribution analysis is appropriate for moderate and large samples, but not for small samples.

The Anderson-Darling method is sometimes suggested as another alternative for distribution analysis. Wes Fulton has added it to SSW options. Further he has developed a new option that will estimate the "power" of these distribution methods. Power is the probability that given samples from the distribution "A," the goodness of fit method will correctly identify distribution "A" as the source. For the Weibull two parameter, pve% is more powerful than likelihood ratio and Anderson-Darling as shown in Appendix D. Note that the power will be less for three-way comparisons than two-way comparisons. For example, [Dumonceaux] considered a famous bearing data set from [Lieblein and Zelen]. His conclusion based on the maximum likelihood ratio (MLR) test was that the lognormal was the best choice but only marginally better than the Weibull three parameter, a two-way comparison.

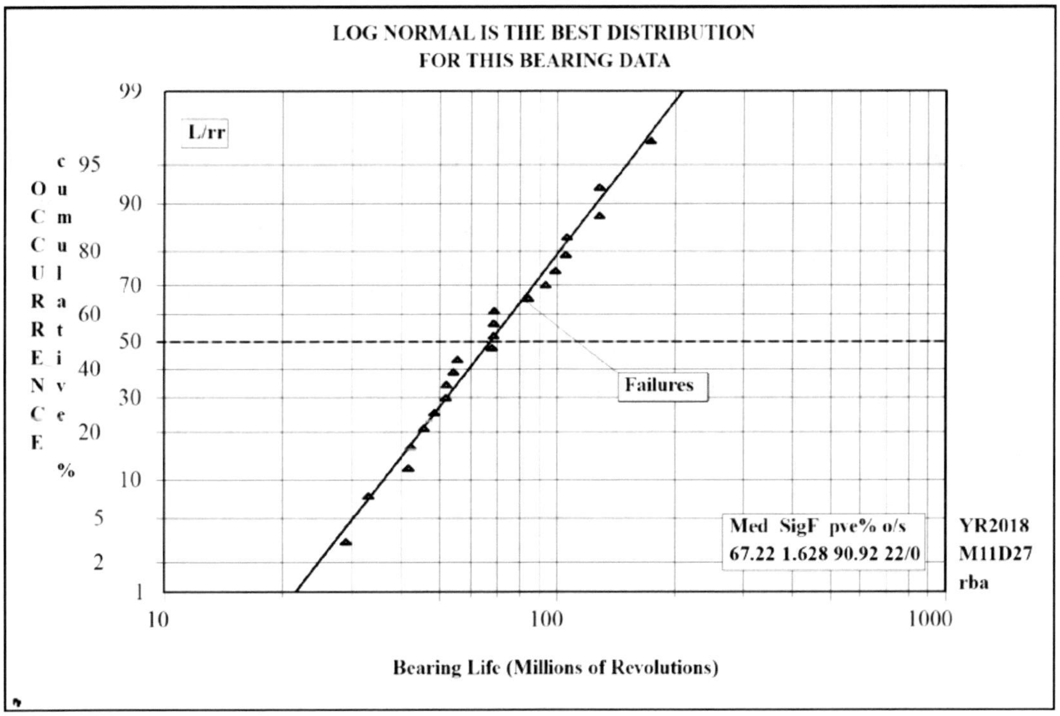

Figure 5-9. Lognormal is the Best Distribution for This Bearing Data

SSW using both tests, MLR and the p-value, agrees with [Dumonceaux]. See Figure 3-7 for the Weibull plot (pve% of 17) and Figure 5-9 for the lognormal plot (pve% above 90). The three parameter Weibull had a pve% of 78. However, the first point in the original data set was deleted as an outlier. The three parameter is the best choice without the deletion. Again, it should be emphasized that the statistical solution is only half the answer; statistics should always be supported by engineering analysis and prior experience. [See Tobias and Trindade]. Appendix D has more on comparisons of goodness of fit methods and their power.

5.10 William S. Gosset

William S Gosset 1876-1937

Gosset, a chemist trained at Oxford joined the Guinness brewery in Dublin in 1899. He was the pioneer of small sample analysis inventing "Student's t," C4, and the "F" test. The author considers him to be the greatest statistician that ever lived. He inspired both Karl Pearson and Ronald Fisher to develop small sample methods. Moreover, among these three peers he was the nice friendly person, the other two being outspoken and arrogant. He was a communication link between these two arch enemies and further, befriended Egon Pearson, Karl Pearson's son, and Jerzy Neyman. When Gosset died unexpectedly in 1937, among the group, his loss was felt most keenly by Egon Pearson, who considered him almost a pseudo father. Egon devoted much of his later life to writing the book ***"Student"*** about Gosset and his relationship with Fisher and Karl Pearson. Gosset became world famous as a statistician but never gave up his pursuit of better beer. Another of his peers was Florence Nightingale David who was also a statistician. She said of the group: "I saw the lot of them. Went fly fishing with Gosset. A nice man. Went to Fisher's seminars with Cochran and that gang. Endured K P. Spent three years with Neyman. Then I was on Egon Pearson's faculty for a year."

Ronald Fisher considered Gosset to be his best friend. Fisher wrote in Gosset's obituary:

> "Student's work has shown that a better course is open to us than that of adopting even the best available estimate of the variance of the population; that, by finding the exact sampling distribution of such an estimate, we may make allowance for its sampling errors, so as to obtain a test of significance which, in spite of these errors, is exact and rigorous. He thus rendered obsolete the restriction that the sample must be "sufficiently large", a restriction vaguely adumbrated by some previous writers, and ignored by others. The claim that "Student's" result is rigorous and exact has, of course, been resented by advocates of "large sample" methods."

And further:

> One immense advantage which "Student" possessed was his concern with, and responsibility for, the practical interpretation of experimental data. If more mathematicians shared this advantage, there can be no doubt that mathematical research would be more fruitfully directed than it often is.

CHAPTER 6. WEIBAYES AND WEIBAYES SUBSTANTIATION TESTING...

Paul Barringer, a close associate, wrote, "The practice of engineering is a combination of art and science. Weibayes analysis is the "glue" that joins engineering art and science together to minimize uncertainties."

6.1 Foreword

Weibayes is defined as Weibull analysis with a given β (slope) parameter. It is a one parameter Weibull. It was developed by the author and other engineers at Pratt & Whitney Aircraft in the seventies to solve problems when traditional Weibull analysis has large uncertainties or cannot be used because there are no failures. Weibayes offers significant improvements in accuracy compared to small sample Weibulls. Therefore Weibayes is **best practice** for small samples and zero failures **if** a reasonable estimate of beta is available. It is helpful to treat data sets with and without failures. Both types of data may have suspensions.

Typical **no-failure situations** include:

1. A redesigned component is under test to demonstrate that a Weibull failure mode has been eliminated or significantly improved. Several units have been tested without failure. Is the testing sufficient or is more testing needed before the redesign is qualified for production?
2. A test plan for a new design or a redesign is needed to show that the design objective has been met or that the new units are significantly better. How many units should be tested for how long?
3. In preliminary design, system reliability models are needed for many purposes, but there is no data available to generate Weibulls. Based on past experience and the corporate Weibull Library, the designer may modify Weibulls from past designs into Weibayes that reflect the new design improvements. These Weibayes distributions may be convoluted to provide the system reliability model.
4. A Weibayes line for a part in-service exceeds the design life providing an increase in design life.

Typical **failure situations**:

1. With small samples Weibayes is more accurate than Weibull if there is prior knowledge of beta. Production acceptance of batches, and research and development tests, are good examples
2. Substantiation tests may be designed with the smallest possible sample sizes if Weibayes is employed.
3. Sudden death tests are also good Weibayes candidates.
4. The age of the units and the number of failures are known, but the failures are not identified. Overhaul depots may generate this kind of data.

The author named this method "Weibayes" when he invented it in the 1970s as it requires judgment when there is minimum prior history on β.In this case Weibayes is regarded by the author as an informal Bayesian procedure. Many Bayesians disagree because the assumed beta is single valued, not a "prior distribution" and therefore not a formal Bayesian method. In many cases beta is essentially known based on a Weibull library of data. Weibull experts agree that the slope, β, tends to be constant with specific or generic failure modes. An experienced Weibull analyst will provide reasonable estimates of β if past Weibull data is not available. Weibayes is more accurate than small sample Weibull analysis if there is good prior information on β.

A library of past Weibull plots is invaluable for Weibayes analysis and many corporations now maintain such libraries. Most have computer search capability. The author highly recommends Weibull libraries as described in Sections 6.20 and 6.21. SSW has capability to make library input almost painless.

A quote from [Hahn 1991]: "The assumption that a parameter of a distribution is known exactly is a strong one, and, if wrong, can result in seriously incorrect inferences. On the other hand, if the assumption is close to being correct, simpler and more precise inferences are possible than if one had to use data to estimate β. In practice, one might compute statistical intervals for different reasonably assumed values of β and, thereby, assess the sensitivity of the resulting intervals to different possible values of β."

6.2 Weibayes Method

In a Weibayes analysis, the slope/shape parameter β is assumed from historical failure data, prior experience, or from engineering knowledge of the physics of the failure. Given β, an equation may be derived using the method of maximum likelihood to determine the characteristic life, η. See Appendix E.

$$\eta = \left[\sum_{i=1}^{N} \frac{t_i^{\beta}}{r} \right]^{1/\beta} \tag{6-1}$$

where: t = time or cycles,
r = number of failed units,
N = total number of failures plus suspension, and
η = maximum likelihood estimate of the characteristic life.

With β assumed and η calculated from Equation 6-1, a Weibull distribution is defined. A Weibayes line is plotted on Weibull probability paper. The Weibayes plot is used exactly like any other Weibull plot. Estimates of B lives, failure forecasts, and reliability are available from Weibayes analysis.

6.3 Weibayes Without Failures

In many Weibayes problems, no failure has occurred. In this case, a second assumption is required. The first failure is assumed to be imminent; i.e., in Equation 6-1, set r = 1.0 (otherwise the denominator in Equation 6-1 would be zero). As no failures have occurred, this is a conservative engineering assumption. The resulting Weibayes line is similarly conservative. Statistically, the Weibayes line, based on assuming one failure, is a lower one-sided confidence estimate. That is, it may be stated with 63.2% confidence that the true Weibull lies to the right of the Weibayes line, if the assumption of β is correct. See Appendix E.

Weibayes lines may be obtained at any level of confidence by employing larger or smaller denominators (assumed imminent failures) in Equation 6-1. For example, 2.3 assumed failures provides 90% confidence. See Appendix E for other values. SSW provides Weibayes lines at any level of confidence.

Historically, Ed Obernesser and Joe Grdenick at Pratt & Whitney did the early development of this analysis in the late 1960s. They were responsible for the original development of the "Weibest" concept. "Weibest" was later modified to "Weibayes" by the author to reflect the fact that β is often an assumption, or a matter of opinion. In the sense that a Bayesian method injects opinion into the quantitative analysis, the author considers Weibayes as an informal Bayesian procedure. If, on the other hand, β is known, Weibayes is better than a Bayesian procedure. Lois Jackson at Gates Rubber and Bhatia at Timken Bearing, independently arrived at the Weibayes concept in the same time period as our development at Pratt & Whitney. [Nelson 1985] provides one-sided confidence bounds for Weibayes with and without failures. See also [Bhatia 1984], [Jackson 1968, 1980], [Lieblein and Zelen 1956], [Medlin 1983], [Harter and Moore 1965].

"Weibest" was an early Pratt & Whitney version based on r = 0.693 assumed failures instead of the usual Weibayes assumption of r = 1.0. 0.693 is less conservative. The "Weibest" line is a 50% confidence bound, a median estimate of the true Weibull characteristic life. Weibest is useful for apples-to-apples, median-to-median, comparisons with Weibull lines. With r = 1.0 assumed (one failure), Weibayes provides a 63.2% lower confidence bound, but Weibayes may be used at any confidence level.

6.4 Weibayes With Failures

When the denominator in Equation 6-1 is based on actual failures, the location parameter, η, is a maximum likelihood estimate (MLE) as shown in Appendix E. A valuable characteristic of maximum likelihood estimates is that they are invariant under transformation. This means that the resulting Weibayes line, B lives, and reliability estimates are all MLE. The Weibayes line is an MLE estimate of the true unknown Weibull, given that beta is known.

Many other statistical estimators are variant under transformation. For example, if the diameters of ten cylinders are measured, the average value will be an unbiased estimate of the true unknown average diameter, but an area calculated from the average diameter would be a biased estimate of the true unknown area. On the other hand, the average diameter is MLE and therefore the biased estimate of the area is also MLE. Confidence interval estimates are another example. As confidence intervals are variant under transformation, rigorously, they may never be employed to calculate other confidence interval estimates.

Weibulls based on samples of 2 or 3, have large uncertainties. See Figure 3-1. If there is good knowledge of β from prior data, significant improvements in accuracy may be obtained with Weibayes. Alternatively, Weibayes may offer cost reductions through reduced testing without loss of accuracy. This is often the case with production acceptance and Sudden Death tests, described later in this chapter. It is also the case when testing a redesigned part in response to a service Weibull failure mode. Confidence intervals will quantify the accuracy improvement with Weibayes compared to Weibull. For example, a company may have employed Weibayes for production acceptance tests for many years and β is extremely well known. To illustrate the Weibayes advantage, assume the original tests required six failures to establish a Weibull B1 life. How would a three failure test plan with Weibayes (β = 2.5, η = 100) compare? The 90% confidence interval for the Weibull six sample B1 interval (2-44) is three and a half times the width of the Weibayes three sample B1 life (9-21). Weibayes provides a significant improvement in precision with half the test time! Of course, it all depends on good knowledge of β.

Many suggest that a Weibull/Weibayes comparison, based on small samples, is a good validity check on the Weibull solution. As Weibayes is more sensitive to suspension times, agreement provides some assurance that there are no abnormalities in the data set. A Weibull library or data bank to provide Weibull slope histories is strongly recommended to obtain the Weibayes advantage.

Weibayes is a one-parameter (η) Weibull. Weibayes will have smaller uncertainties than the two-parameter Weibull. Similarly, the two-parameter (η and β) Weibull has smaller uncertainties than the three-parameter (η, β, and t_0) Weibull. Weibayes uncertainties are reduced by the prior knowledge of the slope, β. For example, compare 90% confidence bounds for B1 life with the one-parameter Weibayes, the two-parameter Weibull and the three-parameter Weibull. For the same conditions, N = 4, β = 3, and η = 100, simulation was used to generate 1000 trials, ranking B1 and produced the following 5% and 95% bounds shown in Table 6-1.

Table 6-1. Ranking B1 Produced 5% and 95% Bounds

True B1 Life = 21.6	*Weibull t_0 3 Parameter*	*Weibull 2 Parameter*	*Weibayes 1 Parameter*
B1 90% Bounds	(-61.8) to 83.0	3.5 to 64.3	15.1 to 26.8
Range	144.8	60.8	11.7

In this case, Weibayes is 12 times more accurate than the three-parameter Weibull and about five times more precise than the two parameter Weibull. This benefit is available **if** β is known. The conclusion is that Weibayes should be considered **best practice** for all small sample Weibulls, 20 failures or less, **if** a reasonable estimate of β is available. With two or more failures, Weibayes solutions using regression analysis are more robust, more consistent and slightly more accurate than MLE solutions as discovered by Todd Marquart and programmed by Wes Fulton.

The distinction between zero failure and one failure Weibayes is worth reviewing. For example, assume five redesigned units have been tested without failure. A Weibayes line is calculated based on the β from the original design. This is a lower one-sided confidence interval for the true unknown Weibull for the redesign. Now assume the same data set includes one failure and four suspensions. The resulting Weibayes is identical to the first zero failure Weibayes, but the interpretation is different. With one failure, the Weibayes is a nominal, MLE, estimate of the true unknown Weibull, not a confidence interval. However, a lower confidence bound for the MLE Weibayes line may be calculated [Nelson 1985] using Chi-Square.

If f is the number of failures (≥1), the lower C% confidence limit for eta is:

$$\eta_c = \eta_{MLE} \left(2f / \chi^2 (C; 2f + 2)\right)^{(1/\beta)} \tag{6-2}$$

Using η_c and β, the lower confidence bound for the true Weibayes line is defined. SuperSMITH will produce Weibayes lower confidence bounds at any level, with or without failures. These bounds are not exact but rather conservative.

6.5 Unknown Failure Times

Sometimes the number of failures is known, but not identified within the data set. Weibayes may provide the solution. For example, an overhaul depot may report that 43 systems were overhauled last month and the times on each are known. Three failed bearings were identified but it is not known which systems they were in. In this case, the number failed and the ages of all 43 bearings are known, but the failures are not identified. A Weibull is impossible, but a Weibayes can be done if a reasonable assumption about β can be made. Use the "LFAIL" option in the SSW Weibayes icon.

If the times on the failed units are known, but the times on the successful units are unknown, a Weibull Dauser shift may be employed. See Chapter 5.

6.6 Weibayes Worries and Concerns

Weibayes is best practice for small samples if beta is known or can be estimated. It is always appropriate to critically question the assumptions required by the Weibayes method since the validity of the results depends on the validity of the assumptions. Typical questions to be raised are:

1. How valid is the assumed slope, β? If this assumption is shaky, a range of slopes may be used to quantify the resulting uncertainty in the results. A library of Weibull experience is invaluable for Weibayes. If there are several relevant Weibull data sets for estimating β, the median value is recommended because β appears to be approximately lognormally distributed. The median value is more typical than the mean value for skewed distributions like the lognormal and most Weibulls. An alternative is to do the analysis twice, at the low and high end of the beta range to bracket the results, and then use the most conservative answer.
2. With a redesign, what is the probability that a new failure mode has been added? A Weibayes test provides insurance against new failure modes that are worse than the mode being demonstrated but no insurance for modes that are better (to the right of the demonstrated mode).

6.7 Weibayes Case Studies

Before looking at three case studies, a word should be said about system models employing Weibayes. In the design phase of a new system, the components may be modeled with a combination of Weibulls and Weibayes from previous designs plus component and specimen tests. Most corporations now maintain a library of Weibull plots from earlier designs (which the author highly recommends in Section 6.19). These are provided to the designer for review before he starts his new design. As the design matures these Weibulls are modified based on new design criteria such as better materials, and better features. The modified Weibulls are Weibayes and form the basis for the preliminary design system reliability model. A new design should not repeat the mistakes of previous designs. Without a systematic method for avoiding past mistakes, "Murphy" says they will be repeated. Failure histories, lessons-learned, and the Weibull library are invaluable for improving the design with Weibayes models. The design engineer should be prepared to defend his new design against all the previous failure modes. Fault trees and failure mode and effect analysis can also help in creating this Weibayes model. As development and production failures occur they are used to update the Weibayes plots for each component. This process continues with service data.

For independent failures modes SSW and SSV may be used to provide failure forecasts for the system based on summing the failure forecasts from the individual Weibulls and Weibayes analysis. See Section 4.6. Most repairable systems have independent failure modes. However, dependant modes such as those that produce secondary damage, catastrophic modes, and benign failures found only at inspection require Monte Carlo modeling. See Section 4.7.

Weibayes Case Study 6-1 Vane and Case. Fifteen vane and case failures have been experienced in a large fleet of engines. Weibull analysis provides a β of approximately 6.0. See Figure 6-1. Three redesigned compressor cases have been tested in engines to 1600, 2900 and 3100 hours without failure. Is this enough testing to substantiate the redesign? Assuming β = 6.0 and given the times on the three redesigned units, Equation 6-1 may be used to calculate the characteristic life for a Weibayes solution.

$$\eta = \left[\frac{(1600)^6 + (2900)^6 + (3100)^6}{1}\right]^{1/6} = 3384 \text{ hours}$$

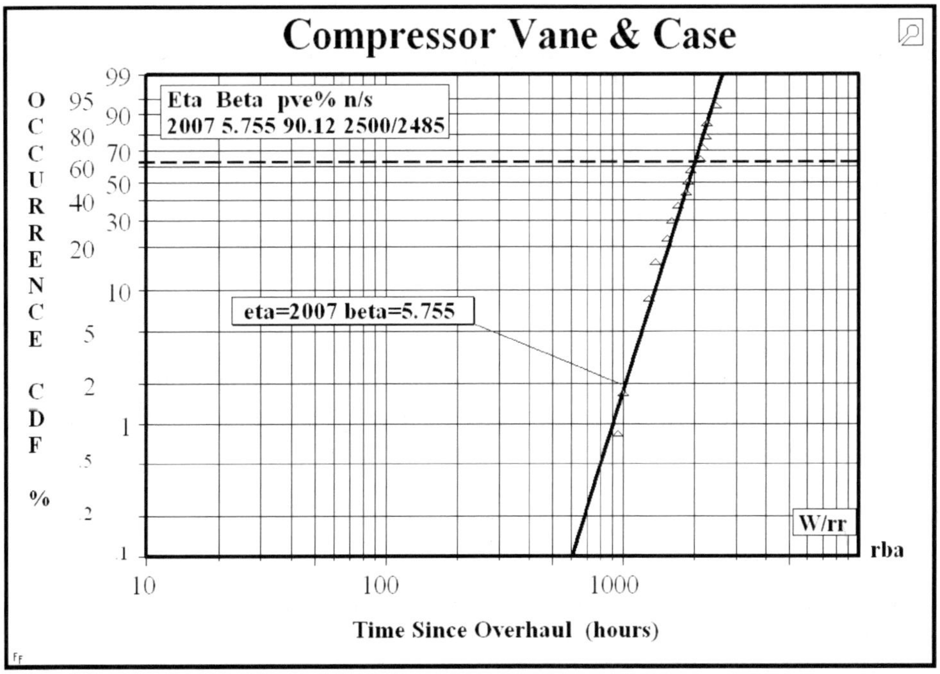

Figure 6-1. Compressor Vane & Case

The null hypothesis for the Weibayes confidence statement is that true unknown Weibull for the redesign is the Weibayes line. Statistically this may be interpreted as the probability of observing the Weibayes data set or better, given the true Weibull is the Weibayes, equals one minus the confidence level.

In this example, if we take $\beta = 6$, and $\eta = 3384$, the probability of surviving to 1600 hours is 0.989, 2900 hours $\approx$ 0.673, and 3100 hours $\approx$ 0.553. The joint probability is the product, approximately 0.368 or 36.8%. This is the α value, the complement of the confidence level of 63.2%. The Weibayes line is plotted in Figure 6-2. We may state with 63% confidence that the Weibull for the redesigned units is to the right of this line and, therefore, significantly better than the bill-of-material vane and case. It is possible that the redesign has eliminated this failure mode but that cannot be proven with this sample of data. As more time is put on these units without failure, the Weibayes line will move further to the right and more assurance will be gained that the failure mode has been eliminated. The assumption of slope, in this case, is based on an established Weibull failure mode and is valid.

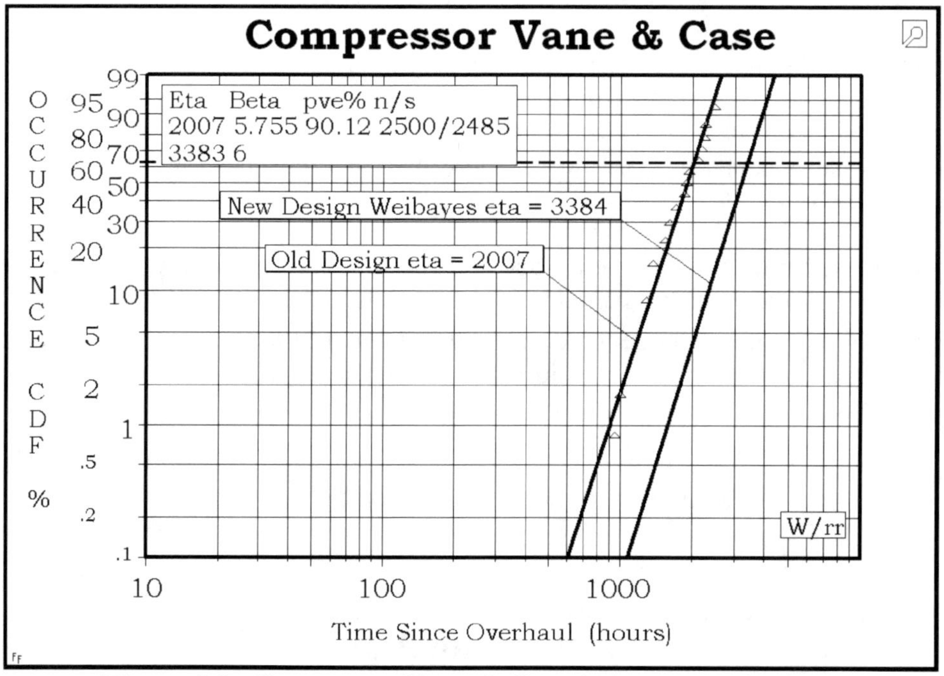

Figure 6-2. Compressor Vane & Case Old and New Designs

Weibayes Case Study 6-2 Turbopumps. [Medlin and Elsaesser 1983] 38 turbopump failures occurred in service on jet engines. See Figure 6-3. Data recreated for this plot is from old records and may be slightly different than original (less than 1 percent). Failure analysis indicated low lubricity of the hydraulic fluid as the cause of the failures. An accelerated bench test was designed using extremely low lubricity fluid. Two more turbopumps failed in a much shorter time, 95 and 155 hours respectively. See Figure 6-4. The almost identical slopes near 2.6 confirmed the capability of the bench test to duplicate the failure mode observed in service. This is an excellent check on the validity of an accelerated test. In addition, the failed parts from the bench test should be compared to the failed parts from service. They should physically look alike. There is always a concern that an accelerated test will produce the wrong failure mode. The ratio of the in-service η to the bench test η is (2255/143) or about 16 to 1; that is ... one hour of test time is equivalent to 16 hours of service time. This is the test acceleration factor. For constant β, the acceleration factor, the ratio of etas is the same as the ratio of MTTFs or any B lives.

The turbo pump was redesigned to fix the problem and two redesigned units were tested on the bench to 500 hours without failure under the same accelerated conditions. Is the redesign successful? What service experience should be expected? Using Equation 6-1 and the slope from Figure 6-3, the Weibayes characteristic life for the redesign is calculated at 653 hours (calculation below), assuming the first failure is imminent. This Weibayes line is plotted on Figure 6-4.

$$\eta = \left[\frac{(500)^{2.59} + (500)^{2.59}}{1}\right]^{\frac{1}{2.59}} = 653 \text{ hours}$$

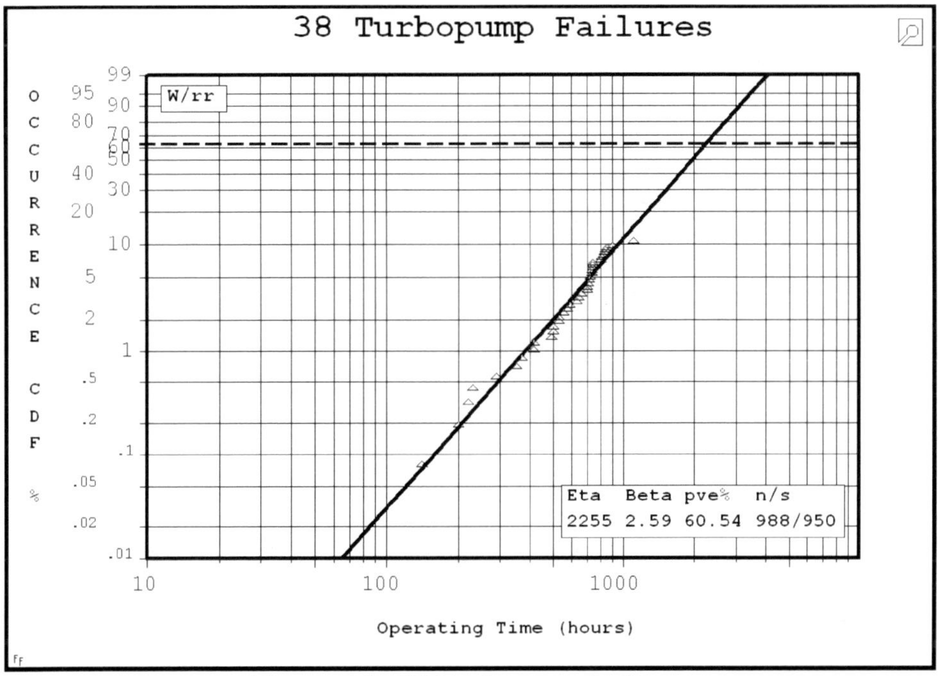

Figure 6-3. 38 Turbopump Failures

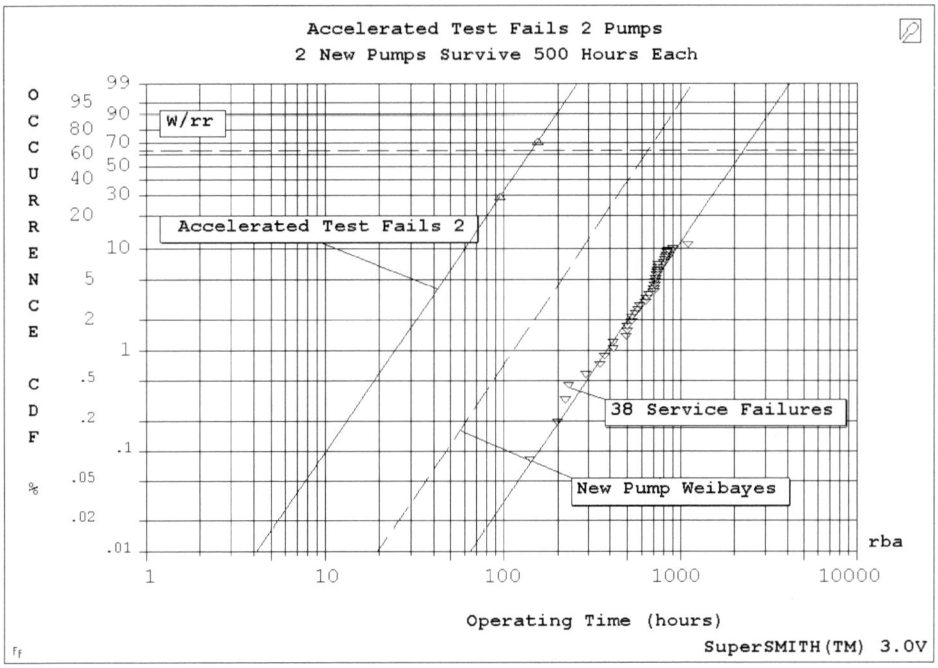

Figure 6-4. Accelerated Test Failures 2 Pumps

A Weibayes line is estimated for the redesigned pump in service by using this ratio. See Figure 6-5. The conclusion is that the new redesigned pump is significantly better (653/143 ≈ 4.5 times better) than the original design. The predicted in-service life is shown on Figure 6-5.

$$\eta_{service} / \eta_{rig} = MTTF_{service} / MTTF_{rig} = B1_s / B1_r$$

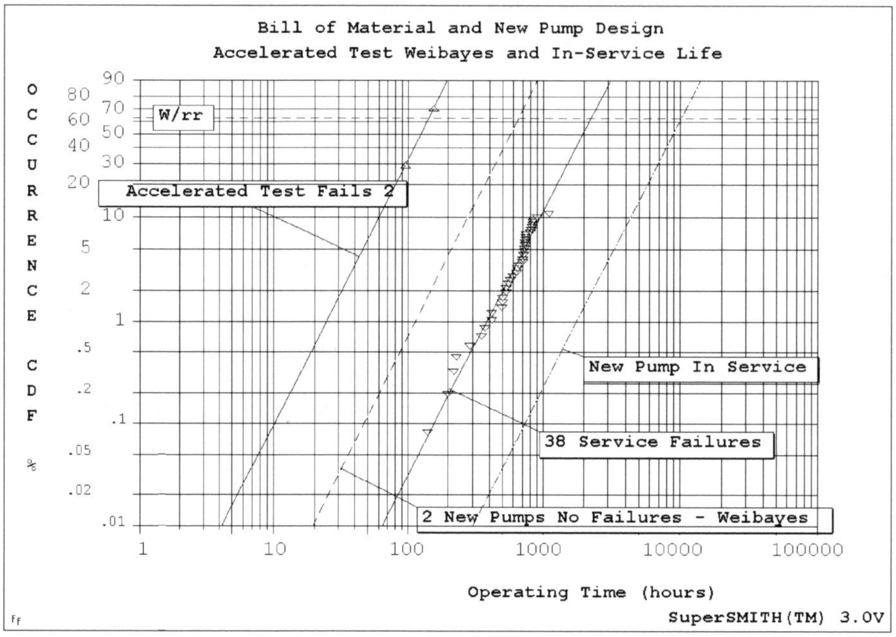

Figure 6-5. Bill of Material and New Pump Design Accelerated Test Weibayes and In-Service Life

Weibayes Case Study 6-3 Bearing. One batch of fighter engine turbopumps (DF3) produced nine service failures involving fire in flight. There were 358 pumps in this batch out of a fleet of 978 units. The original Weibull analysis indicated that it was a batch problem, provided an estimate of the size of the batch, and recommended replacing these pumps after 175 hours of operation. $[Abernethy January 1983] The corrective action was successful. Fifteen of the DF3 replaced pumps were imminent failures for a total of 24 pumps. Two other batches of these pumps, DF1 and DF2, had more service time, but no failures. Teardown and inspection of some of these pumps showed that the failure mode (swelling of the ball bearing plastic cage) was present, but to a lesser degree. There were not enough spare pumps to immediately replace the DF1 and DF2 units. How long can replacement of DF1 and DF2 be safely delayed?

There were no failures in DF1 and DF2 even though symptoms of the failure mode were present. A Weibest analysis, [see Section 6.3], using the existing Weibull slope of $\beta = 4.6$ and assuming the 0.693 failures were imminent, produced the Weibest line shown in Figure 6-6. The 24 failures include original plus imminent. The DF3 retrofit at 175 hours corresponds to a risk level of 7/1000 as indicated in Figure 6-6. The same risk level was applied to the Weibest line and a 700-hour safe period was recommended. DF1 and DF2 pumps were replaced when they reached 700 hours. This did not create a supportability problem as these pumps had acquired very little time. Additional flight time on these units was monitored closely, and the Weibest analysis was updated frequently as more operating flight hours accumulated. Weibest and Weibayes lines move to the right, suggesting longer life, while no failures are observed due to the increase in success time. In this case, the Weibest 7/1000 time eventually exceeded the pump overhaul life of 1000 hours. Therefore, many pumps were utilized to their full life without premature replacement based on Weibest analysis. The Weibest analysis prevented unnecessary down time and needless additional retrofit.

Author's Note: Consider the significant advantages of Weibayes: improved accuracy, reduced costs, and more accurate analysis. See Section 6.19 for more on Weibull libraries.

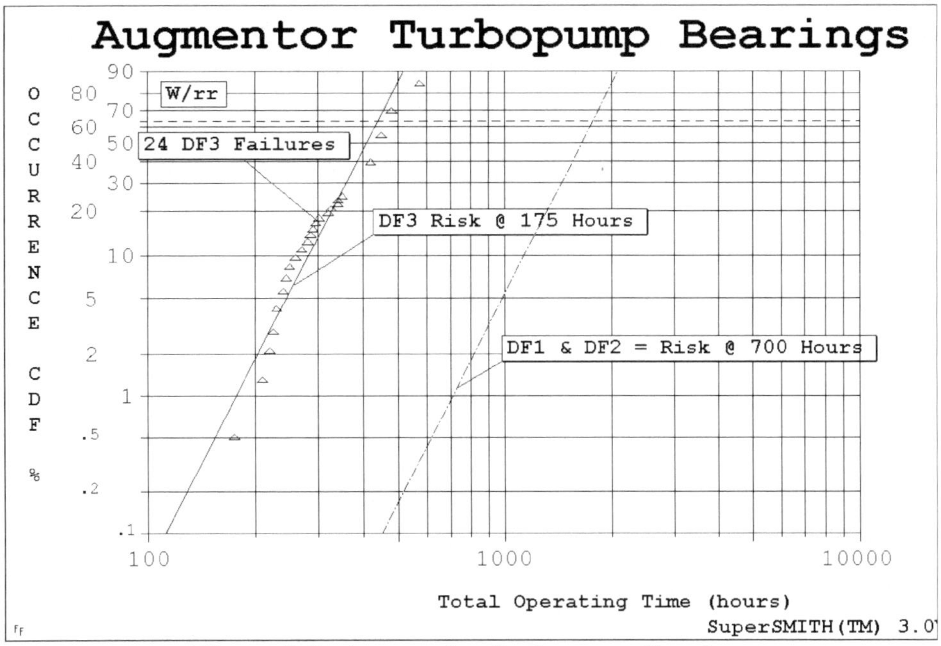

Figure 6-6. Augmentor Turbopump Bearings

6.8 Substantiation and Reliability Testing

In the above examples Weibayes was used to analyze test results. It may also be used proactively, to predetermine test requirements before the test. Test requirements can be reduced based upon knowledge of an underlying distribution such as Weibull.

- Substantiation testing demonstrates that a redesigned part or system has eliminated or significantly improved a known failure mode (β and η are assumed to be known) with some level of confidence greater than 50%.
- Reliability testing demonstrates that a reliability requirement, say 99% reliability at 10,000 cycles with 95% confidence, has been met.
- The scientific method requires the test plan to be documented before the test and that the results are reproducible. Estimates from prior data are valuable but do not "demonstrate" levels of reliability.

Three types of Weibayes are presented herein:

1. Zero failure tests (Section 6.9 and 6.10)
2. Zero or one failure (Section 6.13). (SSW goes further with 0,1,2,...N failures allowed.)
3. Sudden Death (Section 6.14).

Similar normal or lognormal tests are included in the SSW software based on assuming the standard deviation is known, the equivalent of Weibayes.

A **test plan** gives the required number of units to be tested, a stopping rule, and some level of confidence greater than 50%. The stopping rule may be the amount of time to be accumulated on each unit or the number of failures. It also gives a success criterion; the test is passed if the success criterion is met. In a zero-failure test plan the success criterion is **No** failure; the test is passed if every unit runs the prescribed amount of time, and **No** unit fails while on test. In a zero or one failure test plan, the test is passed if there is **one or zero** failures. Test plans that allow one or zero failures trade a higher cost (more testing) for reducing the producers risk, the risk of rejecting a good design.

The **test plan** must be stated *"a priori,"* **before the test**, and agreed to by all parties. A Pre-Test Declaration is used by many organizations for this purpose. It includes the test plan, the test objective, a description of the test articles and instrumentation, test duration, and a hardware comparison of the test articles with the production configuration if there are any differences. It should be signed by the test engineer and the test conductor **before the test.** This engineering discipline was required by NASA for Pratt & Whitney's RL10 rocket engine testing. It proved so useful that it was continued on later Pratt & Whitney projects even when NASA was not involved. Once the test has started the test plan cannot be modified. If you let the results modify the test plan you destroy the statistical confidence characteristics of these plans.

A measure of confidence is built into statistically designed test plans, guaranteeing that if the failure mode in question has not been improved or the reliability requirement has not been achieved, there is low probability that the test will be passed. The zero-failure and zero-one failure test plans provide 90% probability that the test will be failed if the required goal has not been achieved. Thus, a part or system that will have at most a 10% chance of being accepted as satisfactory when in fact it is not.

Prior to the development of these Weibayes plans the popular alternative was to run a small sample of units to failure and plot a Weibull. Usually 5 or 8 units were failed. A one sided 95% confidence bound for B10 life might be compared to the requirement. If the B10 bound exceeded the requirement; a statement was made that the requirement was demonstrated at 95%. For high reliability units this type of Weibull test-to-failure is *expensive, time consuming, and provides erroneous* results in many cases. See Section 6.13. The Weibayes plans presented herein are **best practice**. Another approach was to fail 8 units of the new design and 8 of the old design and compare confidence bounds. This design will be discussed in Chapter 7.

Similar tests for binomial data will be discussed in Chapter 8. Binomial data is counted data, without a time-to-failure scale, such as success-failure rocket firings; go-no go quality data, and destructive tests. If a Weibull test is very short with many units to test, the Weibayes test plan will approach the binomial test.

6.9 Zero-Failure Test Plans for Substantiation Testing

Suppose that ball and roller bearing unbalance test follows a Weibull failure mode, with $\beta = 2$, and $\eta = 500$ hours. This was not adequate life so the system is redesigned and three redesigned units are available for testing. Now the question is: How many hours should each system be tested to demonstrate that this mode of unbalance has been eliminated or significantly improved?

To answer this question Table 6-2 is used. It is entered with the value of β and the number of units to be tested. The corresponding table entry is multiplied by the characteristic life to be demonstrated to find the test time required of each unit. Note that the goal may be to demonstrate an improvement compared to the existing design, or some margin may be added, or the goal may be a new design requirement. In the ball and roller bearing example, Table 6-2 is entered with $\beta = 2.0$ and a sample size of three. The corresponding table entry is 0.876. The characteristic life to be demonstrated is 500 hours. The number of hours that each system should be tested is 0.876 X 500 hours = 438 hours without failure to substantiate the new system is significantly better.

Table 6-2 may be recalculated at any confidence level by employing the following relationship where n = number of units to test and the table entries are k, the ratio of time to η demonstrated:

$$\mathbf{(1 - Confidence) = R^n} \quad (6\text{-}3)$$

$$\mathbf{R = e^{-(t/\eta)^\beta} = e^{-(k)^\beta}}$$

This Equation 6-3 is based on the null hypothesis, that the new design is no better than the old. Given the null hypothesis is true, the probability of passing the test is set equal to one minus the confidence level. Thus, the zero-failure test plan to substantiate the ball and roller bearing system has been fixed is: test three systems for 438 hours each. If all three systems are in balance at the end of the test, then the unbalance mode was either eliminated or significantly improved (with 90% confidence). Note that the Weibayes plot is a good tracking tool during the test; updating the Weibayes plot daily shows what you have demonstrated and when the test will be completed. The amount of improvement is quantified with a Weibayes plot as will be illustrated later. The SSW software will design substantiation and reliability test plans with zero or any number of failures allowed at any confidence level.

Table 6-2. Substantiation Testing: Characteristic Life Multipliers (k)
For Zero-failure Test Plans, 90% confidence, $k = [(-1/n)*\ln(0.1)]^{(1/beta)}$

	Beta (β)									
	0.5	1.0	1.5	2.0	2.5	3.0	3.5	4.0	4.5	5.0
n	Infant Mortality	Random	Early Wear Out					Old Age Rapid Wear Out		
2	1.325	1.151	1.098	1.073	1.058	1.048	1.041	1.036	1.032	1.029
3	0.589	0.768	0.838	0.876	0.900	0.916	0.927	0.936	0.943	0.948
4	0.331	0.576	0.692	0.759	0.802	0.832	0.854	0.871	0.885	0.895
5	0.212	0.461	0.596	0.679	0.733	0.772	0.801	0.824	0.842	0.856
6	0.147	0.384	0.528	0.619	0.682	0.727	0.761	0.787	0.808	0.826
7	0.108	0.329	0.477	0.574	0.641	0.690	0.728	0.757	0.781	0.801
8	0.083	0.288	0.436	0.536	0.608	0.660	0.701	0.732	0.758	0.780
9	0.065	0.256	0.403	0.506	0.580	0.635	0.677	0.711	0.739	0.761
10	0.053	0.230	0.376	0.480	0.556	0.613	0.657	0.693	0.722	0.745
12	0.037	0.192	0.333	0.438	0.517	0.577	0.624	0.662	0.693	0.719
14	0.027	0.164	0.300	0.406	0.486	0.548	0.597	0.637	0.670	0.697
16	0.021	0.144	0.275	0.379	0.461	0.524	0.575	0.616	0.650	0.679
18	0.016	0.128	0.254	0.358	0.439	0.504	0.556	0.598	0.633	0.663
20	0.013	0.115	0.237	0.339	0.421	0.486	0.539	0.583	0.619	0.649
25	0.008	0.092	0.204	0.303	0.385	0.452	0.506	0.551	0.589	0.621
30	0.006	0.077	0.181	0.277	0.358	0.425	0.480	0.526	0.565	0.598
40	0.003	0.058	0.149	0.240	0.319	0.386	0.442	0.490	0.530	0.565
50	0.002	0.046	0.128	0.215	0.292	0.358	0.415	0.463	0.505	0.540

If the amount of test time accumulated on each unit is constrained, Table 6-3 can be used to determine the number of units required for the test. For example, suppose in the previous example that no more than 300 hours could be accumulated on any bearing system. Table 6-3 is entered with the known value of β and the ratio of the test time to the characteristic life being substantiated. In the ball and roller bearing system example, Table 6-3 is entered with β = 2.0 and the ratio:

$$\frac{\textbf{300 test hours per system}}{\textbf{500 hour } \eta} = 0.60 = t/\eta$$

The corresponding entry in Table 6-3 is seven. The resulting test plan is test seven systems for 300 hours each. If all seven systems are in balance at the end of the test, then the unbalance mode was either eliminated or significantly improved (with 90% confidence).

In a no-failure test, if a failure occurs prematurely, the test has been failed. A redesign is in order. It is inappropriate and in fact, *wrong, cheating,* to test another set of the same design or to shift to a one or zero failure test plan. The test plan cannot be modified after the test has started. There may be large economic reasons for these inappropriate responses to failure, but they should be resisted. For example, if enough sets of three roller bearings is allowed to be tested, a set will be found that passes the test even if there is no improvement.

Table 6-3. Substantiation Testing: Required Sample Sizes (N) for Zero-failure Test Plans, 90% confidence, $n = -((\eta/t)^{\beta})*\ln(0.1)$

	Beta (β)									
	0.5	1.0	1.5	2.0	2.5	3.0	3.5	4.0	4.5	5.0
t/η	Infant Mortality	Random	Early Wear Out					Old Age Rapid Wear Out		
0.01	24	231	2303	23026	230259	---	---	---	---	---
0.02	17	116	815	5757	40705	---	---	---	---	---
0.03	14	77	444	2559	14772	85281	492370	---	---	---
0.04	12	58	288	1440	7196	35978	179890	899448	---	---
0.05	11	47	206	922	4119	18421	82380	368414	---	---
0.06	10	39	157	640	2612	10661	43520	177669	725330	---
0.07	9	33	125	470	1777	6714	25374	95902	362473	---
0.08	9	29	102	360	1273	4498	15901	56216	198752	---
0.09	8	26	86	285	948	3159	10529	35096	116984	---
0.10	8	24	73	231	729	2303	7282	23026	72815	---
0.20	6	12	26	58	129	288	644	1440	3218	7196
0.30	5	8	15	26	47	86	156	285	520	948
0.40	4	6	10	15	23	36	57	90	143	225
0.50	4	5	7	10	14	19	27	37	53	74
0.60	3	4	5	7	9	11	14	18	23	30
0.70	3	4	4	5	6	7	9	10	12	14
0.80	3	3	4	4	5	5	6	6	7	8
0.90	3	3	3	3	3	4	4	4	4	4
1.00	3	3	3	3	3	3	3	3	3	3

6.10 Zero-Failure Test Plans for Reliability Testing

This section contains zero-failure test plans for demonstrating a reliability goal when the underlying failure distribution is Weibull with known shape parameter β. First, the reliability goal is re-expressed as a characteristic life goal, and then the test plan is designed.

6.10.1 Re-expression of Reliability Goal to Determine η

Reliability requirements generally assume one of the following forms: [χ is known in all forms].

Form 1: The requirement is stated as a reliability. The reliability of the unit is required to be at least 10,000cycles

Form 2: The requirement is stated as a B life. The B10 life (or B1 life, or B.1 life, etc.) is required to be at least X hours or cycles. By definition, the unit has a 10% chance of failing before reaching its B10 life, a 1% chance of failing before reaching its B1 life, etc.

$$R(t) = e^{-(t/\eta)^{\beta}}$$

Form 3: The requirement is stated as a MTTF. Using the known β, convert the MTTF into an equivalent η using Figure 2-3 or Equation 2-4 or the calculator option in SSW.

Reliability requirements in any of these three forms are transformed into an equivalent characteristic life requirement. Given that the time-to-failure distribution is Weibull, with a known β, reliability at time t is a function of η: This expression can be rearranged algebraically giving:

$$\eta = \frac{t}{\left(-\ln R(t)\right)^{1/\beta}} \tag{6-4}$$

Example: A turbine engine combustor reliability goal is a B1 life of 1800 cycles or longer under service-like conditions. Success is defined as a combustor having no circumferential cracks longer than 20 inches (out of a possible 53 inches). The number of cycles required to reach a 20-inch crack was known to follow a Weibull distribution with β = 3.0. How many combustors must be tested, and how many cycles must each accumulate, to demonstrate this goal with 90% confidence?

Equation 6-4 can be used to express either Form 1 or 2 above into equivalent values of η. If the requirement is, for example, that the reliability of the turbine engine combustor must be at least 0.99 at 1800 cycles (β = 3), then substituting t = 1800 and R(t) = 0.99 into Equation 6-4 gives:

$$\eta = \frac{1800}{\left(-\ln(0.99)\right)^{1/3}} = 8340.9$$

The 0.99 reliability requirement is equivalent to the requirement that η be at least 8340.9 cycles. See Figure 6-8. This is derived from "not more than 1% failures occur before 1800 cycles." Figure 6-9 illustrates an example based on a B10 life requirement.

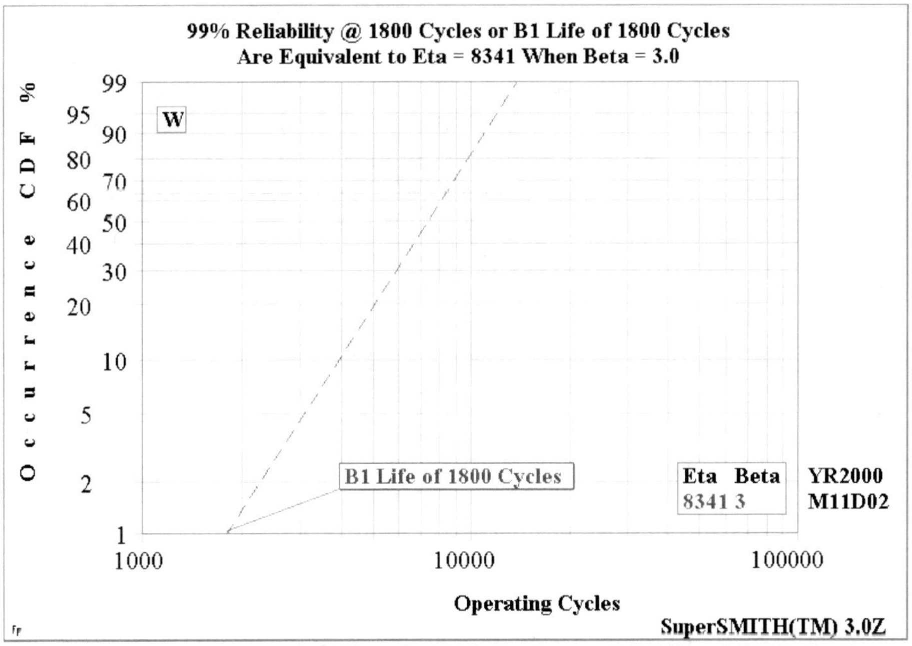

Figure 6-8. 99% Reliability @ 1800 Cycles or B1 Life of 1800 Cycles are Equivalent to Eta = 8341 When Beta = 3.0

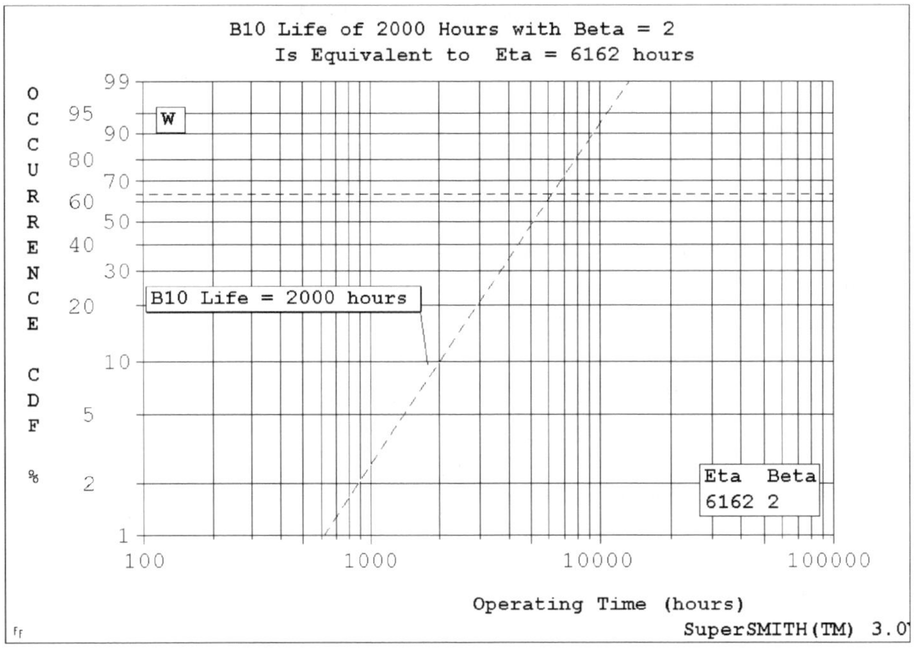

Figure 6-9. B10 Life of 2000 Hours with Beta = 2 is Equivalent to Eta = 6162 Hours

6.10.2 Tailoring and Designing Test Plans

Once the minimum η requirement has been calculated, Tables 6-2 and 6-3 can be used to design the zero-failure test plan. In the combustor reliability example, the 99% reliability goal at 1800 cycles was reexpressed as an 8340.9 cycle characteristic life goal. Ten combustors were available for this reliability demonstration test. To find the test cycles required of each combustor, enter Table 6-2 with β equal to 3.0 and a sample size of 10. The corresponding table entry is 0.613. Multiply the table entry by the characteristic life requirement to find the test time required of each unit. In the combustor example, multiplying the Table 6-2 entry of 0.613 by the characteristic life requirement of 8340.9 cycles gives a test time of: **0.613*8340.9 cycles = 5113.0 cycles**. Thus, the 90% confidence zero-failure test plan to demonstrate 99% reliability at 1800 cycles requires testing 10 combustors for 5113 cycles each. If no combustor develops a circumferential crack longer than 20 inches, then the test is passed.

How many combustors are required if each can accumulate 7500 test cycles? To answer this, enter Table 6-3 with the assumed value of β, the Weibull slope parameter, and the ratio of the test time to the calculated characteristic life requirement. In the combustor example, β is 3.0, and the ratio of the test time to the calculated characteristic life requirement is:

$$\frac{[7500\ \text{test cycles per combustor}]}{8340.9\ \text{cycles}} = 0.90$$

The corresponding entry in Table 6-3 is 4. The resulting test plan requires testing 4 combustors for 7500 cycles each. If no combustor develops a circumferential crack longer than 20 inches, then the test is passed.

Test Units With Dissimilar Test Times:

Julius Wang, Daimler-Chrysler, provided a useful Weibayes equation for zero failure testing that applies when the units are not tested to the same intervals. See [Julius Wang 2004]. The paper contains many examples.

$$R(t_d) = \exp\left[\frac{\ln(1-C)}{\sum_{i=1}^{n}\left(\frac{T_i}{t_d}\right)^{\beta}}\right]$$

where: $R(t_d)$ = reliability at time "t_d,"

t_d = design life or life at a reliability to be demonstrated,

n = sample size,

T = total test time of each part where T_i represents censored times of individual part and n is the total number of sample size.

β = Weibull shape or slope parameter

C = statistical confidence level specified in probability

This equation suggests that each part can be stopped at different times. The usefulness of this equation is for trading off on number of parts and test times when certain parts are no longer available during the testing. This may be due to the fact that certain testing fixtures are broken, certain parts may be required for other purposes and being pulled out of the testing, etc. Note that the basic test plan is not modified.

Example: The plan is to test 14 parts without failure to 1475 hours to demonstrate 95% reliability at 1000 hours (design life in lab testing) at 90% statistical confidence level. The Weibull slope is assumed to be 3.0. Right in the middle of the testing, two parts were stopped at 450 hours due to a broken test stand connector. At this time, another department requires these two parts for other use so you lose these two parts. How long should the other 12 parts be tested to maintain the reliability demonstration requirement?

$$\boldsymbol{R(1000) = 0.95 = \exp\left[\frac{\ln(1-0.9)}{2\left(\frac{450}{1000}\right)^3 + 12\left(\frac{T}{1000}\right)^3}\right]}$$

Solution: Using the above equation, we obtain

Solving for T, we obtain T = 1550 hours. Therefore, the remaining 12 parts will be tested to a total of 1550 hours, rather than 1475 hours as originally planned.

SSW provides a solution to the following similar problem: a number of parts have been tested without failure to dissimilar times. How much additional test time is needed on each part without failure to demonstrate a characteristic life or B life with some level of confidence?

6.11 Total Test Time

Two reliability test plans were constructed in Section 6.10.2 to demonstrate that a characteristic life was at least 8340.9 cycles, with 90% confidence.

Number of Combustors		Test Cycles Per Combustor	Total Test Cycles
Plan 1	4	7500	4 x 7500 = 30,000
Plan 2	10	5113	10 x 5113 = 51,130

Note that, in terms of total test cycles, it is more cost effective to test the smaller number of combustors for a greater number of cycles per combustor. Plan 1 demonstrates the same reliability as Plan 2, but requires fewer total test cycles. This efficiency is realized for all test plans in this section when $\beta > 1.0$. The situation is reversed for $\beta < 1$. In this case, the greater the number of units on test, the lower is the total test time. When $\beta = 1$, the total test time is constant, regardless of the number of items on test.

For wear out modes, if the new design is significantly better, the test plan with a small number of units tested longer will be most cost effective.

6.12 Test-to-Failure Versus Weibayes Zero-Failure Tests

Some manufacturers still **test all units to failure** with small samples using Weibull, not Weibayes. From the resulting plot they compare the lower one-sided bound with the requirement expressed as a B life. For example with the bearing test the procedure might be to demonstrate at 90% confidence a B1 life of 50.1 which is equivalent to eta = 500 and beta = 2.0. The use of these test designs may be caused by a reluctance to assume a constant beta, or lack of information (no Weibull library), or simply ignorance of the great advantages of Weibayes testing. A comparison of the three failure Weibull bearing test with the Weibayes no failure test described above follows.

Comparing **test time**: to fail all three roller bearings would require the average time to failure (MTTF) to be about 880 hours per unit if the new design is twice as good as the old with an eta = 1000. This compares to 438 hours for the zero failure test which provides 90% confidence. Comparing **confidence**: if the failure test of all three bearings employs a one sided Fisher Matrix 90% bound, it provides an effective confidence level of about 75% due to the small sample. See Chapter 7.

Lastly, comparing the probability of passing: if the new units have a characteristic life twice as good as the demonstration value, the **probability of passing** the Weibull failure test is 36.8%, while the Weibayes zero failure test is 57%. If the new units are 50% better, the probability of the failure test passing is 31.9% while the zero failure test is over 35.9%. Here the test time for the failed units would average 660 hours each while the zero failure remains at 438 hours. Clearly the Weibayes zero failure test is more accurate, more cost effective, and quicker.

Sometimes if a manufacturer flunks the all-failure test, he will take a second sample to failure, merge it with the first, and check the lower one-sided 90% bound. This is inappropriate, *wrong*, and will reduce the confidence level to about 70% from the 75% above, a long way from the required 90%.

However, for these marginally better units the probability of passing the test is still too low even for the Weibayes zero failure test. See Figure 6-10. One remedy for this problem is to employ a zero or one failure plan that will be discussed in the next section. Another approach is to reduce the minimum characteristic life requirement enough to guarantee a suitably high probability of acceptance for high reliability units. The curves in Figure 6-10 assist the experimenter in determining how much to reduce the minimum characteristic life requirements to meet high reliability requirements. They give the probability of passing the zero-failure test as a function of the Weibull parameter β and the ratio of the characteristic life of interest to the minimum required characteristic life (Eta demonstrated). For example, the probability of successfully completing a zero-failure test, beta = 2.5, eta = 2000 hours for units whose characteristic life is 4000 hours is 0.67. To see this, enter the $\beta = 2.5$ curve of Figure 6-10 with the x-axis ratio, r, written as:

$$r = (\text{True Eta})/(\text{Eta Demonstrated}) = 4000/2000 = 2.0$$

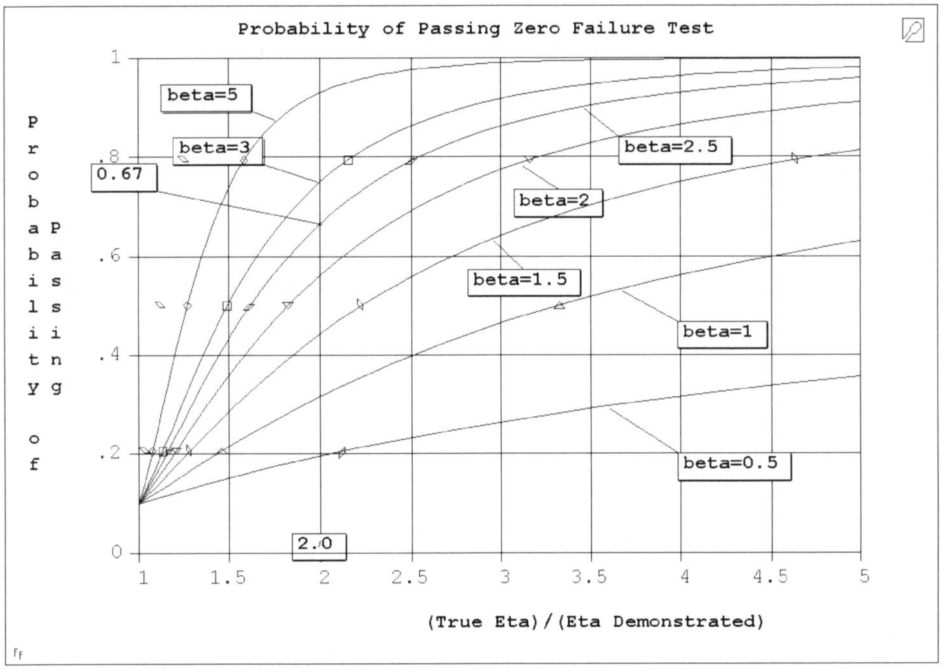

Figure 6-10. Probability of Passing Zero Failure Test

The curves in Figure 6-10 are based on the following equation:

$$\text{Probability of Passing} = R^N = \left(e^{-(k/r)^{\beta}}\right)^N$$

where:

k = Table 6-2 coefficient

r = Ratio of $\eta_{TRUE} / \eta_{\text{DEMONSTRATED}}$

N = Number of units tested

The test time per test unit ("t") is k from above multiplied by the Eta demonstration goal. Note that for a particular β and a particular improvement ratio ("r"), all the test plans have the same probability of passing. If the probability of failing is considered instead of the probability of passing, the curves in Figure 6-10 become Weibull distributions. See Figure 6-11. Using this approach and SSW, the reader may easily calculate the exact probability of failure for his specific test plan and β. The approach is to use the "Marker Line" option, defining the slope and one point on the Weibull. The one point common to all these plans is y = B90 and x = 1.0. The horizontal scale is the ratio of the true η to the demonstrated η. This is the reciprocal of r, the ratio shown on Figure 6-10. (Note Figure 6-10 is obtained by plotting the Figure 6-11 Cumulative Probability Function using SuperSMITH Visual to transform the scales).

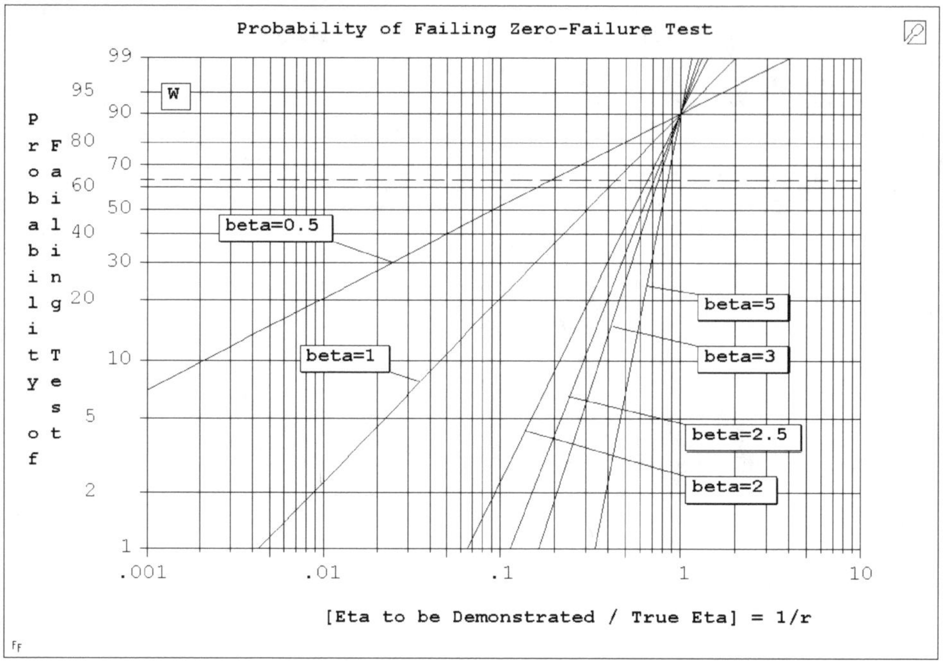

Figure 6-11. Probability of Failing Zero-Failure Test

Reliability demonstration tests that terminate successfully with no failure have one other advantage. Very high reliability often makes a demonstration test-to-failure impractical. In this case, a zero-failure test plan is desirable. The disadvantage is that a Weibayes test will estimate the characteristic life, but it may be grossly conservative, based on zero failures.

6.13 One or Zero Failure Test Plans

Sometimes enough is known about the failure mode to be able to "predict" that the new design will be only marginally better than the old design. This means the new design is likely to fail the zero failure test plans. See Figures 6-10 and 6-11. A one or zero test plan will reduce the probability of failing the test. For this reason a one or zero failure-testing table is presented in Table 6-4.

$$\mathbf{R} = \mathbf{e}^{-(\mathbf{t}/\eta)^{\beta}} = \mathbf{e}^{-(\mathbf{m})^{\beta}}$$

The coefficients in Table 6-4, m, are derived from the binomial distribution as follows:

$$(\mathbf{1} - \mathbf{Confidence}) = \boldsymbol{R^n} + \boldsymbol{nR^{(n-1)}}(\mathbf{1} - \boldsymbol{k}) \tag{6-5}$$

with symbols as defined for Equation 6-3.

Table 6-4 is used in the same way as Table 6-2 to set up one or zero failure tests. Repeating the example of Section 6.9 enter Table 6-4 with $\beta = 2$ and a sample size of three. The corresponding table entry is 1.277. The characteristic life to be demonstrated is 500 hours. In the case of expecting a failure in this test plan the number of hours that each system should be tested is 1.277 x 500 hours = 639 hours.

Therefore, to substantiate the new ball and roller bearing system the one-zero failure test plan is to test three systems for 639 hours each. If no more than one system is out of balance at the end of the test, then the unbalance mode was significantly improved (with 90% confidence). By allowing one or zero failures, the price paid is 201 hours of testing per unit or perhaps, 603 hours of testing total. However, if the bearings are tested sequentially and the first two do not fail, the test is successfully completed, with a reduction in total test time. In this case the zero-one test is a *sequential test plan*. Weibayes plots may be used to track and quantify the improvement with the new design for both zero and zero-one test plans. Figure 6-12 illustrates the

Weibayes lines for the bearing example. The zero-one Weibayes is to the right of the zero Weibayes if more testing is required. The Weibayes line is a lower one-sided confidence bound if it is based on zero failures, but if a failure has occurred it is a maximum likelihood estimate of the Weibull line.

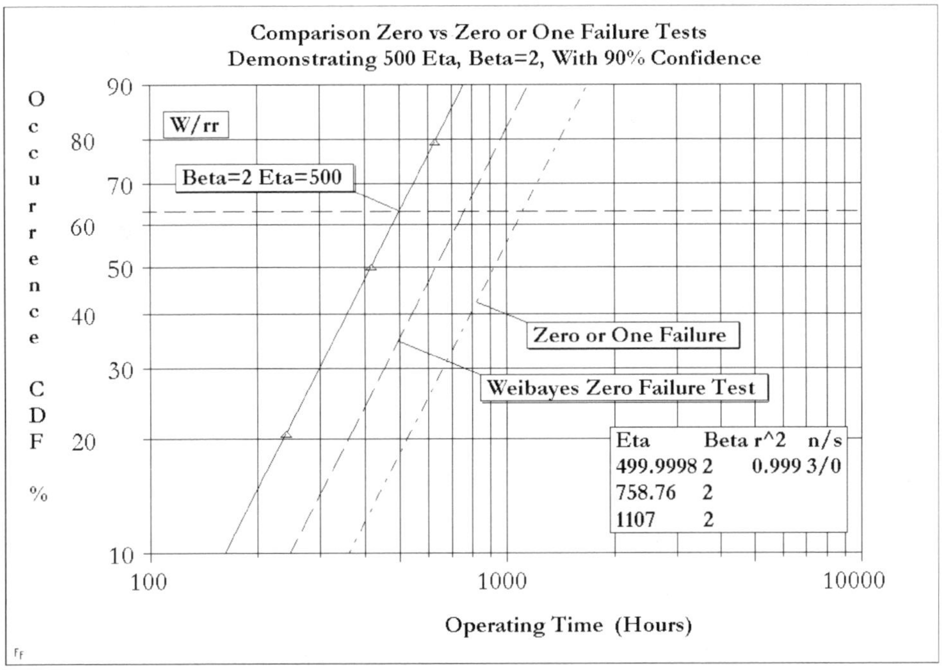

Figure 6-12. Comparison Zero vs. Zero or One Failure Tests Demonstrating 500 Eta, Beta = 2, With 90% Confidence

One advantage of one-failure tests is the results may provide a one-failure Weibayes estimate for the new unit. This will be more accurate than the zero failure Weibayes from zero-failure tests. Another advantage is that the marginally improved redesign will have a better chance of passing a one-zero plan than a zero plan. For the bearing example, the improvement in the probability of passing the test is shown on Figure 6-13. SSW extends this capability to beyond zero-one to zero-one-two and even further, for Weibull, normal and lognormal. The probability of passing the zero-one test is:

$$\textbf{Probability of passing} = R^n + n\left(R^{n-1}\ (1 - R)\right) \qquad (6\text{-}6)$$

In equation (6-7) *m* is the value from Table 6-4 and *r* is the ratio of the new η to the goal η. SSW will calculate the probability of passing the zero or zero-one test plan. In contrast to the Zero Test Plans, the probability of passing the zero-one test increases as the number of units tested increases for a constant β and demonstrated η.

$$R = e^{-(m/r)^{\beta}} \qquad (6\text{-}7)$$

Table 6-4. Characteristic Life Multipliers for One or Zero Failure Test Plans, 90% Confidence, Calculated with Equation 6-5

	Beta (β)									
N	0.5	1.0	1.5	2.0	2.5	3.0	3.5	4.0	4.5	5.0
2	8.819	2.969	2.066	1.723	1.545	1.437	1.365	1.313	1.274	1.243
3	2.659	1.630	1.385	1.277	1.216	1.177	1.150	1.130	1.115	1.103
4	1.295	1.138	1.090	1.067	1.053	1.044	1.037	1.033	1.029	1.026
5	0.769	0.876	0.916	0.936	0.949	0.957	0.963	0.968	0.971	0.974
6	0.510	0.714	0.799	0.845	0.874	0.894	0.908	0.919	0.928	0.935
7	0.363	0.603	0.713	0.776	0.817	0.845	0.865	0.881	0.894	0.904
8	0.272	0.521	0.648	0.722	0.771	0.805	0.830	0.850	0.865	0.878
9	0.211	0.459	0.595	0.678	0.733	0.772	0.801	0.823	0.841	0.856
10	0.169	0.411	0.553	0.641	0.701	0.743	0.776	0.801	0.821	0.837
12	0.115	0.339	0.486	0.582	0.649	0.697	0.734	0.763	0.786	0.805
14	0.083	0.289	0.437	0.537	0.608	0.661	0.701	0.733	0.759	0.780
16	0.063	0.251	0.398	0.501	0.575	0.631	0.674	0.708	0.736	0.759
18	0.049	0.222	0.367	0.472	0.548	0.606	0.651	0.687	0.716	0.740
20	0.040	0.200	0.342	0.447	0.525	0.584	0.631	0.668	0.699	0.725
25	0.025	0.159	0.293	0.399	0.479	0.542	0.591	0.631	0.664	0.692
30	0.017	0.132	0.259	0.363	0.445	0.509	0.561	0.603	0.638	0.667
40	0.010	0.098	0.213	0.314	0.396	0.462	0.515	0.560	0.597	0.629
50	0.006	0.079	0.183	0.280	0.362	0.428	0.483	0.529	0.568	0.601

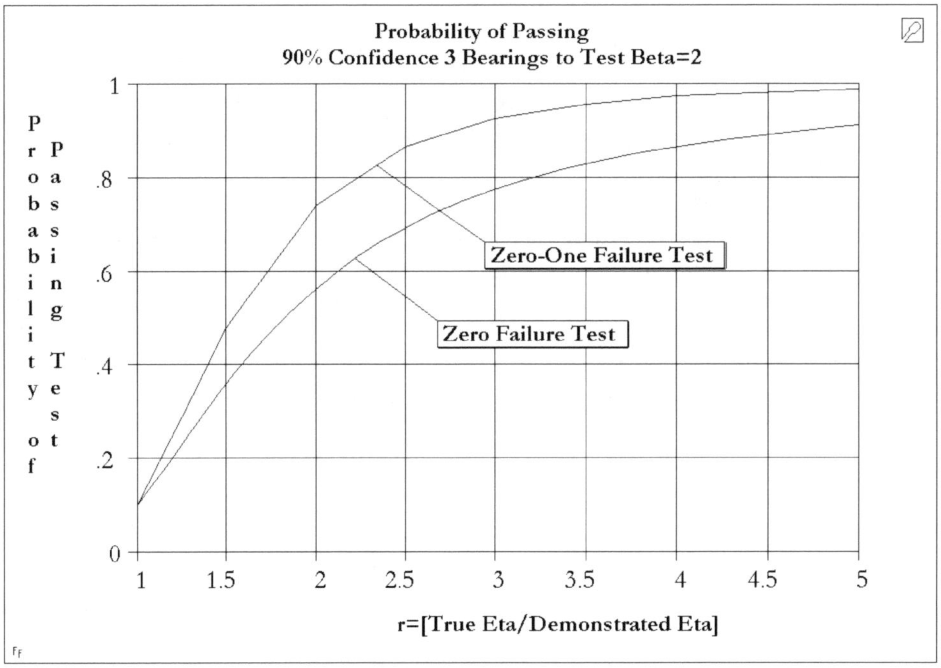

Figure 6-13. Probability of Passing 90% Confidence 3 Bearings to Test Beta = 2

6.14 Sudden Death Tests With Weibull and Weibayes

Sudden Death (SD) testing consists of testing sets of components until the first unit fails. It is widely used in the automotive industry for bearings, wheels, chains, etc. The total number of test specimens is divided randomly into equal groups. Each group of test specimens is tested until the first unit fails, i.e., sudden death. If the cost of testing allows test-to-failure, "Sudden Death" testing is highly recommended. If possible, accelerated testing should also be considered to increase the benefits of SD tests. Compared to the zero and zero-one designs just presented, SD has the advantage of providing the Weibull plot with failures but the test times are longer than a zero-failure Weibayes test. The added information may be useful for warranties and guaranties to show that requirements have been met. After the first failure has occurred, a Weibull plot is made for the failed and suspended units. This plot is updated as more groups are tested.

Leonard Johnson is given credit for inventing Sudden Death. See [Johnson] & [O'Connor]. As conceived decades ago for hand plotting, the method of analysis was quick and easy. (Johnson's procedure may sound a little like the Dauser Shift.)

1- Divide the available number of samples randomly into A subgroups each consisting of B components.
2- SD test rigs are used to test all the units in each subgroup simultaneously.
3- At the first failure in each subgroup the test is ended.
4- Plot the A failures on Weibull probability paper. Draw a best fit line through the data points.
5- Read the median (B50) life from the failure plot. Use this B50 life as the X coordinate and plot a point at the median rank of the first failure of the subgroup size as shown in the table below as the Y coordinate. Lay in a Weibull line through point with the beta of the failures only plot.

Table 6.5 1st Failure Median Rank for Subgroup Sample Sizes

Subgroup size B	***1st Failure Median Rank %***
2	**29.29**
3	**20.63**
4	**15.91**
5	**12.94**
6	**10.91**
7	**9.43**
8	**8.30**
9	**7.41**
10	**6.70**

Now that we have computers and friendly software, we may plot the A failures and A*(B-1) suspensions just like any standard life data. The two methods, with and without suspensions, compare very well. Including the suspensions provides a slightly more conservative answer but the difference is negligible. Almost all bearings throughout the world are tested with SD, usually in subgroups of size four.

Sudden Death produces a reasonably accurate plot in a fraction of the test time that would have been necessary to fail all the units. There will be large cost and test time reductions possible for a small increase in uncertainty compared to failing all the units. If β is known, Weibayes offers additional improvements in accuracy or reductions in test size, since the Weibull and Weibayes methods handle suspensions quite well. Bearing manufacturers typically test ten sets of four bearings as shown in Figure 6-14.

6.15 Case Study: Cost vs. Uncertainty Trades

Mr. Shashank Kolhatkar of Borg Warner Automotive Engine Components, contacted the author to discuss the possibility of using Sudden Death for improving the effectiveness of testing chains. Mr. Kolhatkar provided data on test-to-failure of a dozen chains in cycles.

The Sudden Death plan considered using four groups of three chains shown in the columns in Table 6-6, producing four failures (italics) and eight suspensions. The data in Table 6-6 may be used as sudden death data as follows: the four columns in Table 6-6 represent the four groups. In each group of three, assume the lowest failure time was the failure and the other two were suspensions at the failure time. Figure 6-15 shows the Weibull for all twelve failures versus the Sudden Death results. Using the β of 2.14, simulation may be used to estimate the increase in 90% confidence bounds on the B1 life comparing tests to first failure of three in four groups with testing all 12 to failure. 90% B1 confidence bounds on the 12 failures were 2144 and 56 cycles, while the Sudden Death Weibull gave 2764 and 134. The ratio (2764 - 134)/(2144 - 256) = 1.4, implies that the Sudden Death uncertainty is 40% greater than the full life test.

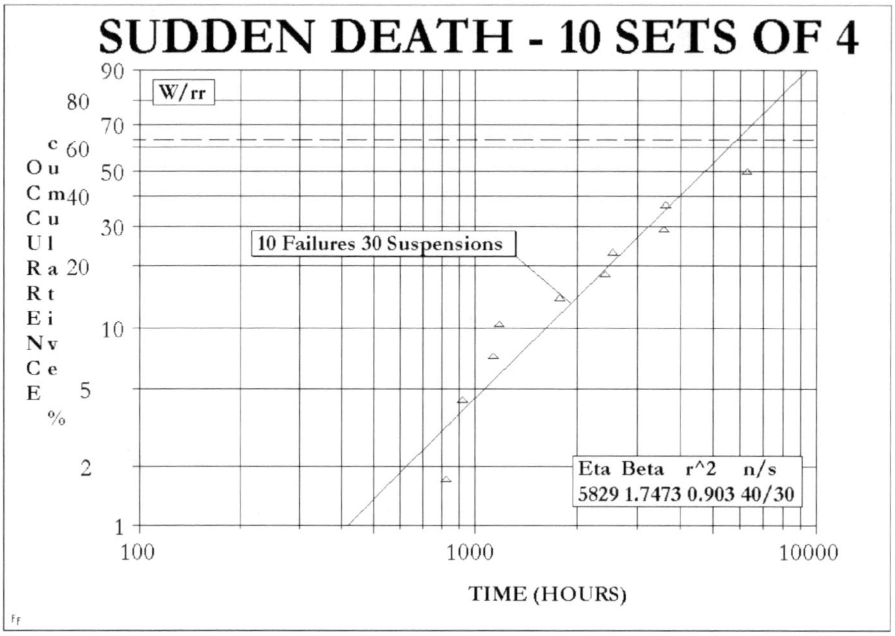

Figure 6-14. Sudden Death - 10 Sets of 4 ... NOTE: r^2 of 0.903 gives pve=24.9%

Table 6-6. Chain Data: Cycles-to-Failure

6169.5	12911.8	12483.1	***2945.6***
3698.1	4789	8274.2	9323.1
7339.7	***4650.6***	***2398.8***	6628.9

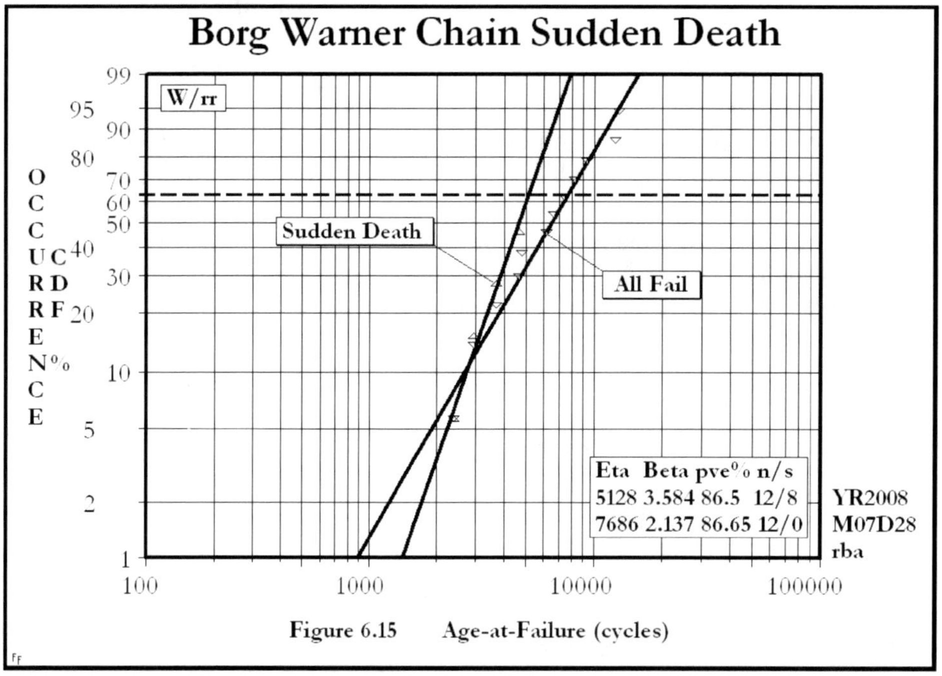

Figure 6-15. Sudden Death

The ratio of test times may be approximated with SSW using the average times-to-failure and the mean rank, (1/(N + 1). Manufacturers that employ Sudden Death reduce the test time by constructing Sudden Death rigs that test each group together, subjected to the same load. For the Sudden Death test with N=3, (1/(N + 1)), for the first is 1/4. The corresponding B life is B25. For a Weibull with $\beta = 2.13$, the B25 life is 0.557η. The ratio of MTTF to η may be read off Figure 2-3 and is 0.8858 for the 12 units tested. Therefore the ratio of test times for Sudden Death would be (0.557 x 4)/(0.8858 x 12) or approximately 0.2, an 80% reduction in test time for Sudden Death compared to testing all to failure.

After some Weibull experience is gained, Weibayes should be considered if β appears to be stable. Weibayes would increase the accuracy for the results from sets of three compared to Weibull if β is stable. For larger sets, Weibayes may be used to reduce the set size for the same accuracy. Again employing simulation, the range of B1 bounds for this Weibayes case is 1207 - 532 = 675. See Table 6-7.

Table 6-7. Summary of Sudden Death Test Designs

Type Test	Failures	Suspensions	90% B1 Bounds	90% B1 Range	Test Time Ratio
Weibull	12	0	2144-256	1888	1.0
Sudden Death Weibull	4	8	2767-134	2633	0.2
Sudden Death Weibayes	4	8	1207-532	675	0.2

The conclusion is that Sudden Death in this case with Weibayes is about <u>three times more accurate</u> than the full Weibull test <u>using 20%</u> of the test time. The number of test setups is usually determined by availability. If you do have a choice in how to group the specimens, it is better to have a smaller number of specimens in each group. For example, it is better to use 5 groups of 3 than 3 groups of 5.

6.16 Normal and Lognormal Tests

This chapter is devoted to Weibull and Weibayes test designs. However, if the data is best fit with the normal or lognormal, similar tests may be designed. With these distributions we assume the standard deviation is known instead of beta. SSW provides test designs for the Weibull, normal and lognormal for zero failures; zero and one; zero, one, and two, etc.

For example, the roller bearing tests described in Sections 6.9 and 6.10 were based on demonstrating $\eta = 500$ hours with a $\beta = 2.0$ with 90% confidence. An equivalent normal distribution with the same B10 life of 162.3 hours and B63.2 life of 500 hours would have a mean, $\mu = 430.16$ and standard deviation, $\sigma = 208.34$ using the SSW predict option. For a zero failure test, Equation 6-3 provides an estimate of R = 0.46416 or F = [1 - R] = 0.53584 for each of the three units under the null hypothesis. Using a table of normal areas or SSW, F = 0.53584 = B53.584 with the above normal distribution corresponds to a test duration of 449.96 hours. Dividing this figure by 500, the demonstration requirement, produces a k value of 0.90. This compares to 0.876 for the Weibull test. Table 6-8 shows the results of repeating this exercise for the lognormal and the normal for zero and zero-one allowed failures for the same roller bearing requirement using SSW.

Table 6-8. SuperSMITH 90% Confidence Test Factors to Demonstrate B63.2 Life

Distribution	Weibull	Normal	Lognormal
Zero Failure Test	0.876 (438 Hours)	0.90 (450)	0.846 (423)
Zero or One Failure	1.277 (638)	1.22 (610)	1.44 (720)

6.17 Accelerated Testing

For high reliability components, test to failure can be very expensive because of long test durations. Accelerated testing is used to shorten total test time required, and minimize costs. Increasing the driving test load accelerates the time to failure. For simple accelerated testing, a relationship is desired between the driving test load value, called the engineering variable, and the life distribution model parameters. The engineering variable (or driving test load) can be any factor affecting the length of lifetime, such as temperature, pressure, stress, vibration, humidity level, and so on. This relationship is estimated based on life data taken at two or more values of the engineering variable.

Case Study: Electronic circuit board switches are tested at 50°C and 60°C with lifetime in number of cycles as shown in Table 6-9. What is the B10 life of the circuit board switch at its normal operating temperature of 30°C?

Table 6-9. Cycles to Failure

50°C	60°C
21045	16551
25077	18935
31407	20996
33812	24363

Analysis: For this small amount of data the Weibull rank regression method is **best practice** if beta is unknown. The solution for both data sets is shown in Figure 6-16 using the SSW accelerated test icon assuming an Arrhenius temperature dependency model to fit the characteristic life, eta. The load values, in this case temperatures, should be converted to absolute temperature (e.g., 50°C = 323K, 60°C = 333K, note the degree symbol is not used with K as K is defined as degree Kelvin). There are two solution possibilities. If the Weibull slope is expected to be the same for all solutions, then the B10 life for 30°C (303K) is about 39250 cycles. This is shown in Figure 6-16. However, if the Weibull slope is allowed to vary as the data indicates (plot not shown for this), then the B10 life for 30°C is approximately 26500 cycles. (Note, for these very small samples the variation in beta may be just statistical scatter.) Certainly, the expected range of B10 values at normal operating temperature would be between these two extremes. With beta constant (Weibayes) the ratio of eta in-service to eta in an accelerated test is the measure of the acceleration of the test. For example, if eta in service is 1000 and eta in the accelerated test is 100, then ten minutes of service aging equals 1 minute of accelerated aging. See Weibayes Case Study 6-2.

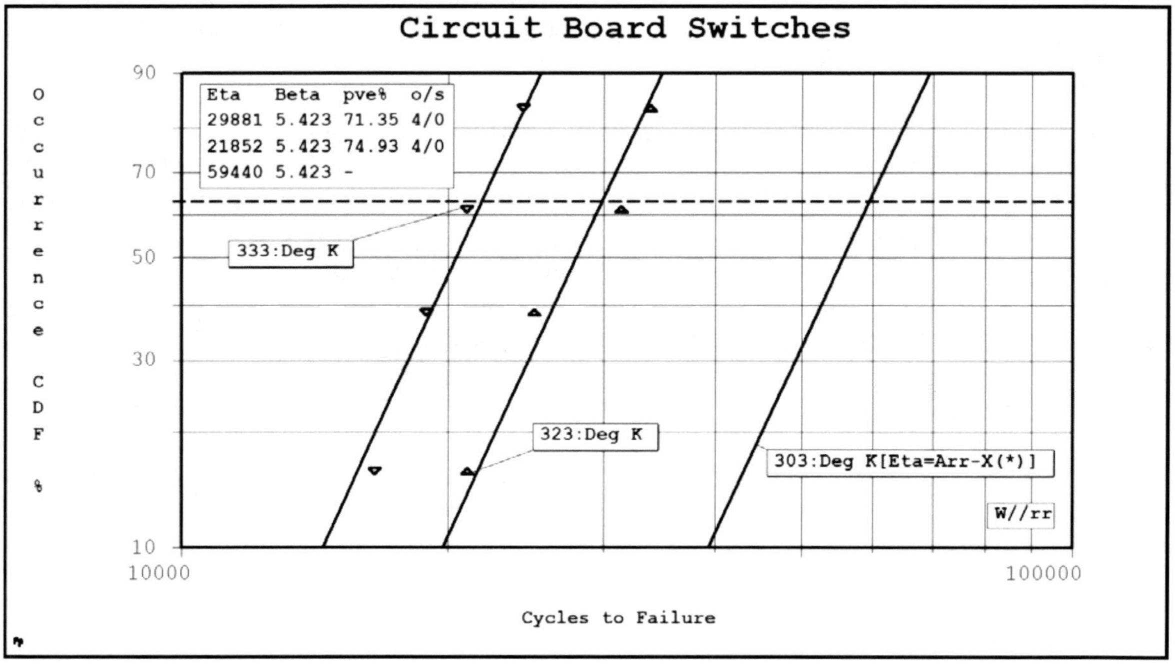

Figure 6-16. 50 Degrees C, 60 Degrees C

Note that with inspection data, time may be used as the load, the accelerating factor. See Figure 6-19. Weibull analysis of accelerated test data also provides some assurance the correct failure mode has been produced if beta on test is close to beta observed from in-service data. [Nelson 1990] is the best reference for accelerated testing. See Case Study 11.15 for temperature effects on the Challenger Space Shuttle.

6.17.1 Accelerated Step-Stress Test Data Analysis* by Wes Fulton (* indicates advanced topic)

One particular form of accelerated testing is "Step-Stress" testing. A typical test plan starts with the test article(s) under a stress that is representative of normal load level for some fraction of one expected lifetime.

Each test article is then subjected to a somewhat higher load for some amount of time (usually shorter in duration for higher stress) and so on until failure. This results in quicker failures in terms of total test time, but it adds significant complexity to the test plan and in the data analysis. The methods described by [Nelson 1990] are used to analyze the resulting information. The benefits of a step-stress test plan include reduced test time plus development of the relationship between stress and product life (the S/N curve).

Many different stress-time schedules are possible. Sometimes the difference between schedules is only the time of testing. Sometimes the difference between schedules is level of stress. Sometimes both may be varied among different groups of the same test article. Analysis capability should be able to solve when more than one schedule is being used on different groups of the same configuration.

Care must be taken to avoid inducing failures that are not representative of what will be experienced in actual service as with all accelerated testing. A common technique is to define an upper limit to the stress that is on the threshold of creating non-representative failure mechanisms (like melting or deformation of material). Then the test planner can subdivide the stress into levels between the normal operating stress and this upper limit in order to get the levels of the stress steps.

The basic step-stress model is based upon some simple assumptions. The Weibull slope (or lognormal SigF value, etc.) is assumed to be relatively unaffected by stress changes within the step-stress test range. Therefore, a common Weibull slope is one of the parameters of the solution. Weibull "Eta" or lognormal "Med" (previously "MuAL") is taken to be in a simple inverse power relationship to stress level. Therefore, the model for the Weibull distribution is: **Eta = (V / x)** ^{p} **where x = stress level**

The solution for Weibull requires finding the following three parameters: (1) Beta = common solution for all test articles, (2) V = stress coefficient, and (3) p = power coefficient.

EXAMPLE 6.1 STEP-STRESS ANALYSIS
Dick Rudy of Daimler-Chrysler provided the data for this example from the automotive industry. An automotive component requires analysis to determine its load capability for use in new applications. The normal operating stress value is 25.6. The useful life at normal operating stress is 800 cycles. The test plan called for seven steps starting at stress value 25.6 and increasing at intervals of 10 up to stress level 85.6. The time at the first load level is 536 cycles (about 2/3 of the unit life at normal operating load). The time for each successive step is half the time at the step before. The resulting step-stress schedule is:

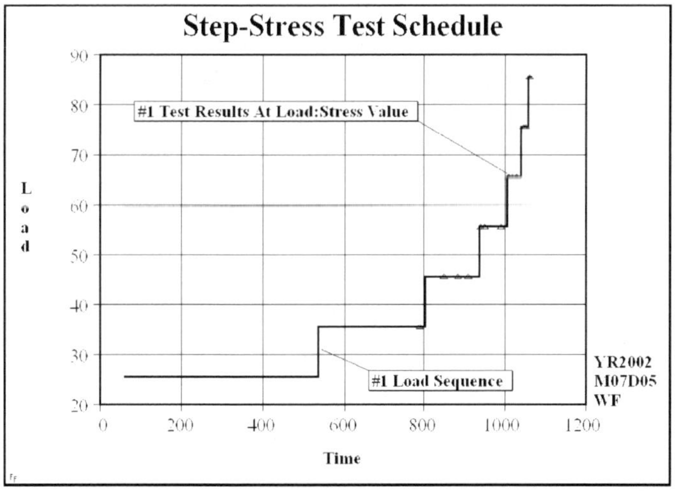

Figure 6-17 - Step Stress Accelerated Test Schedule

Stress Level Value	Cycles At Stress Level Value
25.6	536
35.6	268
45.6	134
55.6	67
65.6	34
75.6	17
85.6	8

Fifteen test articles were subjected to the above step-stress schedule. The resulting age-to-failure in cycles for each unit is: 790, 850, 885, 910, 940, 950, 990, 1008, 1010, 1018, 1027, 1030, 1045, 1050, 1060.

For the normal operating stress of 25.6, reliability can be estimated with different initial Beta guesses. Analysis of this data using an MLE approach gives the following solutions (rounded):

A) Beta=2.435, V=194.4, p=4.256 using the MLE initial guess for Beta

B) Beta=2.181, V=174.9, p=4.596 using the regression initial guess for Beta

Solution likelihood is practically the same with either MLE or regression initial Beta guess for this data. For normal operating stress of 25.6 with either solution (A or B), reliability should be 99.1% for 800 cycles.

6.17.2 Accelerated Testing: A Method for Estimating Test Acceleration Factor with No Existing In-Service Failures* by Wes Fulton

The object is to predict what will happen in service based upon accelerated test results in laboratory. The real benefits of accelerated testing, quicker times from design concept to market, and low-cost design verification, are possible if the laboratory acceleration factor is known.

If there are failures in service, then the acceleration factor of the laboratory test is found with a ratio of the Weibull "Eta" or lognormal "Med" (previously "MuAL") for in-service data to that obtained in the laboratory. The acceleration factor is then available to convert laboratory test times on new designs to in-service capability. However, it is more difficult to get an acceleration factor when no failures exist in service. If no units working in service have failed, and if there is a laboratory test setup for accelerated endurance testing, then there is a possible solution for acceleration factor using a three-parameter (3P) analysis.

A technique to get the acceleration factor without in-service failures is called the Vasan-Fulton former usage to test ratio (FUTR) method. It relies on a linked three-parameter solution for equipment that has experienced some significant operation in service. The procedure includes laboratory testing and analysis of pre-operated equipment. First, take in-service units (non-failures) and group these units by service age. Units in the same group must have similar service ages. It is given here is that all of the in-service units have experienced similar load levels. Test the units to failure in the laboratory at a single higher load (accelerated testing). Twenty or more units total are needed to start, and even more units than this are desirable to improve accuracy. Collect the laboratory failure data grouped into sets by prior service age. Use a three-parameter analysis (negative t-zero) to measure the amount of prior deterioration and link the solutions if there is more than one data set.

The FUTR process involves the premise that all units have the same basic Weibull (or lognormal) model. Also, the requirement here that prior service yields a negative t_0 value must be reasonable. The negative t_0 value is expected to be different for each set and proportional to the amount of prior service. For example, input if there are three groups would include:

A) Equipment with x, y, and z months of prior service, at least some units in each category

B) Three data sets of laboratory cycles-to-failure data (one for each prior service category) at accelerated load

Results are obtained with a linked negative t-zero Weibull (or lognormal solution) (t_0x, t_0y, t_0z) ... where $t_0x = - x / FUTR$; $t_0y = - y / FUTR$; $t_0z = - z / FUTR$. The acceleration factor then is:

Acceleration factor = FUTR [service months/lab hours]

EXAMPLE 6.2 FUTR ANALYSIS

Some units have operated in the field for 12 months, some for 24 months, and some for 30 months. No failures were reported. Units are retrieved from the field for accelerated testing in the laboratory. Lab testing of these in-service units at accelerated load might produce the following (demonstration example data):

- 7 units after 12 months in service (set 1]: Lab failure times [hrs] = 38, 56, 69, 81, 92, 106, 125
- 8 units after 24 months in service (set 2): Lab failure times [hrs] = 28, 45, 57, 67, 78, 88, 101, 119
- 9 units after 30 months in service (set 3): Lab failure times [hrs] = 22.5, 38.5, 50, 59.5, 68.5, 77.5, 87.5, 99.5, 117.5

A linked 3p solution results in the Figure 6-18.

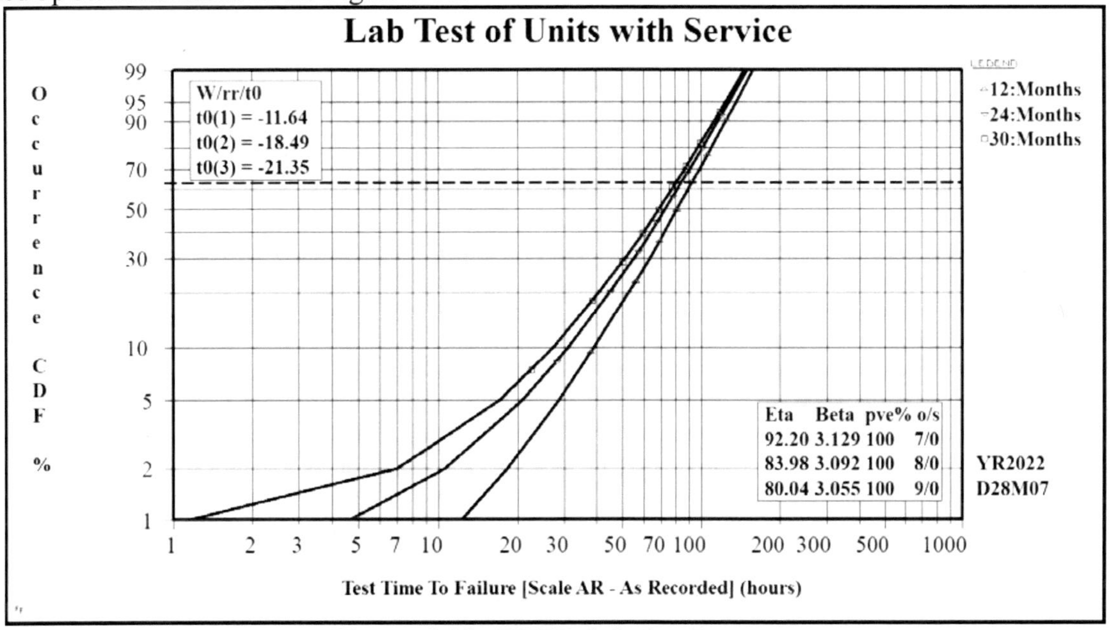

Figure 6-18 - FUTR Solution Process

FUTR Case Study Conclusions:

- Set 1: $t_0(1)$ = - 11.64 lab hours
- Set 2: $t_0(2)$ = - 18.49 lab hours
- Set 3: $t_0(3)$ = - 21.35 lab hours
- Laboratory Weibull: eta = 102.6; common beta = 3.09
- Service Weibull: eta = 126.6; common beta = 3.09
- FUTR = 1.234 [service months / lab hour]

6.18 System Deterioration

There are many system characteristics that deteriorate or degrade as a function of time such as efficiency, fuel consumption, turbine temperature increase, vibration, fuel contamination and so on. All of these characteristics are measurable. There are two different approaches to plotting the data on a Weibull as a function of the data available. Sometimes both approaches may be used.

Periodic Inspection Data: If a number of systems are periodically inspected or calibrated to measure the degradation a good approach with demonstration example data shown in Figure 6-19. The ordinate is the performance characteristic. The acceleration test icon in SuperSMITH Weibull provides the plot with eta as a function of inspection time. Beta may or may not be constant. Note that SSW will interpolate or extrapolate to untested inspection times as shown in Figure 6-19.

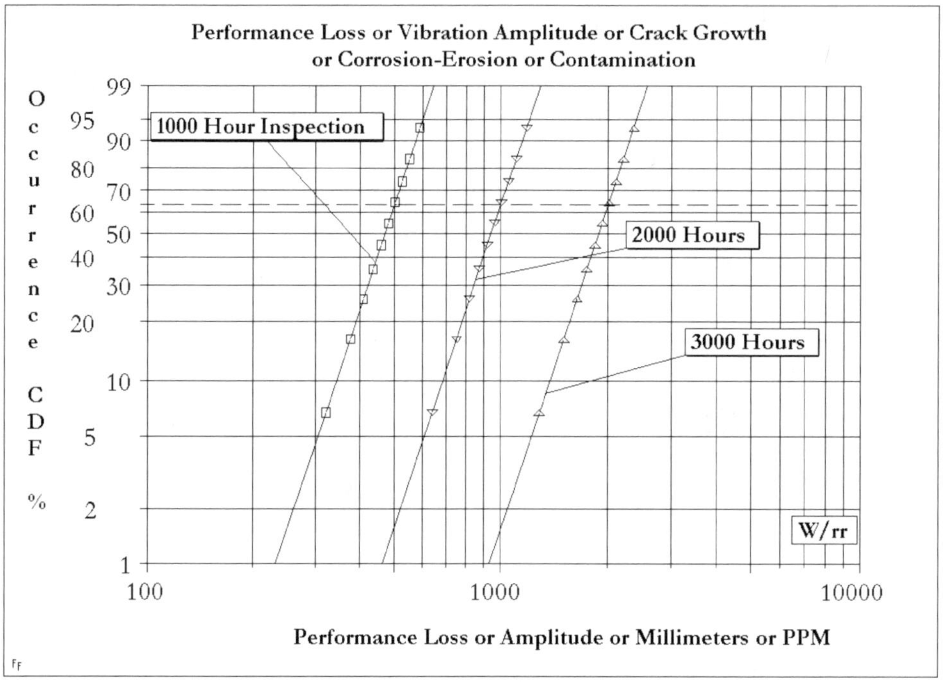

Figure 6-19. Performance Loss or Vibration Amplitude or Crack Growth or Corrosion-Erosion or Contamination

Different Definitions of Failure: The data may allow several definitions of failure in terms of increased degradation. In this case, Figure 6-20 for this type of demonstration example data, the ordinate is time-to-failure. Here again the SSW accelerated test icon provides the plot using the different levels of failure as the "load." Both plots are useful and should be plotted if the data is available.

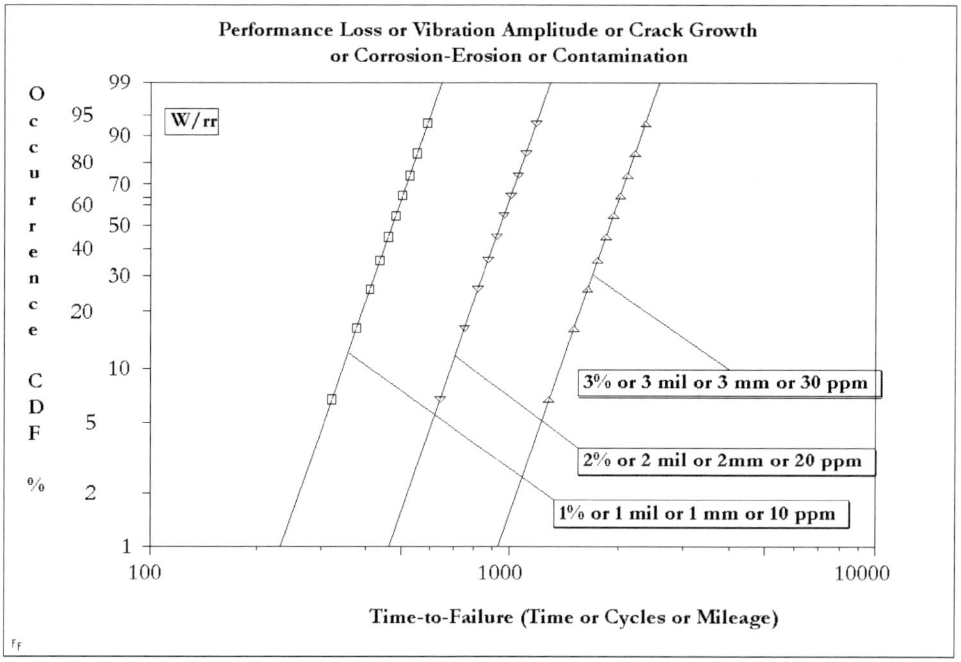

Figure 6-20. Performance Loss or Vibration Amplitude or Crack Growth or Corrosion-Erosion or Contamination

6.19 Weibull Libraries and Lessons Learned

After reviewing the many advantages of Weibayes it should be obvious that a Weibull library is extremely important to every organization. If you don't have one, start one! It can produce enormous cost reductions. There are at least two important reasons for maintaining a Weibull Library: First is to obtain the improvement in accuracy and reduction in test times and cost provided by Weibayes. The ticket to play Weibayes is prior knowledge of beta from the library. The second and equally important reason is to improve new designs.

Without Weibull library designers are free to "invent" the same deficient designs over and over. A much better procedure is to require the designer to review all past failure modes on previous designs before starting his new design. Later, in the design review the designer should be tasked to show how he has improved or eliminated each failure mode in the earlier designs. Without this discipline there may be or will be costly mistakes in the new designs.

It is easy to start a Weibull library as Wes Fulton has provided automatic capability in SSW, now with significant enhancements from Carl Tarum. If you chose to do it, every time you file a SSW data set, the software will take all the data and information from your analysis plus information you want to add such as part name and part number, generic failure mode, test conditions, data source, etc. A text file is produced and can be saved. The library may be searched for key words, edited, formatted, and printed.

A Lessons Learned Library is an extension of a Weibull library. It would include a file or computer folder for every failure mode experienced by your products in test and in service. It should contain the failure report and corrective analysis reports (FRACAS), a root cause analysis, a lessons learned statement that the design manual and process should or should not be changed to avoid the failure mode in the future, the materials laboratory analysis, the FMEA and fault tree analysis, the Weibull library information and plots, performance analysis, the failure forecast, and the corrective action; in other words, all related reports. The lessons learned folder should be identified by a number, a part name, part number, generic failure mode, component and system names, and date.

6.19.1 Honeywell Weibull Library contributed by Randy Pirtle

As a global company with many product groups world-wide we have developed a web application to facilitate data entry to capture relevant Weibull analysis data into a Weibull library. Honeywell uses this information to support risk analysis in two ways, Weibayes analysis and product modeling by aggregation of the product components.

- **A major consideration is ease of entry**. Many times analysis data stored on individual PCs never make it into a common shared database. SSW facilitates this by storing product data with each file and allowing this data to be exported to a text file. As such, the product data need be entered only once at the beginning of an analysis and this is persistent at each re-visitation of the analysis with all parametric data captured automatically by the program.

- **The application supports either manual data entry or txt file import** of the product and analysis parameters. Users are associated with product groups and based on the experience of the analyst, a product group administrator may or may not be notified by email that he needs to review and approve an analysis before the application passes the data into the library. The application provides links to both the SSW file and to a folder containing all failure analysis and design lessons learned information. While we require some standardization we allow some flexibility of field selection by product group, and so, the library supports collection of additional data above that captured automatically by SSW.

- **Quality data entry into the library** is critical. In addition for administrator's review of the analysis, a help facility supports the user SSW PC setup including selected fields, details of field definitions and examples. There are also provisions for administrators to email users on data issues, analysis concerns, and best practices issues.

- **Getting Library data out:** The ease at which data can be accessed from the library application is also important. Many options are provided allowing library data to be viewed and cut and pasted into SSW or our Honeywell modeling tools. Sorting is often by device type, end application and environments.

6.19.2 Weibull Libraries for End Users, Contributed by Paul Barringer, Reliability Consultant

End users of equipment obtain their Weibull libraries from using the equipment, (not necessarily from testing equipment). The Weibull library is important for understanding failure modes, setting the proper maintenance strategies, and modeling systems for new plants or predicting system failures which effect operations.

End users Weibull libraries reflect the grade of equipment purchased, practices employed during installation, effects of maintenance techniques and operating practices over the life of the equipment. You should expect Weibull results to reflect poorer results than Weibull values from equipment suppliers because of usage variation applied to the equipment in the field.

One application for end user libraries is to model existing and future facilities. Strategies must be adopted to avoid replicating former failure patterns based on economics and life cycle costs. This effort requires facts, not opinions. Suppliers acquire much of their data via expensive in-house tests. End users acquire their data as a result of routine operation of the equipment, i.e.; the cost for acquiring the data is minimal. End users need to establish their databases to acquire aging information. Sometimes aging of systems that only wait to perform is more important for failure predictions than for systems that are in continues operation. End users data can be much more valuable than supplier data because "my data" is always better than "your data."

Paul Barringer's website material (see References) contains a sample Weibull Library and a link to David Weber's extensive work on Weibull libraries. Also, there are great Weibull Library resources available from Quanterion Solutions (see References).

6.20 Problems

Problem 6-1: Weibayes Analysis Five redesigned units have been installed on engines in the field. The new units have 800, 1100, 1400, 1600 and 1700 hours on them without failure. The failure mode of the old unit had a slope of 3 with a characteristic life of 500 hours.

a. Construct a Weibayes line for the new units.
b. What is the minimum improvement demonstrated in the characteristic life? Hint: The ratio of etas is the best measure.

Engine Times (t)	**(t)^ β**

Sum =__________________

$(\text{Sum})^{(1/\beta)}$ =______________

Problem 6-2: Design System Calibration The design system predicted B.1 life for the compressor disk is 1000 cycles. Five disks have accumulated 1500 cycles and five have 2000 cycles without failures. If most disk low cycle fatigue failures have a β of 3.0, is this success data sufficient to increase the predicted design life?

Problem 6-3: Failures Not Identified Two bolt failures due to low cycle fatigue have been observed in a flight test fleet of six engines having the following times: 100, 110, 125, 150, 90, and 40 hours. The bolts are not serialized and as failures were discovered after the engines were overhauled, it is not known which engines had the failed parts. If low cycle fatigue failure modes usually have slope parameters between 2 and 5, and after rebuilding the engine will accumulate 100 hours in the next year, predict the number of expected failures. (Take $\beta = 3.0$ and the two new replacement bolts are installed in high time engine rebuilds.)

Problem 6-4: Weibayes Interpretation Plot a Weibayes for two sets of data: (NOTE: ">" in front of the value indicates that the value represents a suspension where it could fail but has not failed so far)
Set # 1 ... >100, >200, >300
Set # 2 ... >100, 200, >300
In the first case there are no failures, only suspensions. In the second case there is one failure and two suspensions. What is the difference between the plots? What is the difference in interpretation?

Problem 6-5: Test Substantiation A new component has been designed. The design requirement is a Weibull with $\beta = 4.0$ and $\eta = 600$ hours. This is twice as good as the existing design which has an eta = 300 hours. How long should the new unit be tested without failure to demonstrate with 90% confidence that it has 2 times the life for 4 units to test versus 8 units to test?

Optional: Plot the Weibayes for the new design and the Weibull for the old design.

Problem 6-6: One or Zero Test How much time should the new unit in Problem 6-5, be tested with one or zero failures, to demonstrate with 90% confidence that it has 2 times the life with 4 units to test?

Problem 6-7: Sudden Death Vs Zero Failure Tests. What are the advantages and disadvantages of these two types of Weibull substantiation tests? With good prior knowledge would you employ Weibayes with Sudden Death tests?

Problem 6-8: (Advanced Problem) Sudden Death versus Zero Failure Tests (SSW recommended) Your company tests 16 randomly selected roller bearings to failure from each batch of 100,000 bearings. You suggest a cost reduction based on Sudden Death bearing rig tests of sets of four simultaneously. Table 6-10 shows the last data set, failures in cycles from 16 roller bearings. For Sudden Death, assume the columns contain the sets and the first bearing failure ends the test for each set. For example, Set 1 would consist of one failure and three suspensions at a value of 47,385 cycles.

Table 6-10. Roller Bearing Times-to-Failure

Set 1	Set 2	Set 3	Set 4
106575	32484	147191	177217
47385	35968	52531	38649
71710	112053	42114	70812
54546	176187	132621	113508

6-8-1. How does the Weibull and Weibayes for Sudden Death, compare to the Weibull for all 16 bearings to failure? Plot both Weibulls and a Weibayes assuming $\beta = 1.5$.

6-8-2. These data were generated by WeibullSMITH Monte Carlo sampling option from a true Weibull with $\beta = 1.5$ and $\eta = 100{,}000$ cycles. Plot the Weibull based on the 16 failures. Which B10 life is the most accurate, Weibull-16F, Sudden Death Weibull 4F-12S, or Sudden Death Weibayes?

6-8-3. Compare the Sudden Death total rig test time to the total life test time?

6-8-4. Compare the Zero-Failure test time for 16 units to test, $\beta = 1.5$ to demonstrate 100,000 cycles.

6.21 Reverend Thomas Bayes

REV. T. BAYES

Reverend Thomas Bayes died in 1761. On his death bed he gave his friend, Richard Price, an unpublished paper, and requested it not be published, but several years later Richard published the paper anyway. [Bayes] The paper produced a branch of statistics based on the Bayes theorem.

Author's View: The Bayesians that follow this school are considered to be the wild-eyed radicals of statistics. What was so controversial about the theorem? Bayes suggested we convolute our personal opinion, the "prior distribution," with the likelihood function to obtain the "posterior distribution," that engineers would call the "results." This process is "fudging" the answer and is repugnant to most engineers. However, when engineers have little or no information or data, they often rely on "engineering judgment" which the author equates to the prior opinion. The only difference is that the formal Bayesian convolution requires a distribution of prior opinion and that is why the author cannot fully endorse the Bayes approach. I may ask an expert to estimate the performance efficiency of an untested design and accept a point estimate, but to ask the expert to describe the distribution of his opinion seems absurd to me. Still I have great respect for the Bayesians, particularly Professor Jeffries with whom I correspond. Except for my problem with the prior distribution I would be a Bayesian.

I named the method in this chapter, "Weibayes" because if there is uncertainty assuming beta, this seems to me to be an informal Bayesian procedure. Jeffries, the foremost Bayesian, does not agree as it is missing the assumed distribution of beta. I would be pleased to assume the distribution is approximately lognormal which it is from statistical scatter but to assume the distribution of the opinion of beta is log normal seems indefensible to me.

A few years ago the Royal Statistical Society moved their headquarters to a new location in London near the home of the executive director. He walks to work each day crossing an old cemetery. One day he noticed an old headstone covered with weeds and pulling them down, found the Bayes' grave. The Royal Statistical Society has decided to permanently maintain the grave. Is this a statistical coincidence?

The two best biographies of Bayes were written by George Barnard and Egon Pearson, neither one of which was a Bayesian.

CHAPTER 7. INTERVAL ESTIMATES

7.1 Interval Estimates

The objective of Weibull analysis is to solve problems by making estimates of parameters such as B1 life, β, η and reliability based on data. The usual estimate is a "point estimate," a single number that might be an average or a median estimate, for example. The point estimate may be determined graphically or analytically, such as a maximum likelihood estimate. Up to this point in the Handbook we have used point estimates, single valued estimates. $\beta = 3.47$ and B1 life = 1200 cycles are point estimates.

An interval estimate is a range within which we think the true value lies. There are at least four types of interval estimates: tolerance intervals, prediction intervals, total uncertainty intervals and confidence intervals. The remainder of this chapter is devoted to confidence and uncertainty intervals. Sometimes confidence intervals have been used erroneously instead of using these other types of intervals. To avoid this error, a clear understanding of the differences is needed.

A **tolerance interval** contains a specified proportion of the units from a population or process. For example, a 95% tolerance interval of 1000 to 1375 cycles-to-failure should include 95% of the specimens in the population with say 90% probability. A 95% tolerance interval at 95% probability would be larger. Instrumentation repeatability is specified with tolerance intervals. Some authors incorrectly use the word "confidence" instead of "probability" for tolerance intervals. There are also unbiased tolerance intervals that on the average contain, say, 95% of the units in the population.

A **prediction interval** is designed to contain one or more *future* observations sampled from the previously sampled population with some probability, say 90%. Our SSW failure forecasts can include predictive interval bounds around the nominal prediction.

A **total uncertainty interval** includes *all the error* associated with the result.

7.2 Confidence Interval Concept

The confidence interval is a range of values, bounded above and below, within which the true unknown value is expected to fall. It measures the *statistical precision* of our estimate. The probability that the true value lies within the interval is either zero or one; it does, or does not, fall within the interval. The fact that we do not know whether or not it lies within the interval does not alter these probabilities. **Confidence is the frequency that similar intervals will contain the true value, assuming the fixed errors are negligible**. If a statistician always selects 90% confidence and makes confidence interval estimates, 90% of the intervals will contain the true value and 10% of the intervals will not contain the true unknown value. *Fixed or systematic* errors are not included in the confidence interval. If the clock measuring time-to-failure has a 1% fixed error, B1 will be off by 1% and so will the confidence interval. The frequency-confidence interval concept was invented in the 1930s by Egon Pearson and Jerzy Neyman. See the last page in this chapter for more on these gentlemen.

Note that the **confidence level must be selected "*a priori*,"** before looking at the data. If the data is employed to maximize the quoted confidence level, the concept is destroyed. Allowing the confidence level to be maximized with the data allows immoral statisticians to prove absolutely anything with some level of confidence. This is the number one abuse of statistics. There are several ways to avoid the abuse. The author always uses 90% confidence to avoid the temptation of allowing the data to influence the confidence level. Wes Fulton suggests for reliability intervals, the confidence level should always equal the reliability level. This concept, called "**reliability assurance index,**" allows the data to influence the confidence level in a predefined, a priori, manner. This concept will be illustrated later in this chapter. The best alternative is to calculate the highest level of confidence the data will support and call that the "P" value instead of confidence. The "P" value interval is not a confidence interval but it is often useful information.

Confidence intervals are estimates of the true unknown parameters, such that they will contain the true unknown value with a given frequency. For example, if the parameter is a Weibull B1 life, 90% confidence intervals for the true unknown B1 life should contain the true value 90% of the time if the fixed errors are negligible. In other words it provides an estimate of the statistical precision error. Repeated samples from the same Weibull will produce various B1 estimates. Interval estimates may be "exact" where the statistical method has a rigorous proof or "approximate" when exact methods do not exist. Approximate intervals are conservative if they are larger than exact intervals.

The author does not recommend confidence interval estimates for use in presentations and reports as they often are misunderstood and misapplied. Confidence intervals contribute more confusion than clarity when communicating with others. Confidence intervals do not measure the total uncertainty or the total error. However, the contract, the customer, or the analyst's superior may demand confidence interval estimates. For your own personal use confidence intervals can be helpful by quantifying the statistical precision. With good data, a good fit to a distribution, and evidence that fixed errors are negligible, most of the uncertainty will be indicated by the width of the confidence interval. In this case confidence interval estimates can be valuable and are recommended for your personal use.

The author has empathy for readers that are confused at this point. It may be comforting to know that his estimate of the proportion of the population of engineers, managers, and statisticians that are confused about interval estimates approaches 100%. Two examples from the Royal Statistical Society "News & Notes":

1. "From a client's test standard," November 1992: "The test specimen shall be machined from the casting. A separately cast bar is not acceptable unless the supplier can show correlation that 99% of the cast test bars show the actual tensile strength of the casting at a 95% confidence level." As this was in the "Forsooth" column, it apparently confused the Royal Statistical Society as well as the author.
2. From the September 1992 Edition, "Letters to the Editor," Dr. W. Edwards Deming wrote: "Degree of belief can not be quantified. It is not 0.99 nor 0.8 nor anything else. A confidence interval is no help."

7.3 Confidence Intervals for B Lives and Reliability

Examples:

- A design requirement specifies test data must be obtained to demonstrate that catastrophic failure modes must be avoided with reliabilities of at least 99% with 95% confidence.
- A new system must be tested to demonstrate 99.85% reliability with 90% confidence.
- A manufacturer requires data to show a B5 life of at least 2000 hours with 95% confidence.
- An engineer reports a lower 90% confidence bound for B1 life.

Note that although the first three above are one-sided confidence bounds, they are still interval estimates, bounded above by 100%. The interval is the estimate of the unknown parameter, not the lower bound.

There are many analytical methods for estimating these confidence intervals. The five most popular, beta-binomial, Fisher's matrix, likelihood ratio, Monte Carlo and pivotal, will be discussed herein. Three other methods for special models are discussed in Chapter 8. All these methods may be applied with Weibull, lognormal, normal and extreme value data. SuperSMITH Weibull provides these eight methods as options.

The distribution fit should be acceptably good before adding confidence intervals as the uncertainty of modeling the data is not included in the confidence intervals.

7.3.1 Beta-Binomial Bounds

The 5% and 95% ranks[1] found in Appendix I are credited to [Johnson 1964]. They have been used for decades by the author and many others. Johnson's derivation is directly related to determining median ranks. The bounds are calculated from the "beta-binomial" distribution, the binomial distribution is used to evaluate the beta distribution as described in the Johnson reference. They also have been employed for data sets with suspensions [Kapur and Lamberson, 1977] although Leonard Johnson did not specifically illustrate this application. Suspensions require interpolation of the 5% and 95% ranks. Beta-binomial intervals are slightly conservative (the width of the interval is large), compared to the Fisher Matrix and Likelihood Ratio methods.

Converting the 5% and 95% ranks into intervals as defined by L. Johnson, provides intervals for the time-to-failure, but several authors incorrectly use them to determine intervals for CDF, the per cent cumulative failure. The author believes Johnson's original method is correct. The following equations relate the 5% and 95% ranks in Appendix I to the Weibull fit line where i is the rank order of the failures:

$$t_{i,0.95} = \eta\left[\ln\left(\frac{1}{(1 - F_{i(0.95)})}\right)\right]^{\left(\frac{1}{\beta}\right)}$$

$$t_{i,0.05} = \eta\left[\ln\left(\frac{1}{(1 - F_{i(0.05)})}\right)\right]^{\left(\frac{1}{\beta}\right)}$$

Example 7.1 Given the Weibull in Figure 7-1, 10 failures with $\eta = 1000$ and $\beta = 3$, calculate the confidence bound for the first failure. $F_{1,10,0.05} = 0.0051$ and $F_{1,10,0.95} = 0.2589$ from Appendix I.

Then:

$$t_{1,0.05} = 1000\left[\ln\left(\frac{1}{(1 - 0.0051)}\right)\right]^{\left(\frac{1}{3}\right)} = 172.3$$

$$t_{1,0.95} = 1000\left[\ln\left(\frac{1}{(1 - 0.2589)}\right)\right]^{\left(\frac{1}{3}\right)} = 669.1$$

These two points anchor the lower end of the confidence intervals in Figure 7-1. The equivalent graphical solution for locating these two points is shown on Figure 7-1. Similarly, each failure point anchors an upper and lower bound calculated and plotted in Figure 7-1. It is wrong to extrapolate the Weibull line and the confidence limits beyond the range of data, but everyone "sins" and extrapolates. The beta-binomial bounds are not easily extrapolated. Rigorously this is an advantage, but practically it is a disadvantage. Further, calculating these limits with large data sets and suspensions is time consuming. Johnson's beta-binomial bounds may be used for other probability distributions such as the normal and lognormal.

[1]The median, 5% and 95% ranks in Appendix I were calculated for the *Handbook* by Mr. Carl Tarum of Delphi. Carl Tarum's Excel formula is listed in the Appendix. SSW provides beta-binomial bounds at all confidence levels.

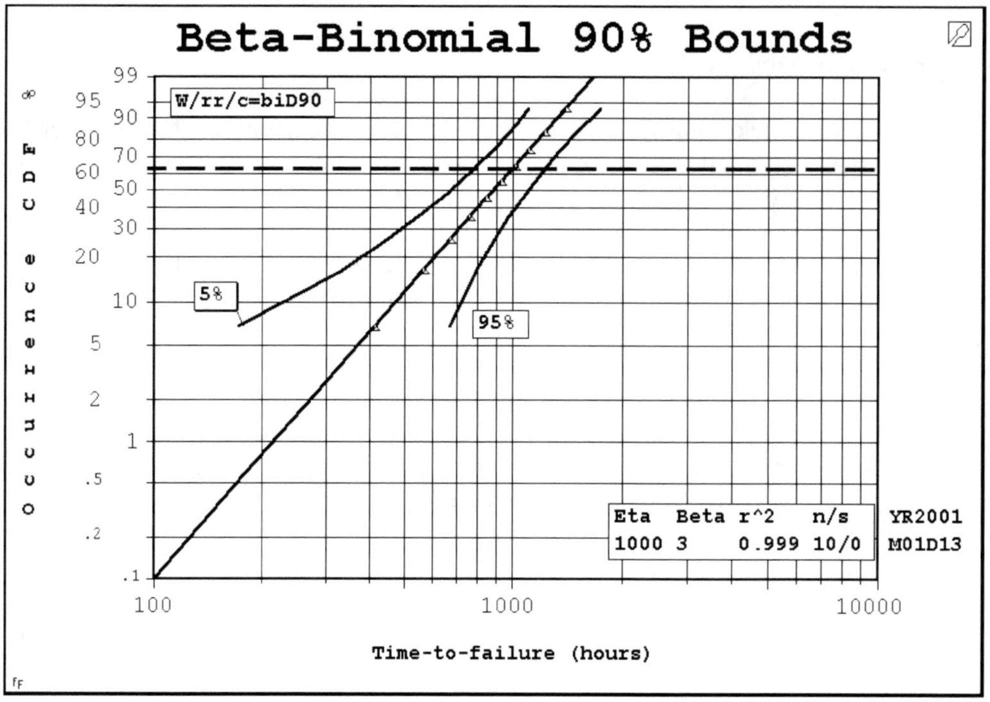

Figure 7-1. Beta-Binomial 90% Bounds

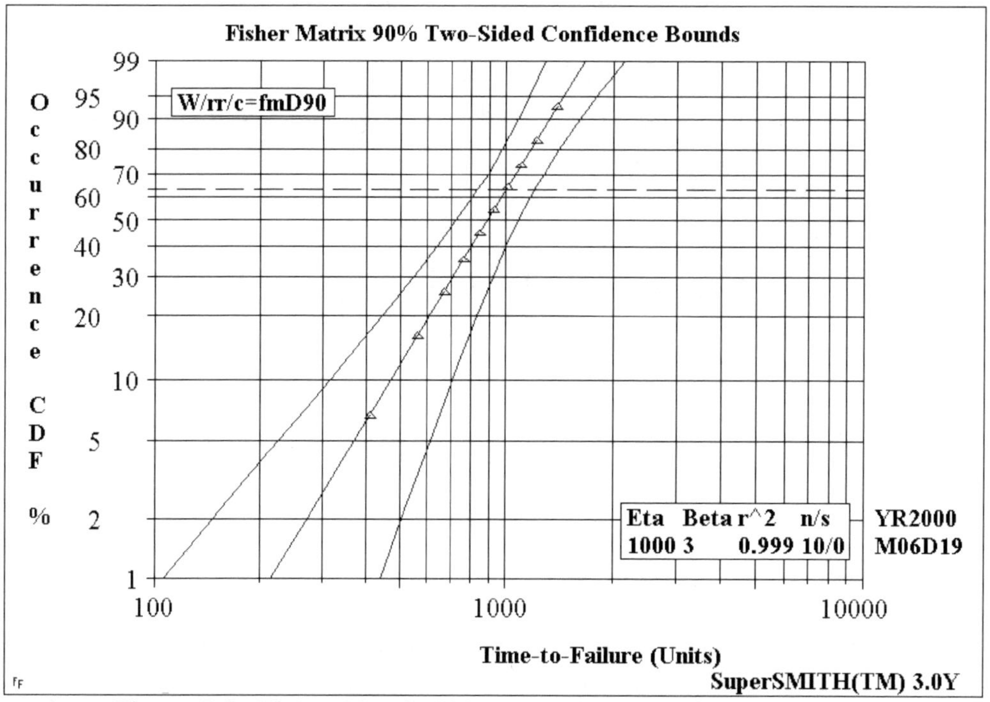

Figure 7-2. Fisher Matrix 90% Two-Sided Confidence Bounds

7.3.2 Fisher's Matrix Bounds

These intervals are described in [Nelson 1982]. The Fisher's matrix solution is the fastest computer solution compared to all the other methods and the bounds are extrapolated to the margins. These are significant advantages. Further, for intermediate and large samples, the apparent confidence level, based on MonteCarloSMITH™ is closer to the requested level, though more optimistic, than beta-binomial. Based on all these considerations, Fisher's matrix confidence intervals are often used. For ten or less failures these bounds are too optimistic. See Figure 7-2.

7.3.3 Likelihood Ratio Bounds

As indicated in Chapter 5, horizontal slices through the likelihood surface produce joint confidence regions for the distribution parameters. The intensive calculations are described in [Lawless 1982]. Computational time is long. Statisticians prefer MLE methods and likelihood ratio confidence bounds follow directly from the MLE calculations. Sir Ronald Fisher and his disciples, such as the author's professor, George Barnard, put great stock in studying the likelihood function. Lawless illustrated many of the advantages of viewing the likelihood function in his keynote address at the SAE Weibull Users Conference in March 1996. Recently, the Royal Statistical Society surveyed the RSS Fellows to identify the "single most valuable statistic." The majority choose the likelihood function. This function has been ignored by engineers, but is now most useful since the author developed MLE-RBA and Wes Fulton developed corrections to remove the likelihood bias for small samples which will be discussed later in this chapter. See Figure 7-3.

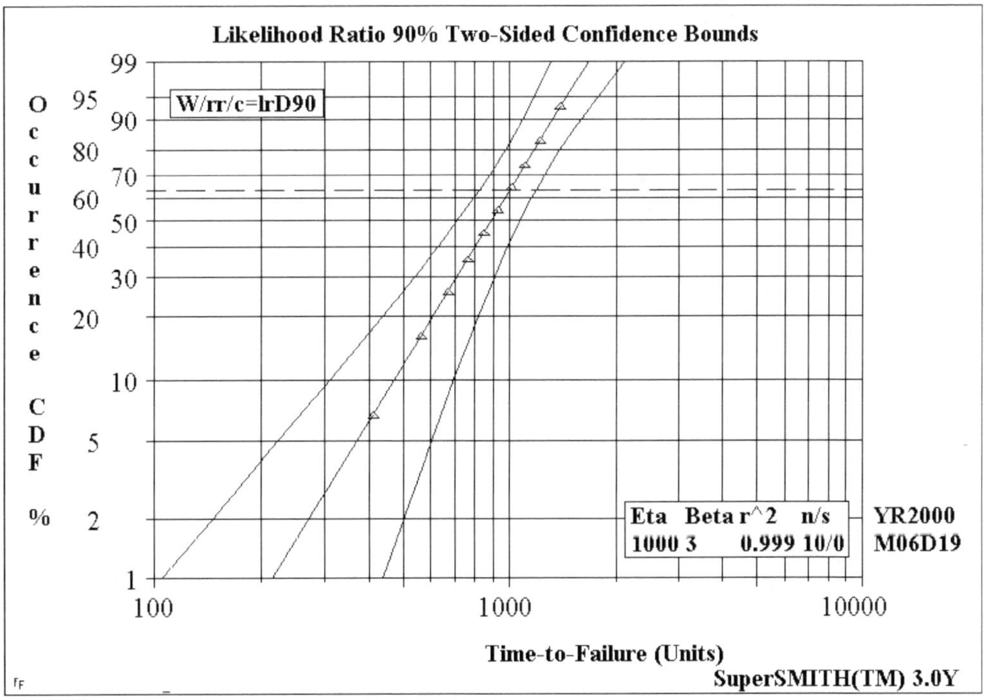

Figure 7-3. Likelihood Ratio 90% Two-Sided Confidence Bounds

To encourage this new concept of looking at the likelihood function, SuperSMITH software provides capability to look at likelihood contours. See Figure 7-4 and Appendix C. For the two parameter Weibull distribution, the contour is a three-dimensional surface resembling a mountain. See also Figure 5-1. Weibull cross sections are elliptical for large samples and triangular for small samples. The MLE are the coordinates of the top of the mountain. Likelihood ratio confidence bounds are determined from contours of equal likelihood. They are also extrapolated in SSW to the margins. For MLE Weibull, likelihood bounds are recommended for very large samples, 100 failures or more. (Fisher Matrix bounds work for very large samples and are much faster.) For smaller sample sizes, use the MLE-RBA method to reduce the bias in β. The likelihood ratio confidence bounds will be automatically adjusted for these sample sizes by the Justified Likelihood Function which will be presented later in Section 7.5. The MLE-RBA value of the likelihood function may be used for distributional analysis. The highest value for the likelihood function at the MLE-RBA among competing distributions with the same number of parameters indicates the best model for the data. (Geometrically the highest likelihood "mountain" is the best distribution.)

When suspensions are present, likelihood ratio bounds are often wider, and thus more conservative, compared to Fisher matrix bounds. However, for complete samples (no suspensions) many will use Fisher's Matrix as it is much faster and provides bounds that are close to the likelihood ratio method as may be seen in Figure 7-5. Likelihood contours also have a valuable new application, testing two or more data sets to see if they are significantly different, described in Section 7.5.

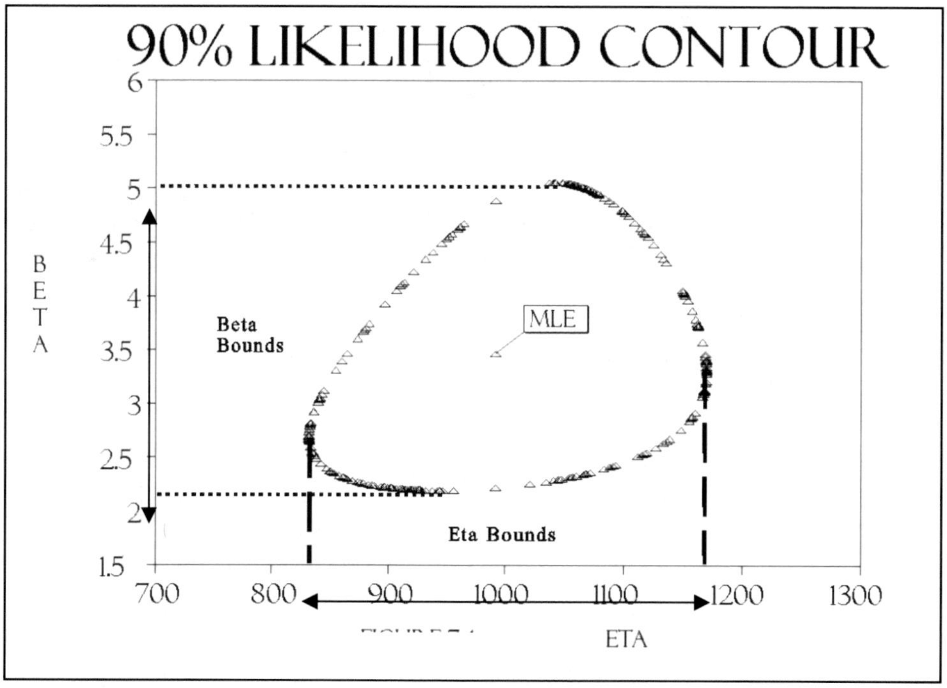

Figure 7-4. 90% Likelihood Contour Bounds

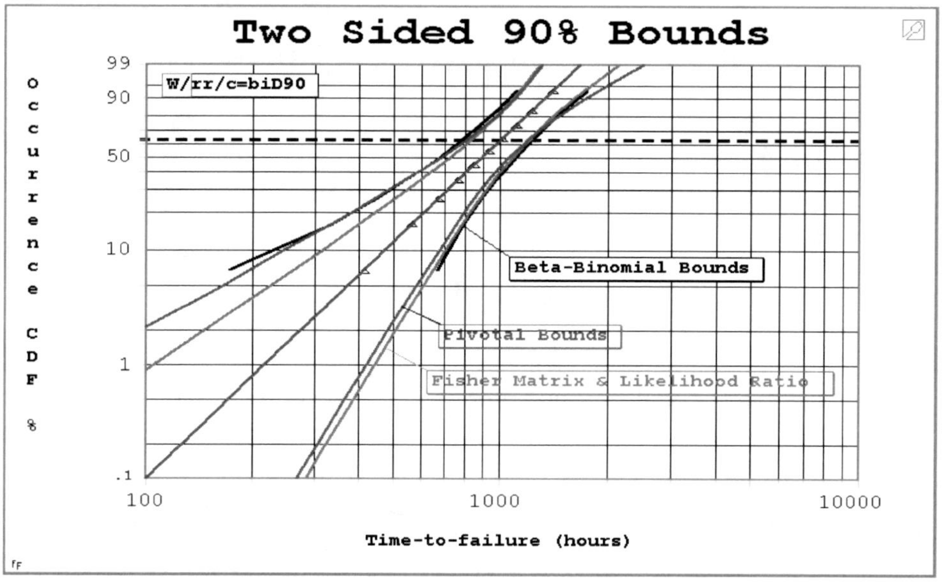

Figure 7-5. Comparison of Confidence Intervals

7.3.4 Pivotal Bounds – Monte Carlo

This method utilizes Monte Carlo (MC) methods based on a "pivotal" statistic. [See Lawless 1982]. A statistic is said to be pivotal if its sampling distribution does not depend on unknown parameters. Many statisticians and corporations recommend the pivotal method. The lower bound is conservative compared to the other methods. See Figure 7-5.

Pivotal (MC) bounds are **best practice** for small and intermediate size samples using median rank regression. For large samples the computing time may be excessive. Recalculating for the same input conditions produces slightly different results each time because the seed number starting the simulation varies. However, if repeatable answers are desired the seed number may be locked and the same data set will produce the same bounds. The pivotal (MC) method, developed by Wes Fulton, properly accounts for the statistical inverse inference. The inference estimates an interval for the true unknown value of B life, rather than the distribution of B values for a given set of distribution parameters, a probability problem. See Table 7-1 for a comparison of these methods.

A major advantage of this method is that it allows the goodness-of-fit p-value ("prr" from a single quick simulation) of the correlation coefficient to be calculated. Using p-value is the second best method for measuring goodness of fit after the likelihood ratio test according to $[Liu]. The highest p-value for the correlation coefficient among competing distributions indicates the best model for the data.

The likelihood ratio confidence bounds are a good alternative to pivotal bounds for small samples if the method chosen is MLE-RBA. These likelihood bounds show reasonably good agreement with pivotal bounds for small and moderate sample sizes.

7.3.5 Reliability Assurance Interval and the "p" Value (not goodness-of-fit p-value)

Confidence intervals for B lives and reliability may be two-sided as illustrated above, or one-sided. The lower one-sided interval is frequently employed with a statement such as the reliability is 99% at 327 hours or less with 90% confidence. The *reliability assurance concept*, proposed by Mr. Wes Fulton is also one sided. With the reliability assurance index the confidence level always equals the reliability demonstrated. SuperSMITH Weibull provides this capability. This concept is relatively unknown because of the *"p" value* concept which is widely used. If we calculate the highest value of confidence the data will support and call it the "p" value rather than confidence, this does not violate the requirement for a priori determination because we do not call it confidence. This confidence-related "p" value is not goodness-of-fit p-value.

7.3.6 Normal Distribution Confidence Bounds With Student's t

Two types of normal data can employ standard normal confidence bounds based on Student's t: (1) complete data, (no suspensions) and (2) non-life data, the typical application of the normal distribution for modeling dimensions and measurements. For these cases, a *modified MLE*, (MMLE), solution that minimizes bias is required for Gosset's Student's t. See Section 5.3 for the interesting history on this modification. The MLE standard deviation is the root sum of the squares of deviations of each data point from the mean divided by the number of data values (n). The MMLE standard deviation is the MLE standard deviation multiplied by the square root of ((n)/(n-1)). It is called the square root of the unbiased estimate of the variance. It has less bias than the MLE standard deviation which is why Gosset elected to use it for his Student t confidence intervals. For example, using simulation, 50,000 sets of sample size ten were generated from a normal distribution with a true mean of 100 and true standard deviation of 10. The mean MLE standard deviation estimate is 9.23 while the MMLE estimate is 9.72. The median rank regression estimate is 10.23, and the unbiased estimate using MLE-RBA described in Section 5.5 is most accurate, 9.98.

Normal regression (nr) confidence bounds are based on the Student t statistic and require the MMLE standard deviation. These bounds are interval estimates for the location of the fit line as well as B values. The standard error of the regression fit is used in conjunction with the student-t statistic to estimate the distribution of confidence about the fit line. This confidence method is only available for the normal distribution fit model and only when no suspensions are allowed. It is generally conservative compared to the other confidence methods.

7.3.7 Summary of Confidence Bounds for B Life and Reliability

The Weibull, normal, and lognormal models were previously presented. Confidence interval estimates for their parameters of interest are presented here. Recommendations for best practice:
For small samples, twenty or less failures, Fisher Matrix and Likelihood Ratio are too narrow, too optimistic. The Pivotal bounds provided in SuperSMITH Weibull using Monte Carlo simulation were developed by Wes Fulton as requested by Daimler-Chrysler to overcome this deficiency. Pivotal bounds are best practice **for small samples for MRR. By comparison, Pivotal bounds generally have the most conservative lower bound. This approach also provides an accurate p-value for the goodness-of-fit correlation coefficient squared.**

- All these confidence methods assume t_0 is known. If t_0 is unknown, the confidence intervals are much larger. The pivotal method has an option to allow uncertainty in t_0 to be included but it is never used as the resulting intervals are huge.
- Use Fisher Matrix bounds for complete samples and quick answers.
- For MLE-RBA solutions **best practice** is to use Likelihood Ratio.
- If your customer or your management has a preferred method, provide their choice.

It is probably confusing to the reader to have all these methods. Why not just use Pivotal bounds for small samples and the fast Fisher Matrix for large samples? The answer is that some customers, organizations, and contracts may require one of the other methods above. Therefore, they are all available in the SSW software. In addition, special methods and models have their own confidence bounds. The Probit data is binomial so binomial bounds are employed. The Kaplan-Meyer model and Crow-AMSAA models also have their own confidence bounds. All of the confidence estimation methods take the uncertainty of the suspensions into account but the pivotal method to a lesser extent. The pivotal and beta-binomial methods do not take the goodness of fit into account.

SSW provides predictive interval estimates for failure forecasts based on Ann Azevedo's suggestion and [Nelson 2000]. Please remember that these estimates often will be optimistic because the real variation is probably much greater than the statistical precision.

7.4 Confidence Intervals for β and η

Confidence intervals for β are useful for comparing data sets to see if β may be considered constant to allow Weibayes solutions. If the β interval estimate includes the other β value the βs are not significantly different. Another application is to see if one (1) is included in the confidence interval for β to test if the data is significantly different from a random (exponential) data. Bounds on η may be used to compare characteristic lives from two data sets.

The B life bounds described in Section 7.3 provide a quick estimate of β and η uncertainties. The B63.2 bounds are the bounds for η. The spread between confidence bounds indicates the statistical uncertainty in β. [Nelson 1982] provides a normal approximation for β and η bounds provided by the SSW report option if Fisher's Matrix bounds are requested. Similarly, if likelihood ratio bounds are requested, the likelihood contours provide these bounds in the SSW report option. They may also be read directly from the contour as the range of the contour, see Figure 7-4. Note that if joint bounds are needed, the degrees of freedom are slightly different from individual bounds. For the example used in Figures 7-1 through 7-3, Table 7-1 compares 90% bounds with all four methods. The B10 life was chosen because the Beta-Binomial B life bounds do not exist at lower values with small samples. Pivotal bounds are **best practice** for small samples.

Table 7-1. A Comparison of Confidence Methods
SuperSMITH Weibull results from precise sample from η = 1000, β = 3.0, N = 10
Bold methods are best practice for small samples

Method	β Bounds	η Bounds	B10 Bounds	B10 Range
Beta-Binomial MRR	1.89-4.11	783-1226	231-721	490
Fishers Matrix MRR	1.80-4.99	832-1201	301-741	440
Pivotal MRR	**1.71-4.94**	**806-1211**	**253-680**	**427**
Likelihood Ratio MLE-RBA	**1.91-4.73**	**817-1190**	**276-669**	**393**

7.5 Are Two Weibull Data Sets Different or From The Same Distribution

The question is are two data sets from the same parent distribution or are they significantly different? The two sets might be from:

1. New design vs. old design
2. Different geographic sources
3. Different fleets or customers
4. Different usage of the product
5. Different production lots
6. Different vendors
7. Different alloys

This question may be answered statistically by testing the two data sets to see if they are significantly different. For example, an automobile failure mode is observed in California and Connecticut. Do the failures come from different populations or the same population? Dichotomies might be United Airlines versus Delta, or new design versus old design. The null hypothesis is that the two sets come from the same parent distribution. This tacitly implies constant slope. The Weibayes method (Chapter 6) is based on a known or assumed slope. If β can be assumed or known, Weibayes provides the **best practice** solution. If the 90% lower bound Weibayes line for the upper set is to the right of the 50% Weibayes line for the lower set, we can state with at least 90% confidence that the two sets are significantly different. It is simple and accurate and provides a good graphic comparison.

If the data sets are significantly different, the information may answer important questions like..."is the new design better than the old?" If the data sets are not significantly different, they may be merged to obtain larger sample size and more accurate results. The null hypothesis is that the two data sets come from the same parent distribution, i.e., they are not significantly different, and therefore, the merged set provides the best estimate of the parent distribution.

Without Weibayes, the problem becomes extraordinarily difficult, particularly if interval suspensions are present. Interval or random suspensions are those that are neither early nor late; they occur within the range of the failures. There are many approaches to this important problem. At the SAE Weibull User's Conference in March 1994, this problem was the hottest topic and the attendees tasked the author and Wes Fulton to review the available methods and recommend the **best practice**.

Wes Fulton and the author spent years on this research and summarized the results in two SAE papers $[Fulton and Abernethy 1996 and 2000]. One major problem is the likelihood function and the MLE estimates for the Weibull and lognormal are biased optimistically for small samples. Therefore, a large part of the research was to develop correction factors for the likelihood function and the Weibull parameter β to provide accurate analysis with small samples. Details of the research are given in the above papers.

There are **three methods recommended as best practice**:

7.5.1 Double confidence bounds do not overlap- when compared at the B10 life
7.5.2 Likelihood Ratio Test- overall comparison
7.5.3 Likelihood Ratio confidence contours do not overlap-overall comparison

Note that the assumption for method (7.5.1) is different than the other two methods. Parametric models were studied since they can be broadly applied and, when appropriate, give much greater accuracy than non-parametric models. The last two methods can make comparisons of more than two sets, i.e., three sets, four sets, etc.

7.5.1 Double Confidence Bounds Do Not Overlap

This method is analogous to the Ford Motor Company method presented at the 1994 SAE Weibull User Conference by Jim Lempke. A simplifying assumption is that the comparison is made at the B10 life. Do the two data sets have significantly different B10 lives? If so, they are considered different. The procedure is to put double sided confidence bounds on the SSW plot say at 80% level. If the upper bound of the lower set is below the lower bound of the upper set at the B10 level, the sets are significantly different at the B10 level with 90% confidence. The B10 level is widely accepted and was recommended by Waloddi Weibull for comparing bearings. Ford Motor Company employs it for comparing vendor new designs with old designs. Based on one million comparisons of same-parent sets with sample size eight using 80% Fisher Matrix double bounds (or single 90% bounds) with rank regression, approximately 90% of the pairs were alike. Double confidence level is displayed on the plot, but the actual confidence level is the single confidence level because *only one bound is used from each set*. The actual percentage is biased somewhat with confidence level, sample size, the choice of B level and Weibull slope differences. So, caution must be used with this method. This method is simple and reasonably accurate. It is a good choice when assuming the comparison at B10 level is acceptable. The other two recommended tests are more general making the overall comparison better. Figure 7-6 shows a significant difference at 95% at the B10 level using double 90% Likelihood Ratio Bounds. As mentioned earlier these bounds are corrected for small sample sizes when MLE-RBA is selected. In this case with eight data points Pivotal bounds would be appropriate if MRR were used.

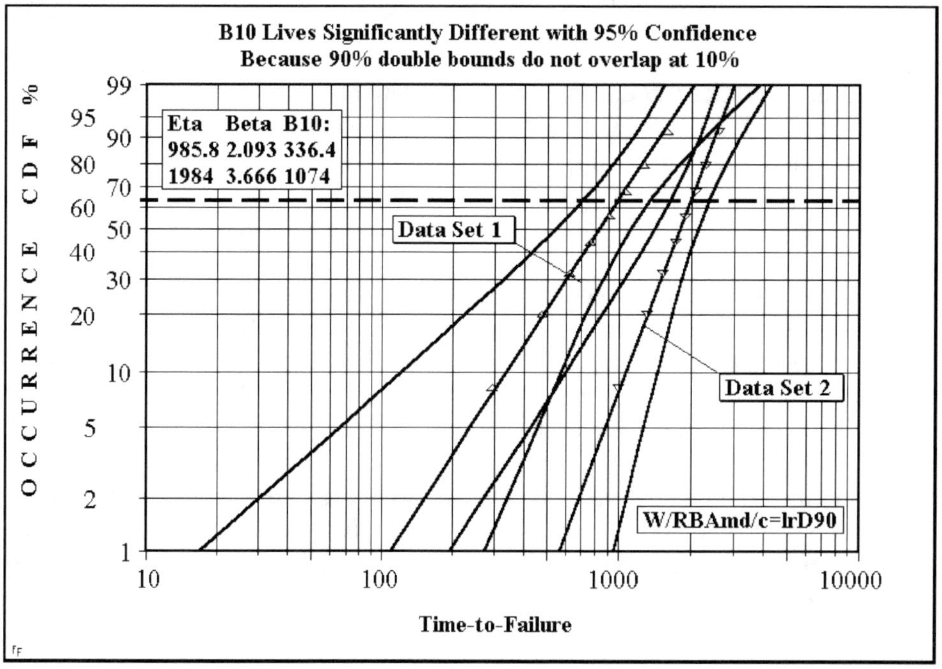

Figure 7-6. Double Confidence Bounds Do Not Overlap at The B10 Level.

7.5.2 Likelihood Ratio Test

This test is described fully in [Nelson 1990]. It is briefly summarized here. The *maximum log likelihood* (MLL) is obtained for each set and the merged set. The sum of the two separate MLLs minus the MLL for the merged data is compared to a chi-squared critical value. This method will test more than two sets at the same time. The standard Likelihood Ratio test does not work well with small sample sizes as it is biased optimistically. The LR test indicates a difference with a "P" value that is too high for all but large sample sizes. Such a restriction would essentially defeat the purpose since it is exactly at small sample sizes where help in measuring the difference is needed most. Wes Fulton developed an adjustment that removes most of this sample size bias making a modified LR Test the most accurate way to compare data sets found so far.

The standard LR Test (biased at small sample sizes) uses inequality Equation 7-1.

There is a significant difference if $\left[\hat{\mathcal{L}}_1 + \hat{\mathcal{L}}_2 - \hat{\mathcal{L}}_m\right] > \chi cd^2/2$ (7-1)

where:

$\hat{\mathcal{L}}$ = log likelihood, see [Nelson 1990] for explanation
$\hat{\mathcal{L}}_1, \hat{\mathcal{L}}_2, \hat{\mathcal{L}}_m$ = maximums for set 1, set 2 and the merged set respectively
χcd^2 = chi^2 at c confidence, d degrees of freedom (in this case d = 2)

The modified LR test inequality includes Wes Fulton's "Fulton Factor" to compensate for sample size bias. Research results indicate essentially no bias. **The modified LR Test uses Equation 7-2.**

There is a significant difference if $\left[\hat{\mathcal{L}}_1 + \hat{\mathcal{L}}_2 - \hat{\mathcal{L}}_m\right](FF) > \chi cd^2/2$ (7-2)

where:

FF = Fulton Factor = (N - 1) / (N + (PHI-1))
N = composite sample size = 2 / ((1 / N1) + (1 / N2))
N1, N2 = sample size of set 1 and set 2 respectively
PHI = golden ratio or divine division = 1.618...

The LR test for the Weibull distribution becomes unbiased with the new factor. For the data sets shown in Figure 7-6, the "p" value is 99.1% that the two sets are significantly different. The likelihood ratio provides the odds that the sets are significantly different compared to not significantly different.

7.5.3 Likelihood Contour Plots

The author hypothesized that if the 90% likelihood contours for the parameters η and β overlap, the data sets are not significantly different with 90% confidence. See Figure 7-7 for likelihood contours for the two data sets illustrated in Figure 7-6. If there is a gap between contours or they are tangent, then the data sets are significantly different. Professor George Barnard of the University of Essex in England was the world's leading authority on likelihood methods when this method was developed. After reviewing the hypothesis with the author, he concurred that this method is rigorous. The use of SSV software provides graphic results and makes this an easy test to perform. However, like the standard likelihood ratio test, the contour test is optimistic without adjustment to correct the bias for small samples. The SSW software uses an adjusted function called the Justified Likelihood Function (JLF) to increase the contour size removing small sample bias. The MLE-RBA is used with the JLF to correct both the likelihood contours and the parameters for small sample bias. The adjustment effects diminish as sample size increases. The JLF requires an unbiased estimate of the true slope on the probability plot (β for Weibull, sigF for lognormal, etc.) produced by the reduced bias adjustment (MLE-RBA) factor described in Chapter 5.

The approximation for the JLF developed by Wes Fulton is:

$$JLF(x, p_1, p_2) = \frac{\{LF[x, p_1, p_2]\}^{FF}}{\{LF[x, \hat{p}_1, (\hat{p}_2/RBA)]\}^{(FF-1)}/e^{(V_{JLLF})}} \tag{7-3}$$

where:

- JLF = Justified Likelihood Function
- x = data values
- p_1 = scale (e.g. eta for Weibull)
- p_2 = shape (e.g. beta for Weibull)
- LF = Likelihood Maximum
- ^ = at Maximum Likelihood
- RBA = Reduced Bias Adjustment per Abernethy
- FF = Fulton Factor = (N - 1) / (N + (PHI-1))
- PHI = golden ratio or divine division = 1.618...
- V_{JLLF} = Likelihood Height (Vertical) adjustment

Note: For adjusting the Likelihood contours with JLF, the Likelihood Height adjustment is not relevant so V_{JLLF} is set = 0, so $e^0 = 1$. Other uses of JLF may use a different value for V_{JLLF}.

The contour test provides a valuable graphic supporting the Likelihood Ratio Test (LRT). For example if the LRT says that five sets are different does this mean one set is different from four, or 2 sets are different from three, or they are all different from each other? The contour test provides the graphic answer, identifying the clusters of related sets.

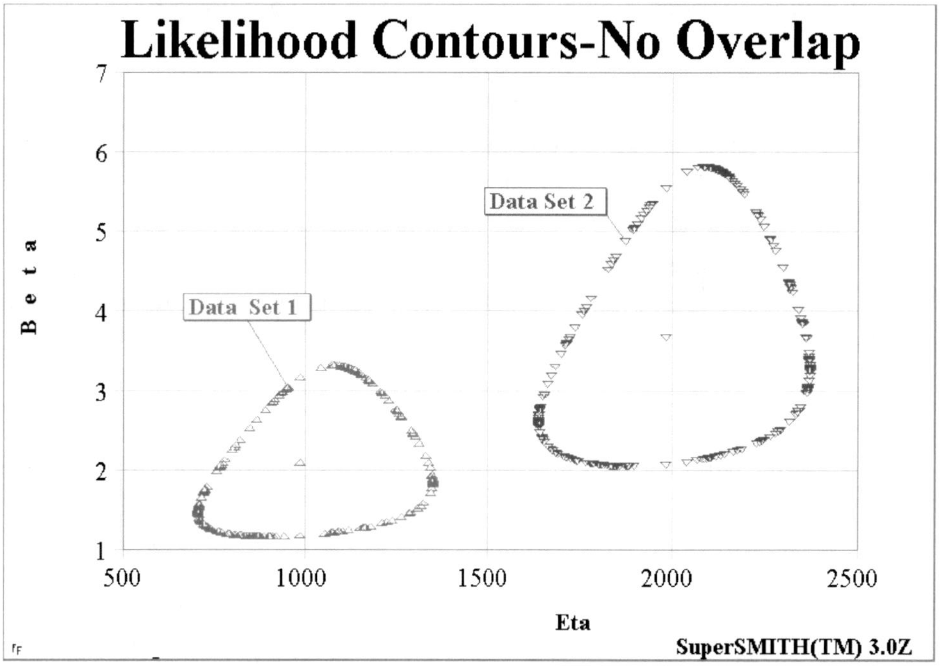

Figure 7-7. Likelihood Contours - No Overlap

Figure 7-8 shows confidence contours using the JLF method compared to those calculated with the unmodified likelihood function (LF). The slope parameter MLE point is adjusted with the RBA factor and the contour size is adjusted due to the JLF.

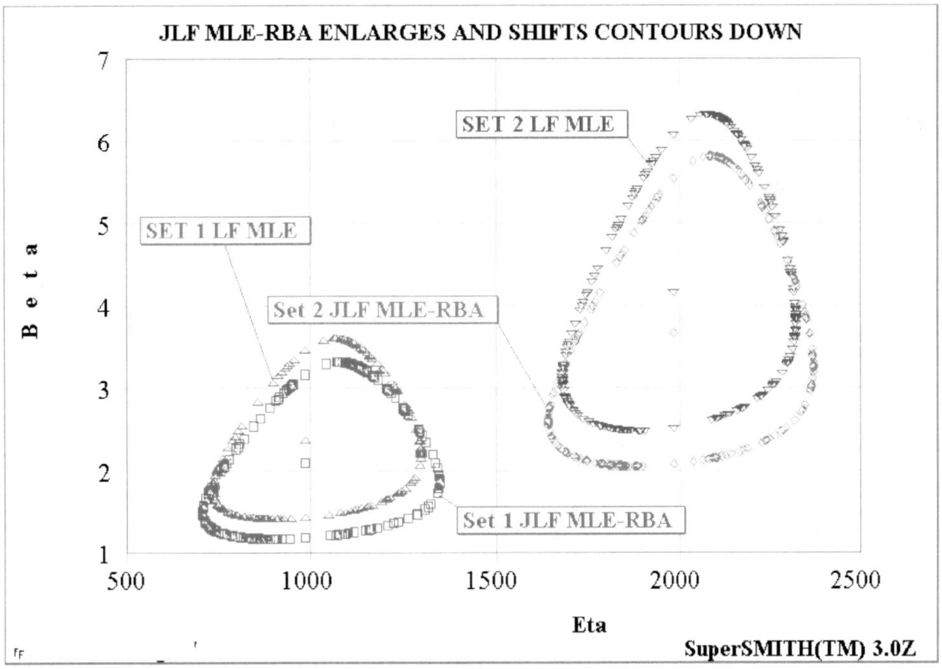

Figure 7-8. JLF MLE-RBA Enlarges and Shifts Contours Down

7.6 Problems - True or False? The answers are in Appendix K.

7-1 The probability that a 90% statistical confidence interval contains the true unknown value is 0.90.

7-2 Statistical confidence intervals include both statistical and engineering uncertainties.

7-3 Data from system tests were used to calculate a lower confidence bound for reliability of 97.3%. This value is an estimate of the system reliability.

7-4 A 99% confidence level for the "reliability assurance index" would have a lower bound of 99% for reliability.

7-5 The simple concept of statistical confidence intervals is readily understood by all managers and engineers.

7-6 Extrapolating the Weibull line and the associated confidence bounds are often done, but it is statistically invalid.

7-7 Johnson's beta-binomial bounds are usually wider, more conservative, than Fisher's Matrix or likelihood bounds.

7-8 With small samples, these Fisher Matrix bounds, are optimistic, but Pivotal bounds will provide acceptable accuracy.

7-9 To compare two Weibull data sets to see if they come from different Weibull distributions, three methods are recommended.

7-10 A component is tested to demonstrate 0.99 reliability with 0.95 confidence. Four of these components are linked in series to make a system. The system reliability is 0.99^4 with 0.95^4 confidence.

7.6 Egon Sharpe Pearson and Jerzy Neyman

Egon Sharpe Pearson 1895-1980 The author was privileged to have attended many Royal Statistical Society meetings in London with Egon Pearson. There was often friendly interaction between Egon, George Barnard, Jeffries, D.R.Cox, G. Jenkins, and visitors like G.E.P Box from America. Solutions from all three schools, Ronald Fisher's, Karl and Egon Pearson's, and Bayesian were compared, sometimes vigorously. Egon collaborated with Jerzy Neyman to invent the *frequency – confidence interval* school of inference, which ended his friendship with Fisher who despised the concept. The concept of tests of significance and the null hypothesis are among Egon's other contributions. He had great admiration for his dear friend, William Gosset, and in his final years he started a book about Gosset, ["Student"] which my professor, George Barnard, finished after Egon passed on. He was a gentleman and a wonderful teacher. Even though the author shares Fisher and Deming's negative views on confidence intervals, I have always admired Egon Pearson and frequently use his Biometrica Tables.

JERZY NEYMAN

Jerzy Neyman 1894- 1981 Jerzy Neyman is considered to be one of great founders of modern statistics. He made large contributions in probability theory, testing hypothesis, confidence intervals, generalized chi-square, and other areas of mathematical statistics. He was born Jerzy Neyman in Bendery, Russia. Neyman was forced to move to Poland due to the war between Poland and Russia. Neyman was 27 at the time. In Poland, Neyman worked with W. Sierpinski before moving to London in 1924. Neyman studied under Karl Pearson while in London. He also made contacts with Egon Pearson, R. A. Fisher, and W. S. Gosset while at University College. By 1934, Karl Pearson had retired and his department was divided between his son Egon and Fisher. Egon invited Neyman to work with him. They worked on the theory of testing hypotheses. They supplied logical foundation and mathematical rigor to the theory that was missing in previous methodology. Their work was disputed by some mathematicians, including Fisher. The Neyman-Pearson ideas eventually spread throughout mathematics. Their ideas made sure that samples were large enough to avoid false representation.

The theory of estimation by confidence sets was Neyman's next topic of choice. He used confidence intervals to guarantee that the probability of covering the true value of the parameter to be estimated was at least equal to a preassigned value called the confidence coefficient. His uses soon appeared in many textbooks and works on statistical methodology.

In 1937 Neyman accepted a position at the University of California-Berkeley. He was asked to start a statistics department at Berkeley. Many people question his decision, but Neyman took the position because he was fearful of Hitler and the start of World War II. It was at Berkeley, were he spent half of his life, that he came up with his BAN (best asymptotically normal) estimation theory. The BAN estimates are now widely used in a manner similar to the use of least squares.

CHAPTER 8. RELATED MATH MODELS

8.1 Introduction

Up to this point, Weibull and Weibayes distributions, good and bad, have been discussed in detail. In Chapter 3, the lognormal distribution was introduced as the most frequent alternative to the Weibull distribution when a Weibull does not seem to fit. In this chapter other math models, related to the Weibull will be introduced. Some are useful for reliability and quality control analysis, either on their own or as part of complex math models. Others are the **best practice** for some types of life data. Some examples:

- One shot devices either work or fail. There is no life analysis (although there may be shelf life).
- The times to failure are unknown. The data are the number of failures and total fleet hours or mileage.
- No failures have been observed and beta is unknown.
- There is an event of interest (outages, in-flight failure, uncontrolled accelerations, etc.) that occurs now and then, at some rate.
- There is an argument that the statistical distribution is unknown.
- There are risks associated with a sampling plan based on a random sample of 20 from a batch of 10,000 at receiving inspection. The batch is accepted if there are 1 or less defective parts found in the sample.
- A system or process is changing, reliability is improving or degrading. Failed parts are replaced. There is a mixture of failure modes. Forecast the future failures.
- Data is available for a period of time, a "snapshot," not including early data.

In this chapter these other models will be discussed, particularly as they relate to the Weibull. These models include the binomial, the Poisson, the exponential, and the Kaplan-Meier survival function. The primary objective is to acquaint the reader with the "situation" that relates to each of these other statistical models and provide simple methods for their application. The Crow-AMSAA reliability growth model has become so important that in this edition of the Handbook it has been updated and moved from Chapter 8 to Chapter 9 even though it is related to the Weibull distribution.

8.2 Binomial and Poisson Distributions ... For Counted Items

8.2.1 Binomial Distribution

In the binomial situation, there are n trials; the trials may be parts, people, missions, tests, systems, etc. In each trial an event may or may not occur. The probability the event occurs in a single trial is p. The probability it does not occur is 1 - p. The binomial has two parameters, n and p. Knowing values for n and p completely defines a specific binomial distribution. The distribution of the events among trials is random. Some examples:

- inspecting 100 parts for good and bad pieces,
- rocket engine successes and failures in 1000 launches,
- finding defects of a given size in 267 systems,
- deaths from cancer in men 76 years old in Florida,
- flipping a coin 100 times or betting on cards.

In each case, there are n trials and r events within those trials. There is probability, p, that an event will occur on any single trial, perhaps the next start, run, card, flight, etc. Each "trial" is independent and has only two possible outcomes; the event occurs or does not occur. The binomial distribution is defined as the probability of exactly r events in n trials as follows:

$$P(r) = \binom{n}{r} p^r (1-p)^{n-r} \tag{8-1}$$

The cumulative probability distribution for binomial is the sum of the above terms for r or less events. If r is summed from zero to n, the cumulative distribution will sum to one:

$$\sum_{r=0}^{r=n} P(r) = 1.0 = \sum_{r=0}^{r=n} \binom{n}{r} p^r (1-p)^{n-r} \tag{8-2}$$

where:

$\binom{n}{r} = \dfrac{n!}{r!(n-r)!}$

Mean $= np = \mu$

Standard deviation Quantity (r) $= \sqrt{np(1-p)} = \sigma_n$

Standard deviation Proportion (p) $= \sqrt{p(1-p)/n} = \sigma_p$

The binomial distribution for n = 20 and p = 0.01 is the mirror image of the binomial for n = 20 and p = 0.99. See Figure 8-1. If p = close to 0.5 the distribution is symmetrical and if n is large, the binomial approximates the normal. However, engineering problems almost always involve p close to zero or one. SSW will solve binomial problems with the calculator option.

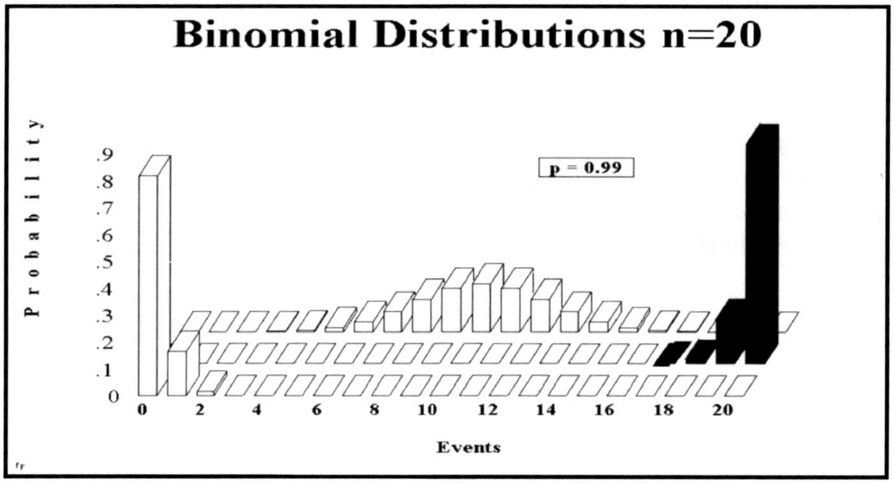

Figure 8-1. Binomial Distributions n = 20 and p = 0.01, 0.99 and 0.5 from front to rear.

For example, based on field data, on the average, one part in 21 will fail per flight. That is, the probability of a failure is 1/21 or 0.048 on any flight. What is the distribution of the successes and failures in 600 flights? Let us assume the process is random, (although in reality it may not be). Random means there is no "memory" from one event to the next, and the failure rate, 0.048 is stable. If so, the distribution of outcomes is Binomial with n = 600 and p = 0.048. The data is discrete, counted, not measured. As such the precision is poor, and accurate estimates require huge samples, measured in the hundreds or thousands.

Example 8.1: Quality control sampling. A manufacturer has agreed with a purchaser to use random sampling to accept or reject the large batches of machine screws that he manufactures. The negotiated sampling plan employs a random sample of 20 screws. If two or more defects are found the batch is rejected. What is the probability of a batch being accepted, if the true percent defective is 1%? Solution: The batch is accepted if there is zero or one defect. Let p be the probability of a defect and r be the number of defects, p = 0.01, (1-p) = 0.99, n = 20. The probability of acceptance is the cumulative probability function for r = 0 and 1, using Equation 8-2. Acceptance of this batch is highly probable.

$$\textbf{Probability of Acceptance} = \sum_{r=0}^{r=1} P(r \le 1) = P(0) + P(1)$$

$$= (1-p)^n + \binom{20}{1} p\,(1-p)^{n-1}$$

$$= (0.99)^{20} + (20)(0.01)(0.99)^{20-1}$$

$$= 0.8179 + 0.1652 = 0.9831$$

Example 8.2: A diesel engine manufacturer agrees to make twenty endurance tests as a one or zero failure reliability demonstration. If the engine reliability is 99%, what is the probability of passing? The solution is the same as Example 8.1.

Example 8.3: Suppose there are 100 successful operations, with no failures. What is the estimate of reliability at this point? A point estimate of the reliability of the system is:

$$R = \frac{\text{Successes}}{\text{Attempts}} = \frac{100}{100} = 1.0$$

While this value is an excellent estimate; unbiased, minimum variance, maximum likelihood, consistent, sufficient, and invariant, it **is** also ridiculous. Most customers will not accept 1.0 as an estimate of reliability. A confidence interval estimate is an alternative. The binomial may be employed to calculate reliability confidence intervals by setting the cumulative probability equal to one minus the confidence level for a lower one-sided bound.

$$\textbf{For Lower 90\% Confidence: } 1 - \text{Confidence} = \sum_{r=0}^{r=f} \binom{n}{r} p^r (1-p)^{n-r}$$

$$= (1-p)^{n-0} = R^n$$

$$\text{so, } 0.10 = R^{100}$$

$$R = 0.977$$

Let p be the probability of failure and (1 - p) is the reliability. r is the number of failures. For this example of n = 100, f = r = 0, we can state with 90% confidence that the true unknown reliability lies between 0.977 and 1.0. Please note that the interval is the estimate, that is, 0.977 is not a reliability estimate, but merely a lower bound of a confidence interval estimate. [Nelson 1982] reports that these are conservative interval estimates.

Note there is new technology using the Crow-AMSAA model that provides much greater precision for success-failure and "go no-go" data when n is large and p is small. This will be illustrated in Chapter 9.

Example 8.4: To illustrate the poor precision of the binomial, calculate the number of zero failure tests required to demonstrate 99.9% reliability, lower bound, with 99% confidence. The equation is:

$$\mathbf{1 - Confidence = R^n} \tag{8-3}$$

$$\mathbf{1 - 0.99 = 0.999^n}$$

$$\mathbf{n = 4603}$$

4603 tests without failure are required! Quality control demonstrations use the same binomial statistics. To demonstrate a defect rate or failure rate of 0.001 or less with 99% confidence, randomly sample 4603 pieces, with no defects or failures. For systems and components, binomial reliability demonstrations are a form of substantiation tests. Similarly for parts, quality demonstrations are substantiation of manufacturing. For both reliability and quality demonstrations, Paul Barringer calculated Table 8-1. For quality demonstrations, remember R = (1 - p).

Table 8-1. Success Testing Tables For Demonstrating Reliability and Quality

Number Of Tests With Zero Failures Allowed (1-confidence) = R^n										
	Note: *B Life %* equals % Defective, with other % values being Confidence Level									
Reliability	*B Life %*	50%	60%	70%	80%	90%	95%	99%	99.9%	99.99%
0.9999	0.01	6932	9163	12040	16094	23025	29956	46050	69075	92099
0.999	0.1	693	916	1204	1609	2302	2995	4603	6905	9206
0.998	0.2	347	458	602	804	1151	1497	2301	3451	4601
0.997	0.3	231	305	401	536	767	998	1533	2300	3066
0.996	0.4	173	229	301	402	575	748	1149	1724	2298
0.995	0.5	139	183	241	322	460	598	919	1379	1838
0.99	1.0	69	92	120	161	230	299	459	688	917
0.95	5.0	14	18	24	32	45	59	90	135	180
0.90	10.0	7	9	12	16	22	29	44	66	88
0.85	15.0	5	6	8	10	15	19	29	43	57
0.80	20.0	4	5	6	8	11	14	21	31	42
0.75	25.0	3	4	5	6	9	11	17	25	33
0.50	50.0	1	2	2	3	4	5	7	10	14
All fractions of the number of tests are rounded upward.										

Number Of Tests With Up To One Failure Allowed (1-confidence) = $R^n + n(R^{(n-1)} * (1-R))$										
	Note: *B Life %* equals % Defective, with other % values being Confidence Level									
Reliability	*B Life %*	50%	60%	70%	80%	90%	95%	99%	99.9%	99.99%
0.9999	0.01	16784	20223	24392	29943	38896	47437	66381	92331	117559
0.999	0.1	1679	2022	2439	2994	3889	4742	6636	9230	11751
0.998	0.2	839	1011	1219	1497	1944	2371	3317	4613	5873
0.997	0.3	560	674	813	998	1296	1580	2210	3074	3914
0.996	0.4	420	506	610	748	971	1185	1657	2305	2934
0.995	0.5	336	404	487	598	777	947	1325	1843	2346
0.99	1.0	168	202	244	299	388	473	662	920	1171
0.95	5.0	34	40	49	59	77	93	130	181	230
0.90	10.0	17	20	24	29	38	46	64	89	113
0.85	15.0	11	13	16	19	25	30	42	58	73
0.80	20.0	9	10	12	14	18	22	31	42	54
0.75	25.0	7	8	10	11	15	18	24	33	42
0.50	50.0	3	4	5	5	7	8	11	14	18

8.2.2 Poisson Distribution

In the Poisson situation, events occur in a continuum: time, rope, plastic, wire production, etc. It is a widely used distribution. It has only one parameter, the mean, μ, the average number of events in an interval of the continuum. Further, it has many applications in engineering and elsewhere. The Poisson has been applied to deaths from horse kicks, absentee data, blood counts, floods, the arrival of telephone calls, turbojet failures, typographical errors, spare parts demands for slow moving parts, nuclear emissions, accidents, and many other events. Like the binomial, the Poisson is a discrete distribution; it is based on counted events and is imprecise compared to distributions of measured quantities like the normal and the Weibull. Poisson events are random in time, that is the average rate of occurrence is stable. Unlike the binomial, there is no counter event. It makes sense to talk about how many aircraft land per hour. It does not make sense to consider how many do not land per hour.

The single parameter of the Poisson is the average or mean value of the number of events. If Z is defined as the expected or average number of Poisson events:

$$\textbf{Expected number (mean)} = \mu = Z = \sigma^2$$

$$\textbf{Standard deviation} = \sigma = \sqrt{(Z)}$$

The first term of the Poisson, Equation 8-4, the probability of zero events, is so important to reliability engineers that it has its own name, the *exponential distribution.* It is identical to a Weibull with $\beta = 1.0$. Thus the Weibull is linked to the Poisson by the exponential. The exponential will be discussed later in this chapter. As the expected number of events increases, the Poisson approaches the normal distribution. In other words, the shape of the Poisson distribution makes a transition from the exponential shape to the normal bell curve shape as the mean increases. See Figure 8-2 from SSV. Poisson problems may be solved using Abernethy-Weber Curves (see Figure 8-3) or with the SSW calculator option.

$$P(0) = e^{-Z} \tag{8-4}$$

$$P(1) = Ze^{-Z}$$

$$P(2) = \frac{Z^2}{2!}e^{-Z}$$

$$\textbf{Generally, } P(C) = \frac{Z^C e^{-Z}}{C!}$$

Example 8.5: The author lived in Palm Beach County, Florida for over three decades, and there were three hurricanes where the eye of the storm passed over the county. (All this was before the hurricanes of 2004 & 2005.Three passed over Palm Beach!) Based on this data, an estimate of the expected number of hurricanes in Palm Beach County is one per decade. How many hurricanes should be expected in a decade? What is the distribution per decade?

$$\textbf{Expectation} = z = 1.0$$

Using the equations above:

C	0	1	2	3	4
P(C)	0.368	0.368	0.184	0.061	0.0153

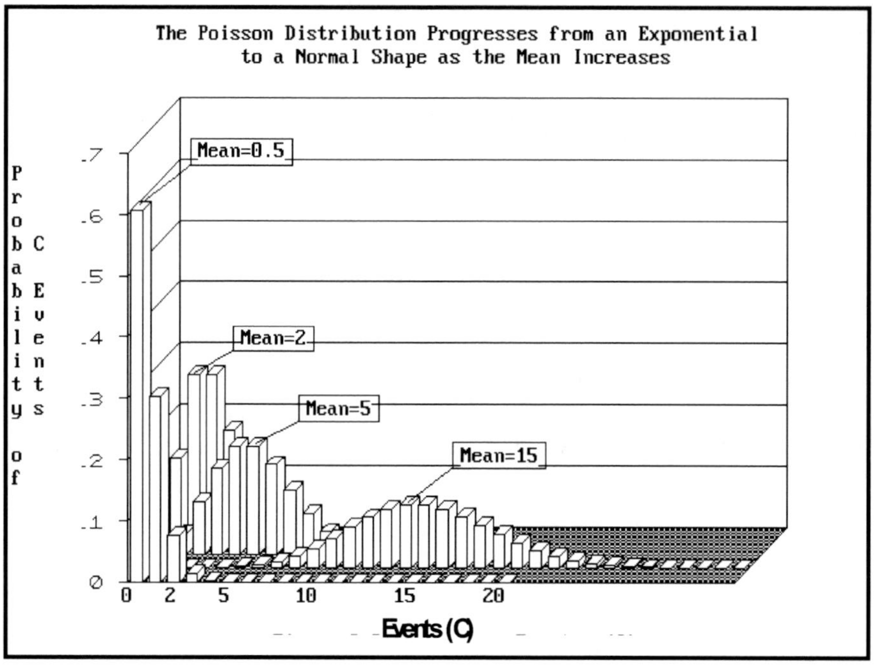

Figure 8-2. The Poisson Distribution Progresses from an Exponential to a Normal Shape as the Mean Increases

An alternative graphical solution was developed decades ago. Statisticians at Bell Labs invented the Thorndike Chart. It depicts the useful range of the Poisson. Although the author used it for years in seminars and workshops, the scales are complex and confusing to many engineers. The Abernethy-Weber plot shown as Figure 8-3 was invented by the author and produced by David Weber to provide an easier presentation of the Poisson using the familiar Weibull plot. The $C \geq 1$ line is the exponential, a Weibull straight line, with $\beta = 1.0$, $\eta = 1.0$. The ordinate is the mean, μ, or average value. The vertical scale is the cumulative probability of C or more events.

For example, the average number of hurricanes is one per decade, read the probabilities of C or more events per decade off the plot using a mean value of one:

C	1	2	3	4	5
P(C or more)	0.632	0.26	0.08	0.02	0.004

The probability of exactly zero events is the difference between the probability of zero or more (which is 1.0) and the probability of one or more (0.632).

$$\mathbf{P(C = 0) = P(C \geq 0) - P(C \geq 1)}$$

$$\mathbf{P(0) = 1.0 - 0.632 = 0.368}$$

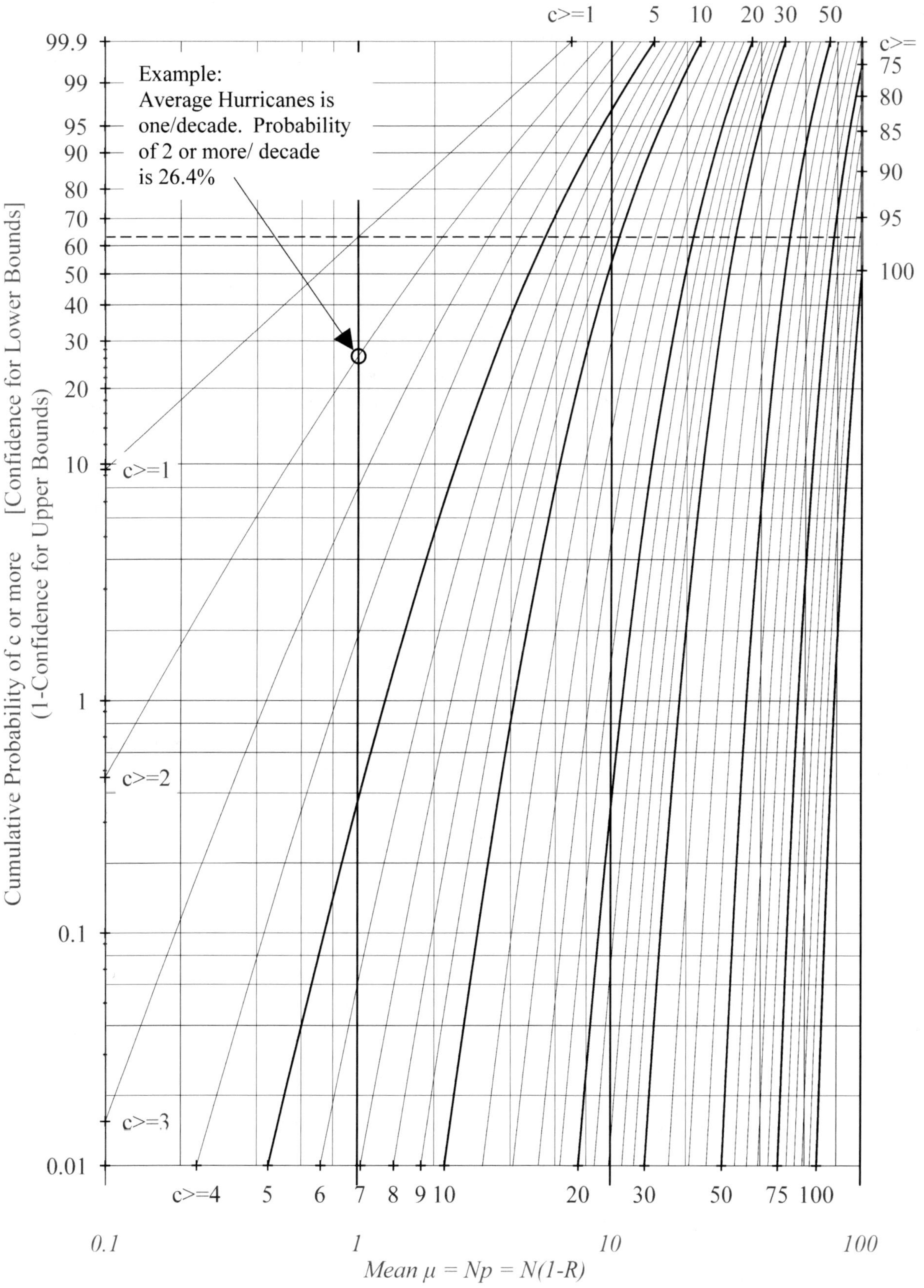

Figure 8-3. Abernethy-Weber Curves for the Poisson Distribution

The probability of exactly one event is the difference between the probabilities of one or more and two or more.

$$\mathbf{P(1) = 0.632 - 0.26 = 0.37}$$

C-1	0	1	2	3	4
P(C-1)	0.368	0.37	0.18	0.06	0.016

Therefore, the space between $C \geq 1$ and $C \geq 2$ is the "one" space.

8.3 Binomial Becomes Poisson...Sometimes

Think of the binomial situation where n is large and p is small. We have a string of many trials and rather rarely, an event occurs. This suggests a continuum where every now and then an event occurs. The Poisson and the binomial situations physically become the same, and asymptotically, so do the mathematical distributions. When n is large and p (or 1 - p) is small, the Poisson is an excellent approximation to the binomial. The rule of thumb is if np (or n[1 - p]) is less than five, the Poisson is a good approximation to the binomial. For engineering problems, p is almost always close to zero or one and usually, n is large. Therefore, most binomial engineering problems may be solved in seconds with the A-W Chart. (Figure 8-3)

For the same reasons, the Weibull with a beta of one and B lives close to one will approximate the binomial. This may be seen by comparing Table 8-1 with Table 6-2. For example, to demonstrate 99% reliability with 90% confidence Table 8-1 indicates 230 tests without failure are required. Table 6-2 for $\beta = 1$, and a ratio $(t/\eta) = 0.01$, which is B1 life or 99% reliability, shows 231 tests are required.

Taking this concept further, the Crow-AMSAA (C-A) model has the Weibull for its foundation and therefore the author suggests success-failure and go no-go data can be more precisely modeled with C-A when N is large and p is small as will be shown in Chapter 9.

This situation, n large, p small, was the inspiration for Poisson to invent the distribution in 1837. For decades it was considered the distribution of rare events until Gosset (penname "Student") proved in 1907 that the Poisson also applies when these conditions are not met.

Three examples of the A-W Chart (Figure 8-3) illustrate the application of the Poisson and the Poisson approximation to the binomial for quality and reliability engineers.

Example 8.6: Quality Control Sampling

A vendor and his customer negotiated a sampling plan to determine the accept/reject decision for large batches of parts. The plan calls for inspection of a random sample of 100 parts. If three or more defects are found in the batch of 10,000 parts, the batch is rejected. If two or less are found the batch is accepted. What is the risk to the customer that a bad batch may be accepted? What is the risk to the vendor that a good batch may be rejected? The standard approach is to plot these risks on an "operating characteristic" or "OC" curve. The expected value is np, which in this case is less than 5, so we can approximate the binomial with the Poisson. If the true percent defective is 1%, the expected value is np = 100 x 0.01 = 1. Read the risk on the Figure 8-3 at np = 1 and C = 3 or more as 8%. Repeat the process for p = 0.5, 2, 3 and 4%. Results:

% Defect	0.5	1.0	2.0	3.0	4
Prob $C \geq 3$	1.5%	8%	32%	58%	76%
Prob $C \leq 2$	98.5%	92%	68%	42%	24%

The reader should verify these values from the A-W Chart (Figure 8-3). The results with some additional values are shown in Figure 8-4, using SuperSMITH® Visual. If it was agreed that 0.5% is "good" quality the vendor's risk of rejecting a good batch is 0.015 (1.0 - .985). The customer's risk of accepting a bad batch, if 4% is "bad" quality, is 0.24.

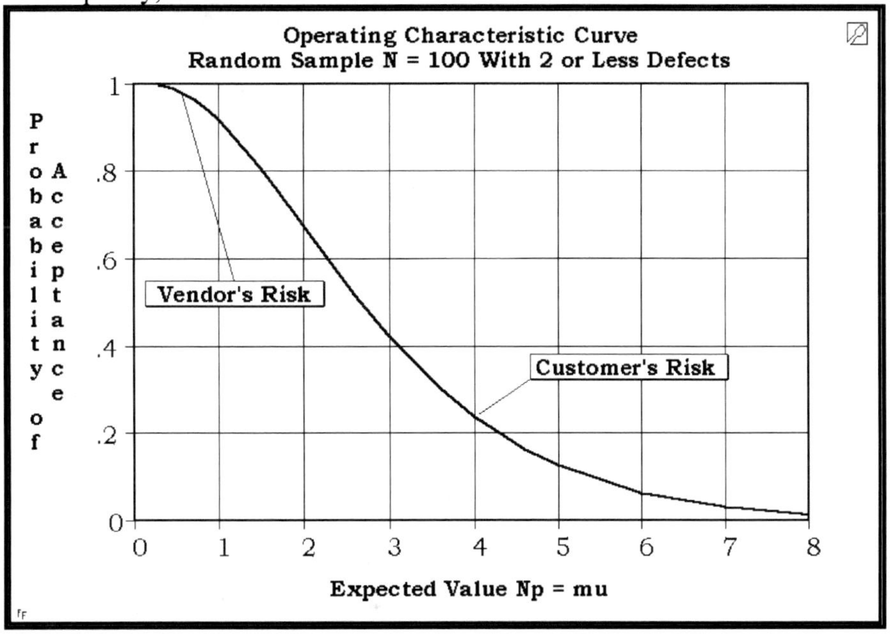

Figure 8-4. OC Curve

Example 8.7: Reliability Demonstration

How many tests without failure are required to demonstrate 99% reliability with 90% confidence? The solution must be somewhere along the $C \geq 1$ curve, as this is a no failure demonstration. The cumulative probability of C or more, vertical scale on the A-W Chart (Figure 8-3), is confidence level for lower bounds. The 90% confidence solution must lie on the A-W 90% line. On Figure 8-3, this line intersects the $C \geq 1$ curve at an expected value of 2.3. Find this point on the A-W chart. For reliability problems the expected value is n(1 - R). R = 0.99. Therefore, 2.3 = n(1 - 0.99) and n = 230 tests are required. Compare with Table 8-1. (Note that np = n(1 - R) = 2.3, which is less than 5, so the Poisson may be used to solve this binomial problem.)

Example 8.8: Poisson Problem

Assume the average value, say for outage rate [or in-flight shutdowns], is 8 per month [or per 1000 hours] based on data from the last five years. How many outages might be expected in a bad month, or what is the upper 90% confidence bound on the number of outages per month if the average is 8.0?

One minus the cumulative probability on the A-W Chart (Figure 8-3) equals the upper bound confidence level [1 - 0.90] = 0.10. The solution is to find the intersection of the 10% horizontal probability line with the vertical mean of 8 line. This is just between the curves for $C \geq 12$ and $C \geq 13$. The space between these two curves represents 12 events, so the upper bound would be 12 outages. (The space between $C \geq 1$ and $C \geq 2$ represents the probability of one event.)

Determine two sided 90% bounds for an average of 8. If we put 5% in each tail, the combination of 95% and 5% bounds will provide 90% two sided bounds. The upper bound is the 5% line. The intersection with the 8 line is between $C \geq 13$ and $C \geq 14$, the 13 space, therefore the upper bound is 13. The lower bound is on the 95% line between $C \geq 4$ and $C \geq 5$, the 4 space. The 90% two sided limits would be 4 and 13 outages per month. If 15 outages were observed in a month would you say that a statistically significant change had occurred? The answer is yes, with 90% confidence.

8.4 The Exponential Distribution

The exponential distribution is identical to a Weibull with a slope, β, of one. It is also the first term of the Poisson and the Gamma distribution with a shape parameter of one. The exponential situation occurs when events, perhaps failures, occur randomly in time. Units that have exponential lives have a constant hazard rate, (the instantaneous failure rate). In other words, of the units that survive to time t, a constant percentage will fail in the next unit of time. Exponential units are said to have no memory. Old units have the same reliability as new units. There are many classes of failures that are random in time. Failures due to nature such as lightning strikes on transformers, foreign object damage in aircraft engines, extreme weather conditions, earthquakes, and woodpecker attacks on power poles, are all exponential. (There is new technology that hints that earthquakes and extreme weather conditions are not random, and further, are predictable, but so far the author is skeptical.) Until about 1980, all electronic failure modes were considered to be random and, perhaps, they were. The newer solid state electronic failure modes tend to be infant mortality, Weibull betas around 0.7. Human failures tend to be random; a mechanic may misadjust an old unit as easily as a new unit.

Before Weibull analysis was widely applied, there were numerous cases of the exponential being employed incorrectly, when it did not apply, even to infant mortality and wear out failure data. The simplicity of the exponential seduced us into employing it and the results were not acceptable. However, there are bona fide random exponential, failure modes. There are also other applications that will be described. Life data tends to collapse into the exponential under certain conditions even when the underlying distributions are not exponential. [O'Connor 1991] describes this phenomena as the central limit theorem of reliability, "...it has the same central limiting relationship to life statistics as the normal distribution has to non-life statistics." Some examples of data that tend toward $\beta = 1$:

- Mixtures of more than 4 or 5 failure modes if the modes have different βs. If the modes have the same β, the mixture will have that β, approximately.
- System or component data when individual modes are not separated are mixtures of modes.
- Intervals between failures, regardless of the underlying mode. Intervals are used when the origin is unknown. For example, intervals between accidents, extreme weather, and earthquakes.
- If Weibull failure times are raised to the β power, they become exponentially distributed.
- If time to failure is uniformly randomly distributed, the reciprocal of the logarithm of the times are exponentially distributed.

The exponential probability function comes from the Weibull with β set to one and MTTF substituted for η as they are equal. MTTF is the mean-time-to-failure, the reciprocal of failure rate, λ.

$$\textbf{Reliability (Time} = \textbf{t)} = e^{-t/MTTF} = e^{-\lambda t} \qquad (8\text{-}5)$$

The author had traveled 1,400,000 miles on Delta Airlines up to 1992 when the first edition of this Handbook was written. If the average number of engines per airplane was three in this period of time, that would be 4,200,000 engine flight miles. If the typical engine has a Mean Time Between In-flight Shutdowns of 25,000 hours or 10,000,000 miles at an average speed of 400 miles per hour, what is the probability that he has been on a plane with an engine shutdown? An exponent of ((4.2M/10M) = 0.42) in Equation 8-5, above, produces a reliability of 0.66 or about one chance in three that he has been on a plane with an in-flight shutdown.

If a set of data is exponentially distributed, the number of failures has a Poisson distribution. The probability distribution may be read off the A-W Chart (Figure 8-3). The expected value is (Time/MTTF). SSW will solve exponential and Poisson problems with the probability calculator option.

Example 8.9: Exponential

If an exponential distribution MTTF is 10 hours, how many failures would be expected in 100 hours? The expected number is the ratio of (Time/MTTF) = (100/10) or 10 failures. The probability of more or less may be taken from the A-W chart (Figure 8-3).

What would the reliability be for one unit operating for one hour? The expected value is 1/10, or 0.1. From the A-W chart (Figure 8-3) at $C \geq 1$ and $\mu = 0.1$, the probability of one or more failures would be about 9.5%, and therefore the reliability would be 0.905. Use a calculator and Equation 8-5, to check this result. This exponential reliability function is displayed in Figure 8-5 using SuperSMITH Visual. SSW provides exponential probability plots.

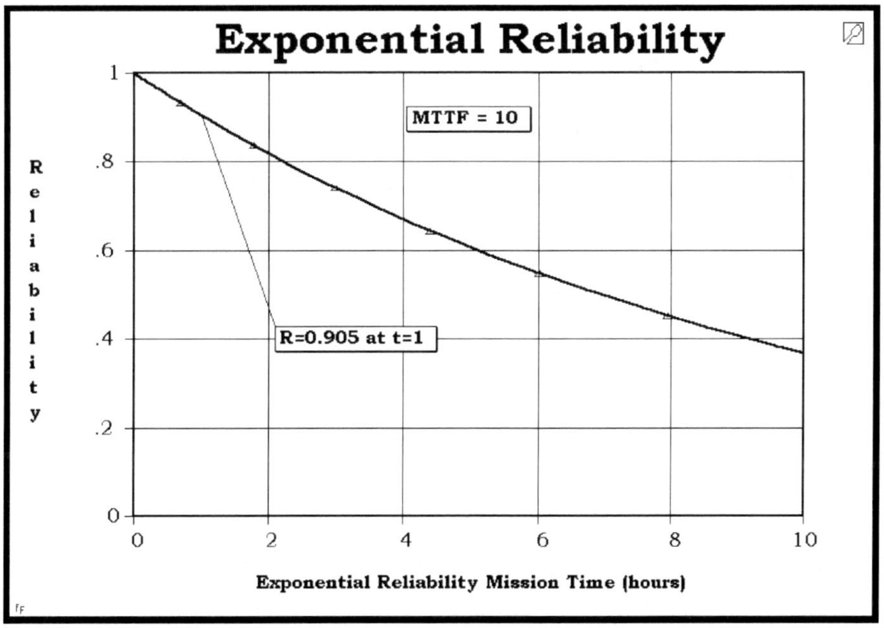

Figure 8-5. Exponential Reliability MTTF = 10

The advantage of the exponential that attracted reliability engineers was the simple system model. If the components in a system are independent and serially arranged such that any failure produces system failure, the sum of the component failure rates is equal to the system failure rate if the individual components have exponential failure distributions. This simple model is useful ***if*** the assumptions are correct. Considering that the reciprocal of the failure rate is MTTF, the reciprocal of the system MTTF equals the sum of the reciprocals of the component MTTF's.

$$1/MTTF_{system} = \sum_{i=1}^{i=n} (1/MTTF_i)$$

8.5 Kaplan-Meier Survival Estimates

This method has been the mainstay of the medical industry for several decades. For life data analysis it has several capabilities that qualify it to be added to your toolbox. The [Kaplan-Meier 1958] survival function is an excellent method to estimate the cumulative survival distribution without making any distribution assumptions. [Cox '84, Nelson '82 and '90, Lawless, Tobias '86.] It is also excellent for large data sets of either or both failures and suspensions. Finally it provides a means to obtain the life distribution for "snapshot" data. (In the medical industry failure implies death so we use survival probabilities rather than failure probabilities.)

Strictly speaking the method only applies when the exact failure times define the intervals. That is, rigorously it does not apply to grouped data but we shall discuss that application later. The number of survivors at the next failure time, n_{i+1}, is equal to the number of survivors at the last failure time, n_i, minus the failure, f_i, plus suspensions, s_i, since the previous failure time.

$$n_{i+1} = n_i - (f_i + s_i) \tag{8-6}$$

$$P(t \geq t_n) = \prod_{i=1}^{n} (1 - (f_i / n_i)) \tag{8-7}$$

$$\text{Greenwood's } \sigma(t_n) = (P(t_n))\sqrt{\sum (f_i / (n_i(n_i - f_i)))} \tag{8-8}$$

The Kaplan-Meier, (K-M), estimate, $P(t_i)$, estimates the probability of surviving to time t_i. The compliment of the K-M survival function, $[1-P(t_i)]$, is the probability of failure. Greenwood's σ may be used to approximate 95% confidence bounds by adding and subtracting two sigma from $P(t_i)$. Strictly speaking the K-M estimates exist only at the failure times but they are often depicted as a bar chart.

For example: Suppose eight systems are placed on life test and failures occur at 100, 300, 400, 500, and 600 hours. In addition, two good units are removed for examination, one at 200 hours and one at 400. These are censored units. The eighth is still on test beyond 600 hours. The probability of surviving to 100 hours is $P(t>100) = 7/8 = 0.875$. The probability of survival is calculated at each failure to obtain the estimate of the cumulative survival function.

The complement of the K-M estimates, the probability of failure, may be plotted on probability papers to test the application of any distribution such as the Weibull or lognormal. SSW software provides K-M analysis and plots, with or without Greenwood's confidence bounds.

$$P(t \succ 100) = (7/8) = 0.875$$
$$P(t \succ 300) = (7/8)(5/6) = 0.729$$
$$P(t \succ 400) = (7/8)(5/6)(4/5) = 0.583$$
$$P(t \succ 500) = (7/8)(5/6)(4/5)(2/3) = 0.389$$

As mentioned earlier K-M does not apply rigorously to grouped or interval data when either the failure times and/or the suspension times fall within intervals. However, [Nelson '90] does apply the method to inspection data when (1) all units are inspected at the same times and (2) suspensions are only removed at inspection times. This is not the general form of grouped data. Nelson suggests plotting the complement of the K-M estimates on probability paper at the inspection times. In this case when the failures and suspensions occur at the same time, the suspension times are arbitrarily considered to follow the failure times. The intervals are not necessarily equal.

In general with **grouped data,** the failures and/or the suspensions occur within intervals. In this case, the **"actuarial correction"** is suggested [Cox '84]. The correction is half the suspensions within each interval and it is subtracted from n_i, the number of units starting the interval. This is based on the assumption that the suspensions are uniformly distributed within the interval. After applying the actuarial correction, the data may be entered into SSW to obtain the K-M analysis. Greenwood's variance estimate still holds based on the modified data. If the intervals are small this correction may be negligible. It is conservative as it increases the probability of failure. Note also that for purposes of failure forecasting the number of suspensions with the actuarial correction is less than the true number obtained without the correction. Therefore K-M is not recommended for failure forecasting when the actuarial correction is used. The Inspection Option or the interval MLE method should be used instead.

One problem with K-M analysis of grouped data is the plotting position. Should the estimates be plotted at the end of the interval, the mid-point, or the beginning? The author recommends the end of the interval. For example, assume there are 100 units at time zero. At each of four later times, t_1, t_2, t_3, t_4, there is one failure and ten suspensions (i.e. removal is 10 good units at the end of each interval). This data provides K-M estimates of $R_1 = 0.99$, $R_2 = 0.9789$, $R_3 = 0.9663$, $R_4 = 0.9519$. This is not grouped data so the actuarial correction is not needed. Now let us assume the same data but the only knowledge comes from an inspection at time t_4. This grouped data would consist of an initial 100 units, with 30 suspensions and 4 failures. The K-M estimates with and without the actuarial correction would be:

$$\textbf{K - M with actuarial} = \frac{\textbf{(100 - 15 - 4)}}{\textbf{(100 - 15)}} = \textbf{0.9529}$$

$$\textbf{K - M without actuarial} = \frac{\textbf{(100 - 4)}}{\textbf{(100)}} = \textbf{0.9600}$$

Clearly this analysis favors the actuarial correction plotted at the end of the interval when the estimates are compared to the ungrouped estimates.

Table 8-2 represents **warranty claims** from six months of production, 1000 units per calendar month. Our objective is to use K-M to predict warranty claims cumulative probability of failure (F(t)) by the age of the unit. For the first month of age there are a total of nine failures among the 6000 units. Four of the failures are in the most recent month's production of 1000 units leaving 996 units suspended. Therefore the actuarial correction is 498 units, half of the first months 996 suspensions. The earlier month's suspensions become failures or suspensions in later calendar months. The SSW data input for the first month is 1 x 9 x 5502. (5502 is the average number of cars at risk during the first month.) In the second service month the population of 5000 was reduced by five that failed in the first service month. Of the 1000 units in the fifth month's production, one failed in the first month and one failed in the second month leaving 998 suspensions. Therefore the average number of systems at risk was (5000-5-(998/2)) = 4496. There were a total of 10 failures in the second service month. The SSW input is 2 x 10 x 4496. The input for the remaining months is 3 x 4 x 3490, 4 x 4 x 2492, and 5 x 5 x 1493. With this input the results are plotted at the end of the interval. Note we have assumed no renewal or replacement of failed parts. Figure 8-6 shows the results from SSW.

Table 8-2. Simulated Warranty Claims, 1000 Units/Month, No Renewal

Service	Month 1	2	3	4	5	6	Suspended
Ages	1000	1000	1000	1000	1000	1000	
Month 1	1	0	0	3	1	4	996
2	4	1	1	3	1		998
3	2	0	1	1			993
4	1	1	2				996
5	1	4					994
6	0						991

There are other issues to consider when using K-M for warranty analysis. Note that warranty data on systems does not fall strictly in the K-M format if there is renewal or replacement of failed parts, as the systems survive even after failure. However, if the probability of repeated failures of the same part is small, the K-M method will probably be adequate and it is in widespread use for this application. For the replacement failed parts, the component age or mileage may be different from the system age or mileage. There is also an issue with subtracting the warranty claims from the number of systems starting the next interval. If all systems are repaired and put back in service, and no failures are catastrophic, it would be better to ignore warranty claims as an adjustment to the number of units starting the next interval. In other words the fleet size is not reduced by claims. In the example above we assumed component data with no renewal. If we had assumed renewal on systems data, the average number of systems at risk in the second month would be 4500, (5000 - (1000/2)). The SSW input would be 2 x 10 x 4500.

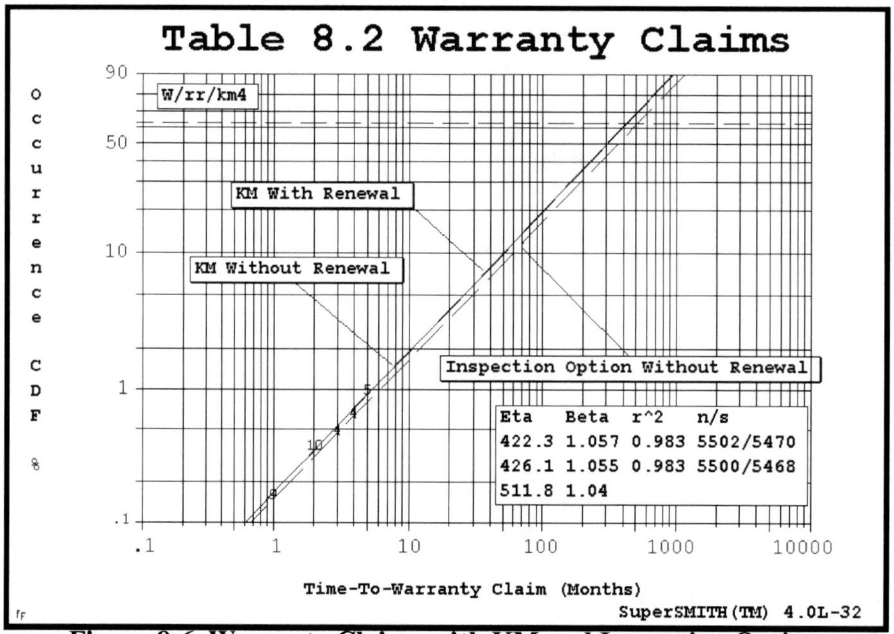

Figure 8-6. Warranty Claims with KM and Inspection Option

Finally, consideration should be given to which way to do the regression. If there is uncertainty about whether all the warranty claims have been received, the regression should use the K-M default, regress Y on X. This was the way it was done above as there was uncertainty. On the other hand, if there is an accurate record of warranty claims, the X on Y special option should be taken. And a last warning not to use K-M for failure forecasting if the actuarial correction is employed as the number of suspensions will be wrong. See Example 9.5 for Crow-AMSAA analysis of this warranty data matrix. See Chapter 9 for much more on warranty claims analysis.

The **Inspection Option** described in Section 5.8 is a competitor with K-M for predicting warranty claims by the age of the units. It locates the suspensions slightly more accurately. As there is no actuarial correction with the Inspection Option, failure forecasting will be more accurate. One new industry study rates the Inspection Option more accurate than the Kaplan-Meier for interval and warranty type data.

Parts that age beyond the warranty period will not generate warranty claims and are treated as suspensions. A more detailed analysis is available with a calculated percent value for each months-in-service value indicating the probability that parts will be outside the warranty limit with respect to some other measure of age such as mileage accumulated. This limit correction information should be located in another row on top of the warranty matrix, above the production quantity values. The percent value above the first production quantity (on the left side) is the largest and corresponds to the maximum months-in-service value possible only with the first batch.

Example 8.10: With the Table 8-2 parts, suppose we know that of the parts with six months-in-service we expect 28% of them to be outside the warranty period. Also, suppose that at five, four, three, two and one month-of-service values we expect 17%, 5%, 1.5%, 0% and 0% respectively to exceed the mileage warranty limit. If the row of values (28, 17, 5, 1.5, 0, 0) is included above the production values in Table 8-2 as a limit correction row this condition can be analyzed with the software. Then the eta becomes 235.5 (was 422) and the beta becomes 1.16 (was 1.057).

Special thanks to M.A. Vasan of Meritor Automotive for this correction. He has contributed to the warranty data analysis methodology with suggestions for new capability and also case studies used for benchmarking.

If failure forecasting is used to predict warranty claims and the warranty limits are input, SSW will eliminate units beyond warranty limits from generating claims. There is much more on warranty analysis in Chapter 9. See Case Study 11.10 for an interesting Kaplan-Meier analysis.

"Snapshot" data can also be analyzed with Kaplan-Meier. By snapshot data we mean that within an interval of time, the snapshot, all the data is available but there is missing data outside the interval. In this case the sample size will not necessarily decrease with age. This data cannot be analyzed with the standard methods but good Weibulls were produced with K-M. This calculation is shown in spreadsheet format (Table 8-3) in Example 8.11 to show intermediate calculations. Regression was Y on X without the actuarial correction and the results were plotted at the end of the interval.

Example 8.11: Table 8-3 shows data on wooden power poles in the Tennessee Valley Authority system, a snapshot of the last ten-year calendar period grouped by age intervals. (Thanks to TVA for their permission to show this data). Weibull is impossible as the failures and suspensions from the earlier periods are unknown but K-M provides an excellent substitute. There are three different failures modes, butt rot, woodpecker attacks, and center rot. The data is treated like inspection data, as if the failures were found at the end of each interval. The woodpecker mode appears to be random (as expected) after a t-zero correction of about 3 years. This might be interpreted as the time needed by the woodpecker to create the damage. (Woodpeckers usually enlarge their nesting hole each year.) The two rot modes need a negative t-zero correction. The negative t-zero implies that some of the life was used up before the power poles were "planted". TVA engineers said the poles were often stored outside for long periods before installation and therefore may have rot from storage. The data is given in columns A-E. Total Samples (column F) is the sum of suspensions (col B) and failures (col C to E). The interval failure estimates for woodpecker attack are in column G = (1- col D/col F). These are converted to K-M survival estimates for probit entry (col H) using Equation 8-7. One minus column H is the K-M cumulative failure probabilities (last column). Figure 8-7 is a SSW plot using the t_0 correction and probit solution.

Table 8-3. TVA Power Pole Data-Kaplan-Meier Analysis

A	B	C	D	E	F	G	H=KM	1-KM
Interval Years	**Suspend**	**Butt Rot**	**Woodpecker**	**Center Rot**	**B+C+D+E**	**1-(D/F)**	**Π(G)**	**(1- (KM))**
0-5y	5469	39	71	7	5586	0.98729	0.98729	0.01271
6-10y	5450	42	152	14	5658	0.97314	0.96077	0.03923
11-15y	6382	70	232	17	6701	0.96538	0.9275	0.0725
16-20	6383	122	448	18	6971	0.93573	0.8679	0.1321
21-25	8637	179	356	27	9199	0.9613	0.83431	0.16569
26-30	8637	270	219	55	9181	0.97615	0.81441	0.18559
31-35	7096	343	214	112	7765	0.97244	0.79196	0.20804
36-40	7097	402	123	84	7706	0.98404	0.77932	0.22068
>40	37779	1573	513	303	40168	0.98723	0.76937	0.23063

SSW Input For Woodpecker Failures shown on Figure 8-7: Time to Failure X # Affected X # Sampled

5 x 71 x 5586	**10 x 152 x 5658**	**15 x 232 x 6701**
20 x 448 x 6971	**25 x 356 x 9199**	**30 x 219 x 9181**
35 x 214 x 7765	**40 x 123 x 7706**	**45 x 513 x 40168**

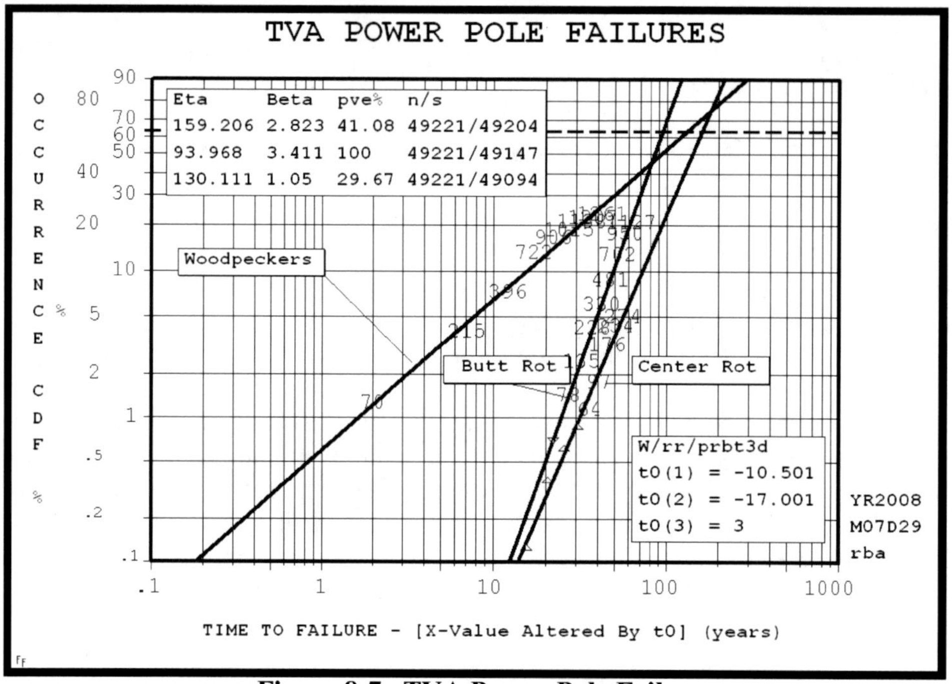

Figure 8-7. TVA Power Pole Failures

8.6 Probabilistic Design

Significant progress has been made by design engineers applying probability analysis to the design function. This practice is encouraged and contributes to better designs. The US Air Force Structural Analysis Teams (SATs) have been successful applying these methods to aircraft and engine designs to identify critical areas. NASA Jet Propulsion Lab has a team developing rocket engine design improvements based on probabilistic design. The Society of Automotive Engineers (SAE) is actively encouraging the application of this technology. Several of the most popular applications involve complex statistics. These solutions are easy with Monte Carlo methods, but they may be extraordinarily difficult, sometimes impossible, using analytical methods, There are usually two distributions that must be convoluted to obtain the third or product distribution. Solutions to these problems have been eased by incorporating the Monte Carlo methods in the SSW software.

8.6.1 Strength-Load and Life-Usage Interactions

Strength is generally equated with a single value representing the highest load capability. However, realistically it varies from sample to sample and should be represented by a distribution. Using fracture mechanics and material characteristics the designer may predict the **strength** distribution and using customer specifications, the load distribution may be estimated for a given design life. The **strength** distribution interacts with the **load** distribution. Failures occur when load exceeds strength. See Figure 8-8.

Alternatively a **life** prediction may be made as a function of age for a fixed stress based on material characteristics from specimen tests. If the **usage** distribution is estimated from the customer's conditions, a failure will occur when the usage as a function of age exceeds the life value. In both cases the proportion failed may be estimated by comparing random samples in pairs from both distributions. The SSW "Interaction/Life-Usage" option is under the Calculator icon. The proportion failed is estimated from a random sample of 30,000 pairs. The precision is given as two-sigma (binomial), approximately 95% confidence. If the precision is unsatisfactory the sampling may be increased up to two billion pairs.

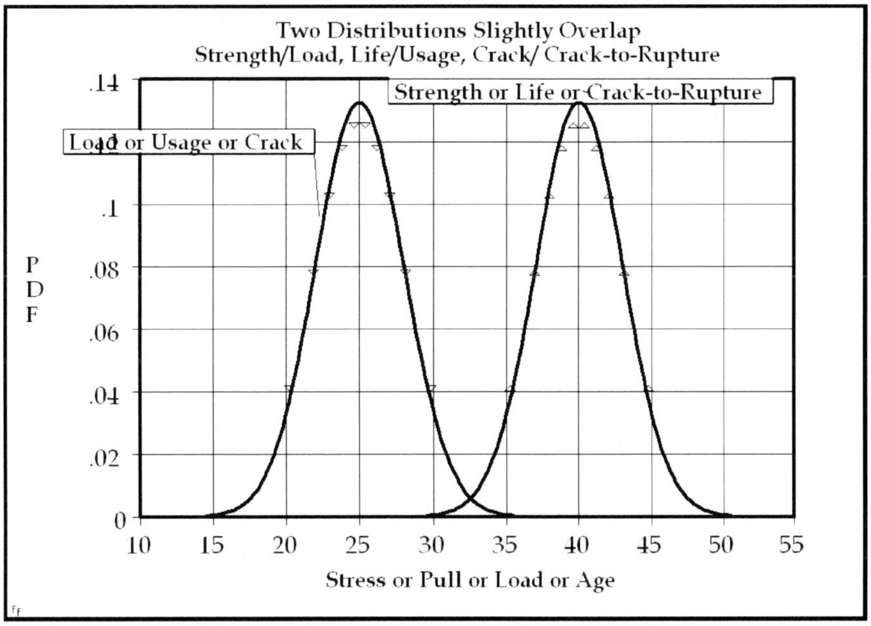

Figure 8-8. Two Distributions Slightly Overlap Strength/Load, Life/Usage, Crack/Crack-to-Rupture, Mode A/ Mode B

Example 8.12: Assume the strength distribution is Weibull, eta = 1000 lbs, beta = 5.0 and the load distribution is normal, mu = 500 lbs., sigma = 20 lbs. SSW estimates the proportion failed will be 3.1% based on a random sample of 100,000 pairs. This estimate is precise to ±0.11%. Note: Monte Carlo simulations with different starting values, (seed numbers,) produces slightly different results.

Example 8.13: If the life distribution is Weibull, eta = 1,000,000 cycles at 1000 kpsi, and beta = 5.0 and the usage distribution is Weibull, eta = 500,000 and beta = 3.0, the proportion failed is estimated to be 4.3% (Reliability 95.7%). The 2 sigma precision is ±0.04% based on one million trials.

8.6.2 Total Life = Crack Life + Crack-to-Rupture Life

The rupture of a disk in a turbine is catastrophic. Turbine manufacturers may identify the time to crack incidence as a Weibull and the crack-to-rupture life as a lognormal based on separate tests. The total life is the parameter of interest. In the aerospace industry it was standard practice for commercial airlines to remove disks at their B.1 life, discarding 99.9% of their life. The total life is the sum of the time to crack incidence (say 5 mil cracks) and the crack-to-rupture life. This convolvement is done with SSW. The two distributions are entered and a sample of say 1000 random pairs are summed and plotted with SSW. A distribution analysis will indicate which distribution is best. Confidence bounds and median estimates are available for the total life distribution parameters and B lives.

Example 8.14: The time to incidence of a 5 mil crack is a Weibull with eta = 1000 hours and beta equal to 4.0. The crack-to-rupture distribution is lognormal with Med ("median") = 100 and SigF ("sigma factor") = 2.0. What is the B.1 life (reliability of 99.9%) for the total life distribution? A sample of 6000 sums of pairs transferred to SSW indicates the best fit distribution is the two parameter Weibull. The median B.1 life is 237 hours with 90% confidence bounds of 231-244 hours.

8.6.3 Does Failure Mode A Cover Mode B?

If there are two competitive failure modes will one completely cover the other or will there be a proportional division of failures? What is the proportion of "A" failures? This is another problem with probabilistic design. The solution is to randomly sample pairs of times-to-failure from the two distributions. If we subtract the "A" time from the "B" time, the proportion of negative answers will be the proportion of "B" failures.

Example 8.15 A quality problem has produced some defective parts, a batch mode B, with a Weibull eta = 10,000 and a beta = 0.4. The same part wears out with mode A, a Weibull eta =100 and a beta = 3.0. What is the batch proportion? Answer 13.8%. 86.2% of the failures are mode A. These results are based on simulation with one million trials with precision, 2 sigma = 0.069%.

8.7 Production Process Reliability by Paul Barringer, Reliability Consultant

Pareto principle: A few contributors are responsible for the bulk of the effects--the 80/20 rule where by 10 to 20 % of the things are responsible for 60 to 80% of the impact. Named for the Italian economist Vilfredo Pareto (1848-1923) who studied the unequal distribution of wealth in the world and by Dr. Juran who described the Pareto concept as separating the vital few issues from the trivial many issues.

Name plate capacity: The maximum production capacity of the factory under theoretically ideal operation and control. It is usually provided by the site contractor that designs and constructs the factory. It is rarely measurable as it is impossible to achieve the ideal conditions.

Efficiency/utilization losses: The difference between the name plate capacity and the demonstrated Weibull line is generally a result of efficiency losses or by utilization of the facility.

Production losses: The difference between the demonstrated Weibull line and the actual production data point associated with the same % CDF.

Crash and burn output: A euphemism for seriously deficient production quantities during periods of substantial process upsets or deteriorations.

Cutbacks: A production quantity recorded during a period when output is restricted by partial failures resulting in a slowdown from the intended/scheduled production rate. The reliability of a manufacturing process is seldom quantified because failures are not defined. For most manufacturing efforts, the objective is to make the "most you can"--the emphasis is on the large quantities and small quantities are shrugged-off as bad luck. When you measure how you're doing, then you'll understand some symptoms of why problems exist, and Weibull analysis of production data provide evidence needed for root cause analysis. The data for Figure 8-9 is one year of daily production output. This figure shows reliability of the process, defines the demonstrated process capacity, depicts

the normal range of production output, gives clues about nameplate capacity, and quantifies losses from both reliability issues and from efficiency/utilization problems. Weibull analysis helps build a Pareto improvement list.

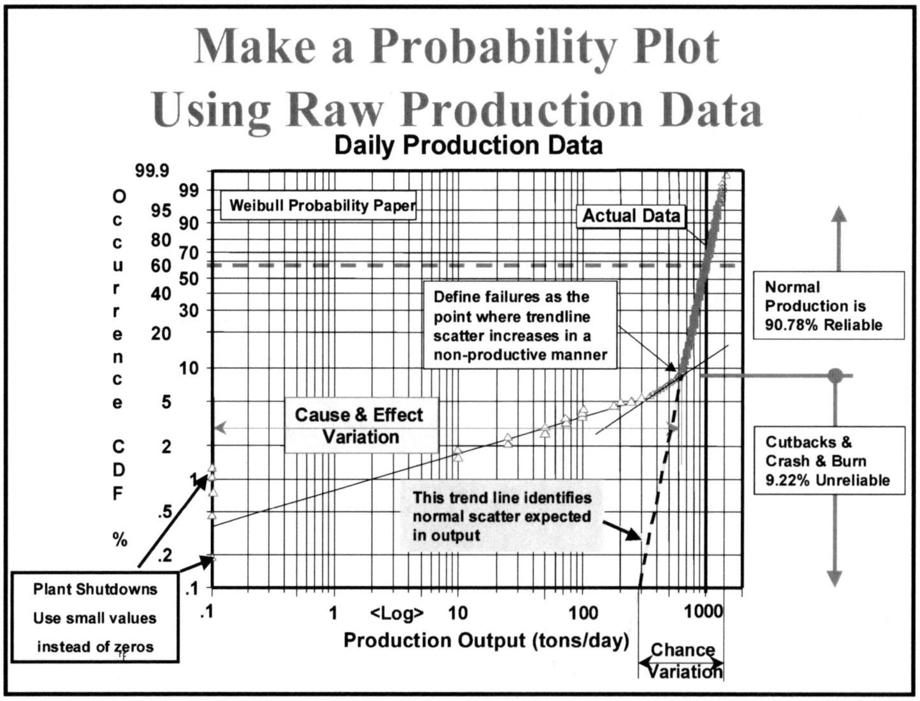

Figure 8-9. Shows Process Reliability = 90.78% At the Point Where Consistency Is Lost

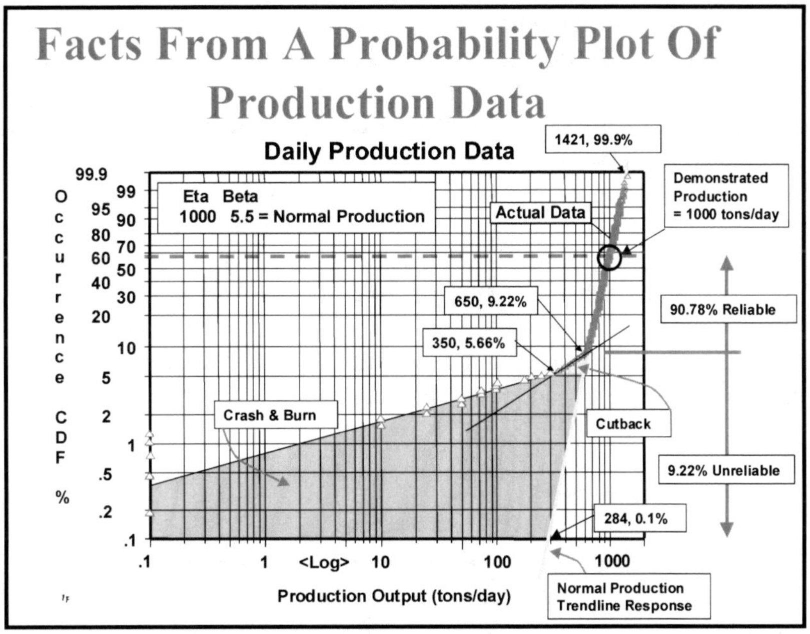

Figure 8-10. Facts from a Probability Plot of Production Data
(See files pr-Raw-Production.w or Fig8-9.w in Smithdat directory for raw data for these figures.)

Consider the manufacturing process as a black box. Take the daily production output numbers (pieces, tons, etc.) from the black box, and make a Weibull probability plot. Most plots will show three distinct groups of data:

1. A straight line trend in upper reaches of the plot defines **"normal"** production when all is well--as quantities deviate from this segment, failures occur (by definition) because the process loses predictability.
2. A middle zone exists where production fails to meet steep portions of the curve and this defines a **cutback** segment--usually with a cusp at each end shown in Figure 8-10 above.
3. A **crash and burn zone** where production quantities are substantially deficient. Figure 8-10 shows these three zones from a production process. The bottom cusp of the steep curve defines process unreliability (9.22%) and the complement is reliability for this process at 90.78%. Inherent features of the Weibull plot magnify the bottom end of the curve to accentuate problem areas of most concern.
4. Another important point identified on the Weibull plot is demonstrated production capacity. Demonstrated capacity is the single "talk about" number at 1000 tons/day shown at 63.2% which "best" represents production output. Many production processes have zero outputs which cannot be plotted on Weibull probability paper; if these data are changed from zeros to about two orders magnitude less than the smallest production number they may be plotted on Weibull probability paper.

Normal production, in Weibull terms, has an eta = 1000 and beta = 5.5. See Figure 8-10. Loss occurs when data lies to the left of the normal trend lines. Loss for the cutback area is about 1,397 tons (1.4 days of normal production) and about 8,786 tons for the crash and burn zone (8.9 days of normal production). For example, if N = 365 days of actual production output values in rank order from least to greatest, for the tenth point recorded output is about 50 at CDF = 2.81% by Benard's median rank. The Production line (Eta=1000, Beta = 5.5) gives calculated demonstrated production of about 500 tons/day at this same 2.81% CDF. Actual production was only about 50 tons. Loss for that day is then 450 tons (=500-50). The total for the year's production loss from cutback, crash-and-burn, and zero days can be similarly estimated. Costs from lost production may then be calculated. Further, the daily losses may be accumulated by cause, ranked, and plotted on Pareto chart to identify the priorities for correcting the problems.

If the nameplate capacity is a valid number, losses may be calculated against nameplate or against the nameplate distribution. However, in many cases the nameplate number may be outdated by later production improvements.

8.8 Extreme Value Statistics*

The probability of rare events may be estimated with extreme value statistics. Examples include the probability of "Upper" maximum values like extreme floods, earthquakes, dams breaking, and maximum gust loads. Minimum events of "Lower" values include examples like wall thickness for boiler tubes and pipelines, material properties, time-to-failure, and lowest temperature.

The largest (upper, maximum) or smallest (lower, minimum) value in some interval of time or within a sample of data is the objective of the analysis. If the distribution of the data is known, the exact distributions of the smallest and largest may be computed. Otherwise, the data will converge asymptotically to one of three distributions, depending on the limits of the data. As each distribution may deal with either maximum or minimum events, there are six versions of the distributions, Type I, II and III, maximum and minimum. Type III minimum or smallest is the Weibull distribution.

In their simplest form, "normalized," the distributions are:

$$\mathbf{F(t) = 1 - e^{-e^{x}}, \text{ where } -\infty < x < \infty \text{ (Gumbel Type I)}}$$

$$\mathbf{F(t) = 1 - e^{-(-x)^{-\alpha}}, \; x \le 0 \text{ (Type II)}}$$

$$\mathbf{F(t) = 1 - e^{-x^{\alpha}}, \; x \ge 0 \text{ (Weibull) (Type III)}}$$

For life data, Type I and III are most applicable and have an interesting relationship. If we take the natural logarithms of Weibull times-to-failure, they will be distributed as Type I smallest. In other words the relationship between normal and lognormal is the same as Type I smallest (Gumbel) and Type III smallest (Weibull). [Tobias] says, "So, if the Weibull had not been named in honor of its chief popularizer, it probably would have been called the log-extreme value distribution." Some experts like Wayne Nelson prefer to work life data analysis in the Type I domain and transform back and forth. The author believes this makes it difficult for the reader, but technically there is no objection as the same answer will be obtained with either distribution. For example, r and r^2 will be the same using median rank regression with either Gumbel Type I Minimum or Weibull. Nelson develops MLE methods for standard and interval Type I data.

Let *t* be a Weibull random variable. The Type III Weibull equation is:

$$F(t) = 1 - e^{-(t/\eta)^{\beta}}$$

Then the random variable, $\boldsymbol{X = \ln t}$ has the Type I extreme value CDF with $\boldsymbol{\psi = \ln \eta}$ and $\boldsymbol{\delta = 1/\beta}$.

* Advanced topic.

$$F(x) = 1 - e^{-e^{(x-\psi)/\delta}}$$

Type I-smallest probability paper has a linear scale for the ordinate but a Weibull scale for F(t). For Type I-largest the vertical scale is 1-F(t). Type I largest may be useful for the largest value in an interval of time, gust loads, turbine temperature, pressure ratio, etc. SSW provides both Type I smallest and largest for probability plots and distribution analysis. See Figures 8-11, 8-12, and 8-13 for examples of Type I largest.

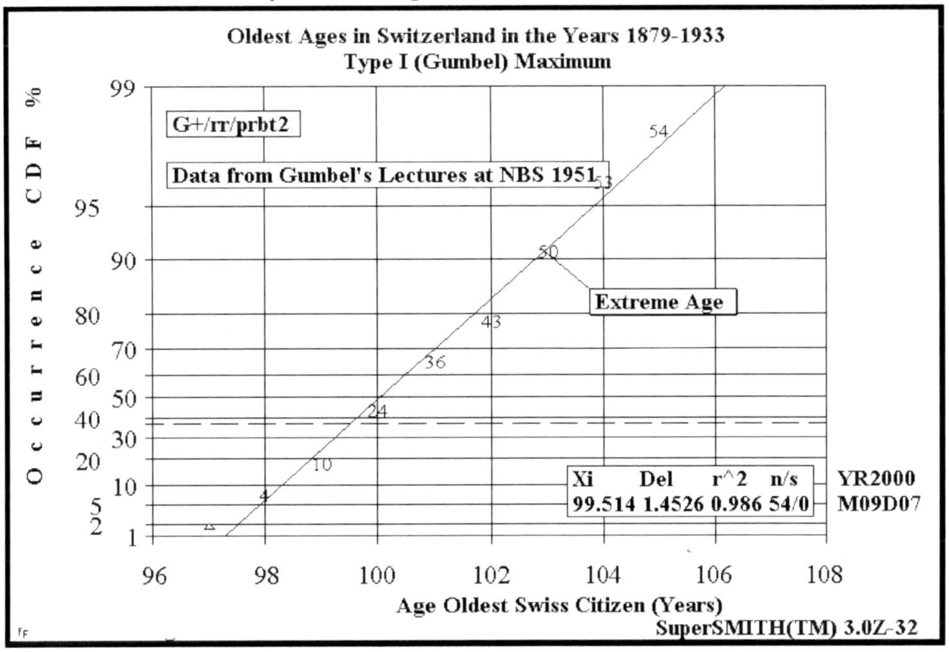

Figure 8-11. Oldest Ages in Switzerland in the Years 1879-1933 Type I (Gumbel) Maximum

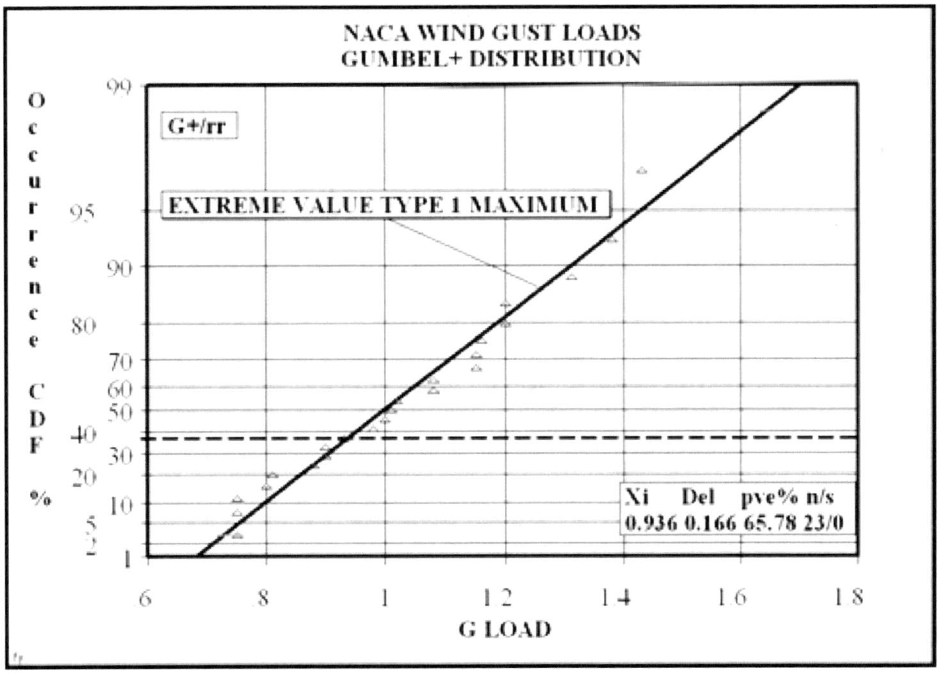

Figure 8-12. NACA Wind Gust Loads Gumbel + Distribution Type 1 (Gumbel) Maximum (Repeated from Chapter 1, Figure 1-5)

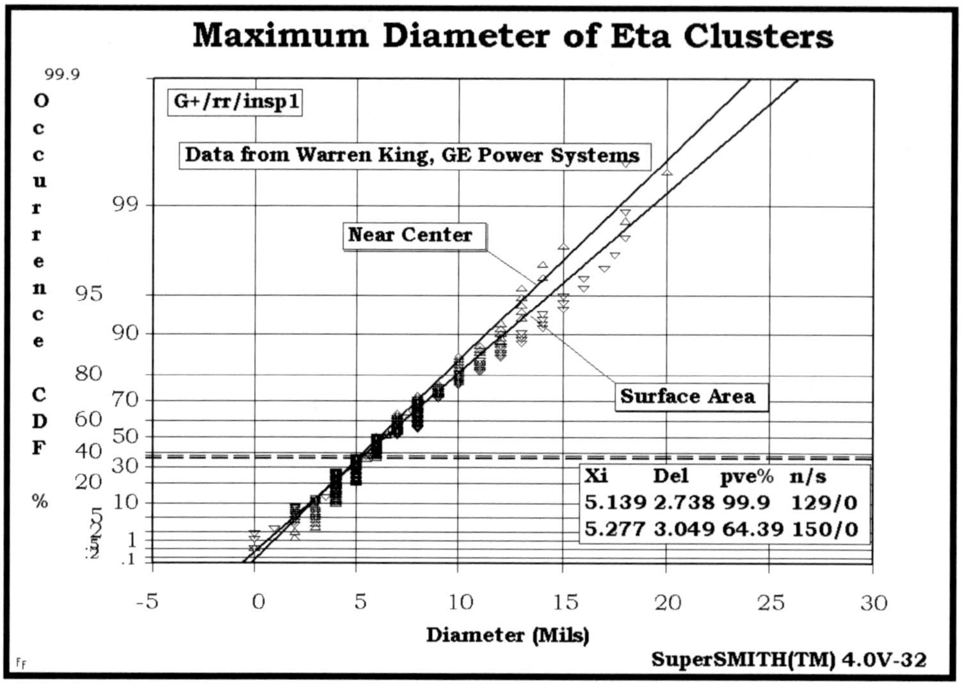

Figure 8-13. Maximum Diameter - Eta Phase Clusters in Metal Surfaces Type 1 (Gumbel) Maximum

8.9 Batch Effects

In several places in earlier chapters batch problems were discussed. As batch problems are numerous it is appropriate to summarize batch detection clues and methods. A batch is a subset of the fleet of units in service that are susceptible to a failure mode that the rest of the fleet is immune to or at least more resistant to. The batch failure mode may or may not be mixed with other failure modes.

Batch Clues:

1. Early failures of new units with many *older units unfailed*, late suspensions. (Section 3.7)
2. The failed units have *close serial numbers* indicating they come from the same production run. Similarly a batch is indicated if vendor A's units fail and vendor B's do not, or client A versus client B, etc. (Section 3.7)
3. The *MLE beta is shallow* compared to the MRR beta. (Section 5.3.3)
4. The Weibull plot pattern *steep line followed by shallow line* indicates a batch problem, but there may be batch problems with Weibulls with good straight line fits. (Figure 3-20 and Section 3.7.3)
5. If there is only one failure mode and some of the units will never fail, these safe units are called *"perpetual survivors."* The data will plot as steep slope line followed by a shallow gently curve to a shallower slope.

Batch Analysis:

1. The Now Risk is much higher than the observed number of failures. Use the 90% lower Poisson Bound on the Now Risk. If it is greater than the observed number of failures there is evidence of a batch problem. The ratio of the observed number of failures to the Now Risk is an approximate batch proportion.

2. The Aggregate Cumulative Hazard shows a batch problem. See Appendix F for an explanation of this excellent graphic approach. It works for complete samples which the Now Risk method cannot handle. It also works for samples with suspensions. See also Appendix J.
3. The Weibull failure forecast will be too high if there is a batch problem. If the batch units are randomly distributed, (probably not true), the ratio of the observed number of failures to the Now Risk is approximately the batch proportion. See Case Study 11.14 for a method to estimate the batch proportion of failures. The ACH plot may provide a similar ratio. This approximate batch proportion may be entered into the Abernethy Risk menu in SSW to obtain a more accurate failure forecast.
4. The Crow-AMSAA failure forecast will be more accurate than the unadjusted Abernethy Risk forecast.

Recommended Action: The number one priority is to identify the batch, identify the root cause and fix the problem if possible. The Weibull analysis should be based on batch units only. If the non-batch units cannot be purged out of the data set, the Crow-AMSAA model should be used.

8.10 Problems

Problem 8-1: Binomial SQC. A sampling plan requires one or less defects in 100 pieces. What is the probability of passing if the percent defective is 1%?...3%? Hint: May we use the Poisson approximation to solve this binomial problem?

Problem 8-2: Binomial Reliability. A reliability demonstration requires 231 tests without failure. What reliability is demonstrated at the 90% (lower one sided) level? See above Hint.

Problem 8-3: Exponential Reliability. A system has three components with the following MTBFs: 300,600, and 900. Assuming constant failure rate, what is the system MTBF?

Problem 8-4: Poisson. An aircraft engine produces 6 in-flight shutdowns per 100,000 hours (or uncontrolled automobile accelerations, or forced outages, or reactor scrams). What is the upper one-sided 90% confidence bound?

Problem 8-5: Which Distribution? Select the most appropriate distribution from 2 and 3 parameter Weibulls, lognormal, normal, exponential, Crow, Poisson, rectangular, or binomial:

a. Receiving inspection reports 2 defects in a sample of 100.
b. North Palm Beach averages 13.7 long power outages per year.
c. Transformer failure data by calendar time caused by lightning strikes.
d. Bearing times-to-failure
e. Professor Mischke says a specimen will rupture under fixed conditions as described by a Weibull, but in service, the time to failure is the product of the Weibull time-to-failure and factors for temperature, vibration, acceleration, corrosion, etc.
f. IQ scores
g. Dr. Von Ohain, the German inventor of the first turbojet engine to fly, told me that the MTBF of the Jumo engine was 25 hours. What was the reliability of the Jumo for the typical 20 minute mission?
h. A reliability demonstration requires 2300 tests without failure.
i. A combined environmental test to force failures and improve reliability, a CERT test.
j. Mechanics sometimes make mistakes....so do engineers.
k. The duration of forced outages.
l. Performance estimates.
m. Foreign object damage as a function of operating time.

n. Uniform random numbers are the basis for Monte Carlo simulation.

o. The reciprocals of uniform random numbers are useful.

p. Weibull times to failure raised to the beta power.

q. Wire cable is made of multiple twisted strands of wire. Under load what is the expected distribution of first strand time to rupture?.. Last strand?...total strands rupture?

r. Accident data for a plant or facility.

s. The probabilistic statement that if two events are equally likely, the worst will happen is derived from who's law?

Problem 8-6: Exponential. All the failures lumped together from a system like an automobile will have an exponential distribution. If we model the time to failure as Time = MTBF(-ln random number), what is the mean time to failure for the first ten units. Let MTBF = 1000. Use the random numbers shown.

0.329	0.604	0.615	0.300	0.070	0.845	0.494	0.624	0.085	0.194

Average__________________

Note that the Crow-AMSAA model is the best way of plotting mixtures of failures like system data.

Refer to Appendix K for answers to the problems.

Chapter 9 – Crow-AMSAA Modeling, Warranty Analysis, & Life Cycle Costs

Contributed by David P. Weber, Reliability Consultant 1993, Updated 2006 by the author

9.0 The Crow-AMSAA Reliability Growth Model

It is remarkable how the Crow-AMSAA (CA) model has gained popularity since the first edition of this Handbook. Most industries are routinely doing CA analysis for all the applications listed below. In most cases both Weibull and CA are employed. Reliability growth analysis originally was developed to track the progress of reliability growth in a development process to determine when the configuration was good enough to put in production. Now there are a number of new applications. The Crow-AMSAA, (CA), model is **best practice** for reliability growth and the following newer applications:

1. Tracking **development reliability growth** during component development, the original application.
2. Tracking significant **events for management**.
3. Tracking **repairable systems** in service for reliability, maintainability.
4. Utilizing **dirty data**. The data deficiencies include changing reliability levels, mixed failure modes, batch problems, and missing portions of data. CA modeling is the most "robust" method in this handbook, providing reasonable accurate results with terrible data. ("Robust" to a statistician means a method that works well even when the assumptions and data are deficient.)
5. Predicting **warranty claims** and failures by calendar month.
6. Predicting how many **<u>new</u> failure modes** will be found with additional testing.
7. Failure forecasting with data that includes a **batch effect**.

Crow-AMSAA benefits include:

1. Trend charts make reliability more visible and manageable.
2. More accurate program decisions are made such as when to go into production, program scheduling, increasing testing, funding additional effort, reallocating program resources, and delaying the program until interim reliability goals have been demonstrated.
3. The progress of the reliability improvement program, the effects of design and quality changes, is measured and displayed.
4. Reliability predictions are available months or years early, to compare against program requirements, allowing corrective action. Therefore, CA is a pro-active method.
5. Adverse system reliability trends are indicated much sooner and more accurately than moving averages provide.
6. System level (or component level) reliability trends can be measured even with mixtures of failure modes. (whereas Weibull analyzes one failure mode at a time).
7. Trend charts verify engineering predictions of the effectiveness of development fixes for observed problems.
8. The rate of incurring new engineering problems can be determined by projecting trends of first time problem occurrences.
9. The method can be used to analyze reliability growth with intervals of missing or suspect data. For example, electric utilities have data on systems operating for decades before reliability data was collected. The method provides good risk predictions even with these deficiencies.

9.1 Background History

James T. Duane at GE Motors Division pioneered growth analysis. He observed that failure rates plot as straight lines when placed on a cumulative test time scale on logarithmic paper. (Duane 1964). The AMSAA model, a major improvement in Duane's approach, was developed by Dr. Larry H. Crow while he was at the Army Material Systems Analysis Activity (AMSAA). The AMSAA model and the Duane Postulate describe the same growth pattern. However, the Crow-AMSAA (CA) model provides statistical analysis of the growth process. This model has previously been called the Duane-AMSAA model, but recently the IEC changed the name to Crow-AMSAA recognizing Dr. Crow's significant contributions.

Reliability growth results from an iterative design-development process which includes: (1) detection of failure sources; (2) feedback of problems identified; (3) redesign effort based on problems identified; (4) fabrication of changed hardware; and (5) verification of redesign effect by retesting and iterating the process. These elements require a failure analysis and corrective action system (FRACAS). FRACAS is a formal closed loop system established to identify and track failure events, determine corrective actions and track the effectiveness of incorporated corrective actions. [SAE] (FRACAS, Fault Tree Analysis (FTA), and Failure Mode and Effects Analysis (FMEA) are qualitative methods highly recommended by the author, although they are not described in this Handbook on quantitative methods.) *[O'Connor]* is a good reference for these methods.

Four decades ago twenty-seven different growth models were in use. The Aerospace Industries Association Technical Management Committee studied this array of methods as applied to mechanical components and concluded that the CA model was the best. (The author was the United Technology Corporate member.) The U.S. Air Force redid the study, including both mechanical and electronic controls, [McGlone], reached the same conclusion. (It was conducted by Dr. Abernethy's RMS Group at Pratt & Whitney Aircraft.)

The original hallmark text was MIL-HDBK-189. Today the new standards are *IEC 61164, IEC 61710,* and an update of MIL-HDBK-189, published as *AMSAA TR-652.* These documents introduce new and better methods for CA modeling. All were developed or influenced by Larry Crow. IEC 61710 emphasizes tracking repairable and non-repairable systems while the other two standards are devoted to the component development reliability growth. *[$AMSAA TR-652]* may be downloaded from the internet, see References.

9.2 CA Methods

This section presents a framework of the features, applications, and modeling methods for the CA model. Specific methods will be illustrated for:

1. Simple graphical and regression solutions - Section 9.2.1,
2. IEC Unbiased estimates for time and failure terminated test data sets- Section 9.2.2, and
3. IEC maximum likelihood solutions for interval or grouped data - Section 9.2.3.

9.2.1 Simple Graphical and Regression Solution

The logarithm (log) of cumulative failure events **n(t)** versus log cumulative time **(t)** is a linear plot if the CA model applies:

$$\mathbf{n(t) = \lambda t^{\beta}} \tag{9-1}$$

Taking natural logarithms of Equation 9-1 yields:

$$\mathbf{Ln\ n(t) = Ln\ \lambda + \beta\ Ln\ t} \tag{9-2}$$

which implies that Equation 9-1 is a straight line on logarithmic paper. Figure 9-1 is a plot of this function on logarithmic paper.

The model intensity function $\rho(t)$ measures the instantaneous failure rate at each cumulative test time point. The intensity function is:

$$\rho(t) = \lambda\beta t^{\beta-1} \tag{9-3}$$

Mathematically the intensity function, Equation (9-3), is the first derivative of cumulative events, $n(t) = \lambda t^{\beta}$, Equation (9-1). For repairable systems $\rho(t)$ is called the rate of occurrence (ROC). The reciprocal of the $\rho(t)$ is the instantaneous MTBF.

The scale parameter, λ, is the intercept on the y axis of n(t) at t = 1 unit. (The log of 1 is zero). With the data plotted in Figure 9-1, the value read off of the y-axis is λ = 0.698 at t = 1 on the Cumulative Test Time x-axis. The slope β can be measured graphically as linear values (6.41 cm/10.0 cm = 0.641) if 1 x 1 scale, standard log-log paper is used. This plot and the parameters were calculated with *SuperSMITH Visual*. The IEC methods are the most accurate method for n>5. Alternatively, the line could be drawn in by eye or regression analysis. Both methods are available in SSV. The regression solution provides estimates of beta = 0.716 and lambda = 0.510. The plot is approximately 1 x 1 scale, but this is not generally true with computer plots.

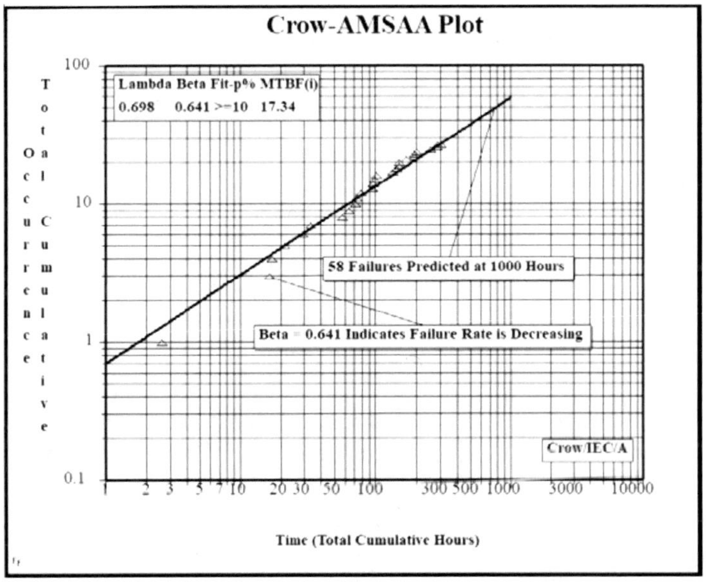

Figure 9-1. Crow-AMSAA Plot

The slope, β, is interpreted in a similar manner as on a Weibull plot. If the slope β is greater than one the failure rate is increasing. In a development process the slope should be less than one, indicating a decreasing failure rate, i.e., reliability growth. If the slope is constant, the failure rate is constant (β = 1) and the process is called a homogenous Poisson process. When the slope is not one, it is a non-homogenous Poisson process. If $\beta < 1$, the failures come more slowly; If $\beta > 1$, the failures come more rapidly. Typical component development is illustrated in Figure 9-1, with a slope of 0.641. This data set is from MIL-HDBK-189A, section 6.3.1.6.4 Example, Table IX. See Logbook in SSV for Fig9-1.v for more information.

Plotting the data on logarithmic paper is the first step in reliability growth analysis. Some authors skip the plot of cumulative failures versus cumulative time and go directly to plots of failure rate or MTBF. The author strongly disagrees as the basic plot is useful and informative. For example, failure forecasting is based on a linear extrapolation of the plot. One thousand hours of testing is budgeted for the units shown in Figure 9-1. If 27 failures have been observed to date, how many more failures will be observed? [Answer: 58-27=31.] Goodness of fit is indicated by the proximity of the points to a straight line and the eyeball usually is a good judge. This is part of the CA analysis. It is important to understand that both the process and the system are modeled, not just the system. For example, if a development process is being tracked and the number of engineers is doubled, the reliability growth should accelerate. That is, the slope should decrease. A cusp or corner should appear on the plot at the point the change was made. The straight line should be fit from the cusp to the right to model the latest process.

Cumulative failure rate, C(t), equals cumulative failures, n(t), divided by the cumulative test time, t. C(t) is estimated using Equation (9-4):

$$C(t) = \lambda t^{\beta-1} \tag{9-4}$$

The instantaneous failure rate estimate, ρ, is Equation (9-5):

$$\rho(t) = \lambda \beta t^{\beta-1} \tag{9-5}$$

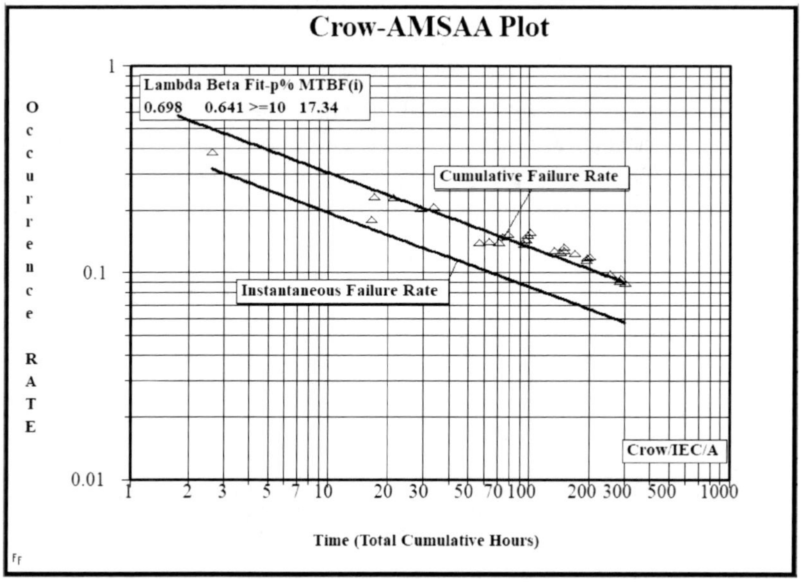

Figure 9-2. Failure Rate Plot

The log of the instantaneous failure rate and its reciprocal, the instantaneous MTBF, are both linear functions of the log of cumulative test time. See Figure 9-2. If the units are repaired after failure, the instantaneous failure rate, ρ(t), becomes the rate of occurrence (ROC). For component development, the instantaneous failure rate, ρ(t), is the failure rate of the latest configuration including all the fixes implemented. In this case extrapolating to 1000 hours, ρ(t) = 0.04 failures per hour. If the latest configuration is put into production, ρ(t) = 0.06 is an estimate of the in-service failure rate. It is not a very good estimate, (+/-20%), [McGlone] but better than any other method. For tracking in-service system reliability 12 months ahead, the same study showed better results, uncertainties of +/-10%.

SuperSMITH Visual provides the CA solution and plots including all three methods, regression, IEC unbiased and IEC MLE. The regression solution is not as accurate as the newer IEC unbiased and MLE methods except for very small samples, five events or less. Comparisons will be provided later. The IEC unbiased method is used for standard data, when exact times to failure are known; the MLE method is used for grouped or interval data.

If the process is stable as indicated by a good fit to a straight line, both Figures 9-1 and 9-2 may be extrapolated to make predictions for six months or a year ahead or to the end of the test program. On Figure 9-1, the predicted number of failures is the difference between the current value, n(t) and the future value, n(t + k), on a straight line extrapolation as illustrated earlier. (k) is the expected number of hours in the future period. Similarly, the predicted instantaneous failure rate is a straight-line extrapolation for (k) hours on Figure 9-2. At 1000 hours of testing, the instantaneous failure rate is predicted to be 0.04 failures per hour. The same method also applies to MTBF plots. Thus, failures, ρ(t), and MTBF predictions are simple, straight line extrapolations, even with missing portions of data, *as long as the process remains stable.* In other words, the instantaneous line can be extrapolated based on the improvement rate *demonstrated with the current process.*

Larry Crow established the relationship between the CA model and the Weibull distribution; the foundation of CA is Weibull. Crow refers to the CA model as a "Weibull Power Process," (WPP). The interpretation of the slope, β, is similar. However, the time scales are different. The Weibull distribution uses individual times-to-failure, but CA uses cumulative times. Weibull is one failure mode at a time: CA handles mixtures of modes. The single Weibull failure mode is considered static, unchanging; but CA models a process where overall reliability may be changing with cumulative time.

9.2.2 IEC Solutions for Time and Failure Terminated Data

Earlier editions of this Handbook recommended maximum likelihood solutions for time and failure terminated data. The new IEC standards provide more accurate MLE methods which now are recommended as **best practice**. Much of the following is from IEC 61164. These methods utilize system times-to-failure data to estimate reliability growth, the instantaneous failure rate or MTBF, and predicted failure rate at the end of component development. In this section we assume the times-to-failure are known. In the next section we will deal with grouped or interval data where the times are not known exactly.

Type I tests (time terminated) are concluded at T*, which is not a failure time, and type II tests (failure terminated), are concluded at failure time T_N, use slightly different formula as indicated below. It is assumed that time data is available for every failure with these methods. Statistic "U" estimates reliability change.

Test statistics are calculated:

$$U = \left(\left(\sum_{i=1}^{N} T_i\right) - N\left(\frac{T^*}{2}\right)\right) / \left(T^*\sqrt{\frac{N}{12}}\right) \qquad \text{(Type I)} \quad (9\text{-}6)$$

or

$$U = \left(\left(\sum_{i=1}^{(N-1)} T_i\right) - (N-1)\left(\frac{T_N}{2}\right)\right) / \left(T_N\sqrt{\frac{(N-1)}{12}}\right) \qquad \text{(Type II)} \quad (9\text{-}7)$$

Where: N is the total number of failures;
T* is the total accumulated test times for type I test (Time Terminated);
T_N is the total accumulated test times for type II test (Failure Terminated);
Ti is the total accumulated test time at the i-th failure.

Under the hypothesis of zero growth, the failure times follow a homogenous Poisson process, and the statistic U is approximately distributed as a standard normal random variable with mean 0 and standard deviation 1. This allows a normal distribution significance test at an alpha significance level for positive or negative reliability growth. This test is included in SSV. If significant positive or negative growth is indicated at the alpha level the IEC standard recommends proceeding with the analysis shown in the next step below. If significant growth is not indicated the growth analysis is terminated and the hypothesis of exponential times between events is accepted, a homogenous Poisson process. At a 20% significance level, the critical values for a two sided test are +/–1.28. If $-1.28 < U < +1.28$ then there is no significant change in reliability growth. (Ref IEC61164, section 9.2.1)

The next step is a summation:

$$S_1 = \sum_{i=1}^{N} \ln\left(T^* / T_i\right) \qquad \text{(Type I)} \quad (9\text{-}8)$$

or

$$S_1 = \sum_{i=1}^{N} \ln\left(T_N / T_i\right) \qquad \text{(Type II)} \quad (9\text{-}9)$$

Next calculate an *unbiased* estimate of β:

$$\hat{\beta} = (N-1)/S_1 \qquad \text{(Type I)} \quad (9\text{-}10)$$

Or

$$\hat{\beta} = (N-2)/S_1 \qquad \text{(Type II)} \quad (9\text{-}11)$$

Finally calculate an *MLE* estimate of λ;

$$\hat{\lambda} = N/(T^*)^{\hat{\beta}} \qquad \text{(Type I)} \quad (9\text{-}12)$$

Or

$$\hat{\lambda} = N/(T_N)^{\hat{\beta}} \qquad \text{(Type II)} \quad (9\text{-}13)$$

The instantaneous *MLE failure intensity* is similar to (9-5);

$$\hat{\rho}(t) = \hat{\lambda}\,\hat{\beta}\,t^{\hat{\beta}-1}$$

The reciprocal is instantaneous *MLE MTBF(t)*;

$$M\hat{T}BF = 1/\hat{\rho}(t)$$

The instantaneous failure intensity and MTBF are estimates for the "current" configuration over the range of the data. At T^* or T_N they are estimates for the latest configuration.

Example 9.1: Time vs. Failure Terminated Testing

This problem demonstrates using the MLE equations above. Data are in Table 9-1 below. For the failure terminated test, eleven units were tested to failure. Testing was halted after the last failure at 83.6 hours. The cumulative test time is recorded at each failure point. Natural logarithms are calculated and summed for N = 11 cumulative test times. ***U* = – 1.027**. The point estimate of the slope β is then calculated from equations 9-9 & 9-11 as:

$$S_1 = \sum_{i=1}^{N} \ln(T_N/T_i) = 13.35551 \qquad \hat{\beta} = (N-2)/S_1 = 0.674$$

and the scale parameter (Equation 9-13) is:

$$\hat{\lambda} = N/(T_N)^{\hat{\beta}} = 0.557$$

In Case 1 "Time Terminated," we have more than 11 units and we continue testing beyond the last failure at 83.6 hours until we terminate the test at 90.0 hours with no additional failures. ***U* = – 0.833**.The point estimate of the slope β is then calculated (equations 9-8 & 9-10) as follows: **S_1 = 14.16693**

β = (10/14.16693) = 0.7059 and the scale parameter (Eq 9-12) is: **λ = 0.4591**

These data are plotted in Figure 9-3. The time terminated results were obtained by assuming the testing continued to 90.0 hours without an additional failure. Using IEC 61164 methods the Cramer-Von Mises statistic accepts the goodness of fit at an alpha of 10% as indicated on the plot. Reliability growth is not significant as ABS(U) < 1.28155. SSV shows Reliability Growth significance at 79.8%, which is less than the 90% usually targeted for significance. These equations are the ones to use when precise cumulative times are known for each failure point. Otherwise the interval or group equations should be used. These are discussed in the next section.

Table 9-1. Failure Data

Cumulative (t)	Cumulative N(t)
2.5	1
5.8	2
10.7	3
16.2	4
23.0	5
36.2	6
45.3	7
49.5	8
70.1	9
80.3	10
83.6	11

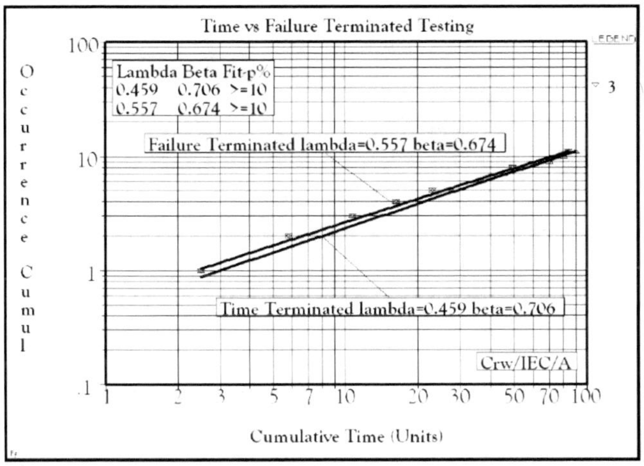

Figure 9-3. Time Versus Failure Terminated Tests

9.2.3 IEC MLE Solutions for Interval or Grouped Data

Grouped or interval data requires a more complex maximum likelihood equation to estimate the shape parameter, β. The solution is to find β which satisfies the top equation below. This method is especially appropriate where data are gathered by inspection at intervals or with coarse data collection, i.e., monthly reports. This type of data is very common in industry, so this method has wide application. The intervals do not have to be equal in time or numbers of events. Several intervals are needed, however, to have appropriate degrees of freedom for the analysis. This method recommended in IEC61164 is the same solution recommended in earlier versions of the Handbook for grouped data.

Note that t(0) and ln(t(0)) are defined as zero in the following equation, per IEC61164. An iterative method is needed to solve for the MLE β:

$$\sum_{i-1}^{K} n_i \left[\frac{t_i^{\hat{\beta}} \ln t_i - t_{i-1}^{\hat{\beta}} \ln t_{i-1}}{t_i^{\hat{\beta}} - t_{i-1}^{\hat{\beta}}} - \ln t_k \right] = 0 \quad (9\text{-}17)$$

$$\lambda = \frac{\sum_{i=1}^{K} n_i}{t_K^{\hat{\beta}}} \quad (9\text{-}18)$$

Figure 9-4 presents Example 3 from IEC 61164. The calculations and plots were produced with SSV software. The Chi-squared test indicates an acceptable fit. Another Chi-squared test indicates significant positive reliability growth. 90% confidence intervals for beta are (0.320-0.836). These methods are described in IEC 61164. If *"IEC"* is selected in SSV the correct IEC method will be selected automatically, type I or II or MLE for grouped data. IEC is the recommended choice for Crow-AMSAA solution at least to start. Regression is an alternative but is not as accurate except for five or less failures.

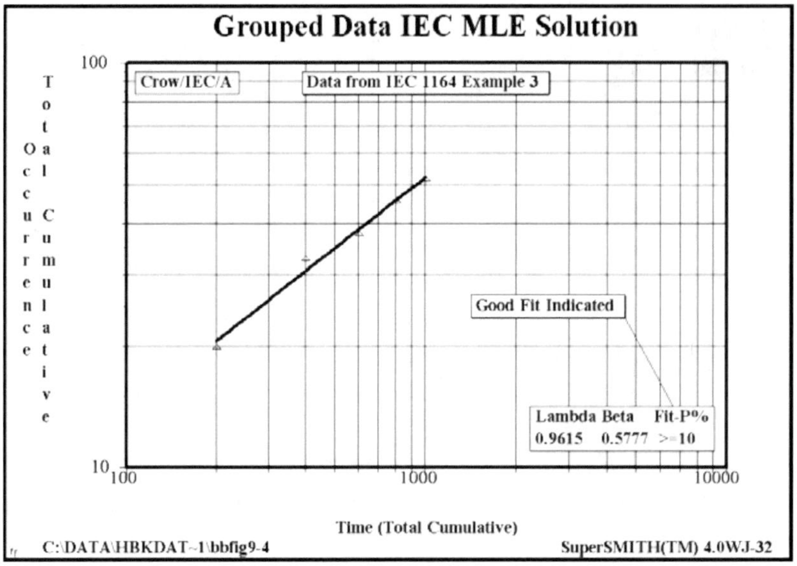

Figure 9-4 Grouped or Interval MLE Solution

Example 9.2: Printed Circuit Board Solder Joint Failures – Grouped Data

This problem demonstrates using the maximum likelihood estimator equation shown above to estimate the slope β from interval or grouped data. It also illustrates the relationship of data treated as Weibull data and also plotted as cumulative CA. A set of 12 printed circuit solder joints were selected for mechanical fatigue testing to simulate mechanical and thermocyclic solder joint fatigue. Each of the test specimens was made alike with the same materials for solder joints, leads and printed circuit boards. Twelve units were mechanically cycled (in a rig for the set) by pulling on all the leads with the same force, at the same time, until each failed. None were replaced. This is *nonrepairable* data.

The testing yielded Weibull data. Weibull data were analyzed (MLE) and a Weibull slope $\beta = 2.0722$ and characteristic life $\eta = 4{,}698.5$ cycles were computed. The data for CA is grouped into k = 10 intervals with the first interval containing the first 3 failures. The first interval cumulative cycles is the sum of 2231 cycles on 10 units plus the cycles-to-failure on two previous failures. These CA cumulative cycles are obviously not the same as in the Weibull cycles. The next step is to cumulatively plot these data. (see Figure 9-5)

Example of Cumulative Cycles:

25607 = 10*2231+1123+2174

29612 = 9*2676+1123+2174+2231

Table 9-2. Solder Fatigue Test Data

Weibull Failure Cycles	CA Cumulative Cycles t_i	CA Cumulative Failures
1123		
2174		
2231	25607	3
2676	29612	4
3269	34356	5
3377	35112	6
4060	39210	7
4079	39305	8
4562	41237	9
6172	46067	10
7480	48683	11
8527	49730	12

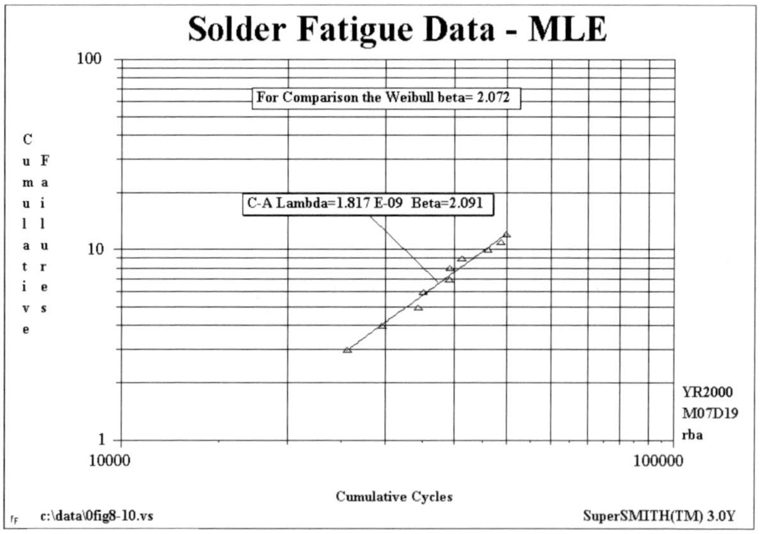

Figure 9-5. Solder Fatigue Data – MLE

These data are used in the grouped data MLE equations illustrated above, iteratively trying values of the slope β using a Newton-Raphson method, until convergence. SSV will do this for you. The slope $\beta = 2.09$ was computed. This is close to the Weibull slope computed, as it should be for *single failure modes*. The scale CA parameter $\lambda = 1.82$ E-9 is computed from the slope estimate.

Analyzing the same data by the two different methods, CA and Weibull, has revealed their differences and similarities; they are not equivalent, but their slopes are approximately the same. This shows that the Crow-AMSAA model can analyze data with an adverse trend, as well as for reliability growth. Here it is used with wear-out mode data with no improvement in design introduced during testing and with no replacement of failed units during testing.

Example 9.3: Bomber Unscheduled Maintenance

This problem also demonstrates the use of the maximum likelihood estimator equation to estimate the slope β from grouped data. These data were gathered from an operational fleet of bomber engines. The reliability index is Unscheduled Maintenance Events which include Unscheduled Engine Removals (UERs), Unscheduled Line Replaceable Units (LRUs), and maintenance adjustment events made at the flight line. This is repairable systems data, where engines are put back in service after repair or replacement of component parts and with staggered starting times. This is interval data. The monthly reports include the number of events and the cumulative engine flight hours. The precise cumulative engine flight hours at each maintenance event point is unknown. Data are shown in the left section of Table 9.3.

First, the cumulative CA plot is examined. Several cusps or discontinuities are noted. An initial reliability improvement trend is indicated by the shallow slope up to 800 hours, followed by a constant failure rate trend ($\beta = 1$) up to 30,000 hours, finally giving way to a moderate improvement trend after 34,009 cumulative engine flight hours. The last 8 intervals of data are used to determine the latest trend; there were 257 events at the start of this last trend. Results are shown in Figure 9-6. This trend predicts the prescribed goal at 100,000 cumulative engine flight hours (900 or less events) is going to be met (which it did with subsequent data).

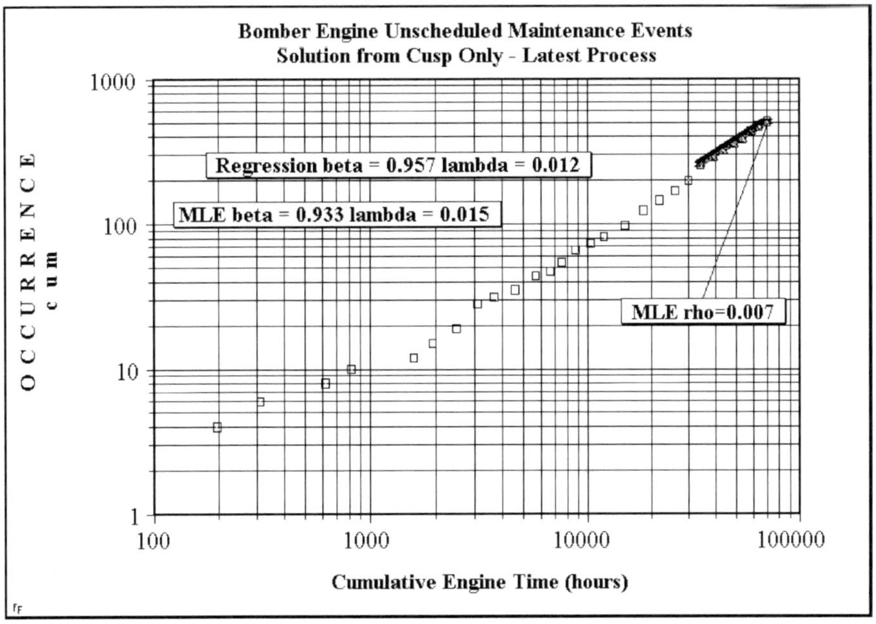

Figure 9-6. Bomber Engine Unscheduled Maintenance Events Solution from last Cusp Only - Latest Process

Table 9.3 Bomber and Warranty Data

Bomber Data Example 9.3		Warranty Data Example 9.4 With Renewal		Warranty Data Example 9.4 Without Renewal	
Cum EFH	Cum Failure	Cum. System Months	Cum Claims	Cum. System Months	Cum Claims
34009	257	1000	1	1000	1
39029	289	3000	5	2999	5
43181	321	6000	8	5994	8
48485	356	10000	13	9986	13
53470	385	15000	20	14973	20
58457	435	21000	32	20953	32
63436	470				
70081	505				

Example 9.4. Warranty Claims Predictions

To forecast warranty claims by calendar month the Crow-AMSAA model is excellent. Using the data shown above, there is one claim the first month, 4 in the second month, three in the third month, etc. These sums are diagonals in Table 8-2 in Chapter 8. Each month a thousand units are produced which contribute 1,000 system months each calendar month if there is renewal or replacement of failed parts. The data are entered in SSV by block, copy, and paste the matrix in Table 8-2. The plot is shown in Figure 9-7. Extrapolating the line to the next month's cumulative system-months, (28,000), the forecast would be 44 claims less 32 or 12 more claims in the next month. The SSV report option predicts the time to the next claim, 2 claims, 5 claims, etc. CA modeling is **best practice** for predicting warranty claims by calendar time for repairable systems.

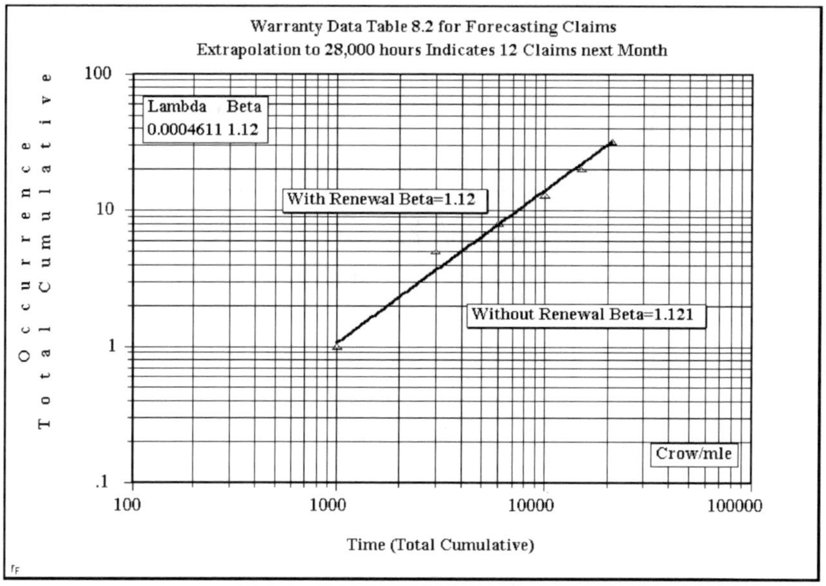

Figure 9-7. Warranty Data Table 9-2 for Forecasting Claims Extrapolation to 28,000 hours Indicates 12 Claims Next Month

The other related warranty problem, predicting events by the age of the system, is usually modeled with the inspection option, interval MLE or the Kaplan-Meier method. However, this may not be rigorous as repeated events of the same repair will be improperly plotted according to some experts like Wayne Nelson. Most analysts ignore these considerations. When you have recurrent failures/repairs that are recorded by product age, consider using Wayne Nelson's rigorous approach with his **graphical repair analysis,** described in Appendix M. Here he has contributed an excellent writeup with examples. His new text on the subject, [Nelson 2003] is recommended. The data required *consists of the age to each repair and the current ages of* *all systems along with an identification number or name relating each data input value to a particular system.* The methods apply to costs as well as the number of repairs.

Example 9.5 Measuring Maintenance Improvements

CA provides a quantitative measurement of improvements in maintenance planning, reliability centered maintenance, improved inspection methods and frequency, and operations of the system. Figure 9-8 illustrates the effect of maintenance plan improvement with a significant decrease in observed failures. The effect of the maintenance plan was to avoid 131 – 29 = 102 failures. If the average price of failure is known and is an input, then this plot becomes a measurement of the cost reduction.

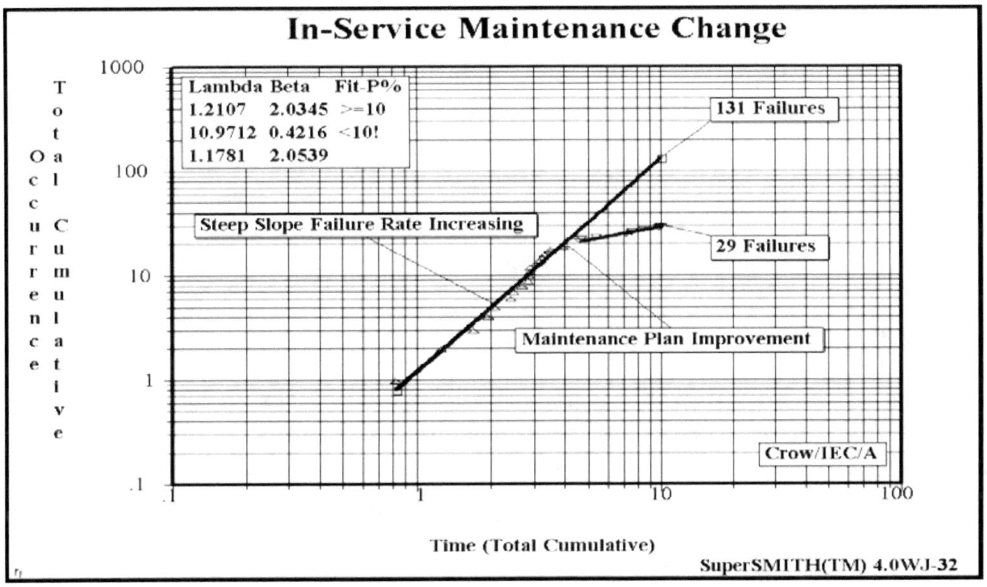

Figure 9-8. Measuring Improved Maintenance

Example 9.6 Success Failure Data Ariane Rocket Launch Reliability

An analyst ensuring satellite launches in London asked if there is any way to improve on the poor precision of the binomial distribution to estimate the reliability of the Ariane Rocket system. The author's solution is related to the fact that if the binomial parameters are large N and small p, the Poisson distribution accurately approximates the binomial probabilities. As the Poisson is related to the Weibull and the statistical foundation of the Crow-AMSAA is the Weibull, the author hypothesized that the CA may offer a much more precise solution. Using the instantaneous failure rate it is easy to convert to reliability. This option is called "SURV" in the SSV Crow-AMSAA menu. See Figure 9-8A.

Could this method be used to estimate the Space Shuttle reliability after two failures? Try it.

Example 9.7 Estimating Availability with Crow AMSAA

Wes Fulton has added another option to SSV using Mean Time to Repair, (MTTR). Availability is estimated as (MTBF)/ (MTTR + MTBF). Therefore, a CA MTBF plot may be modified to plot instantaneous availability by inputting MTTR to the Crow-AMSAA option in SSV.

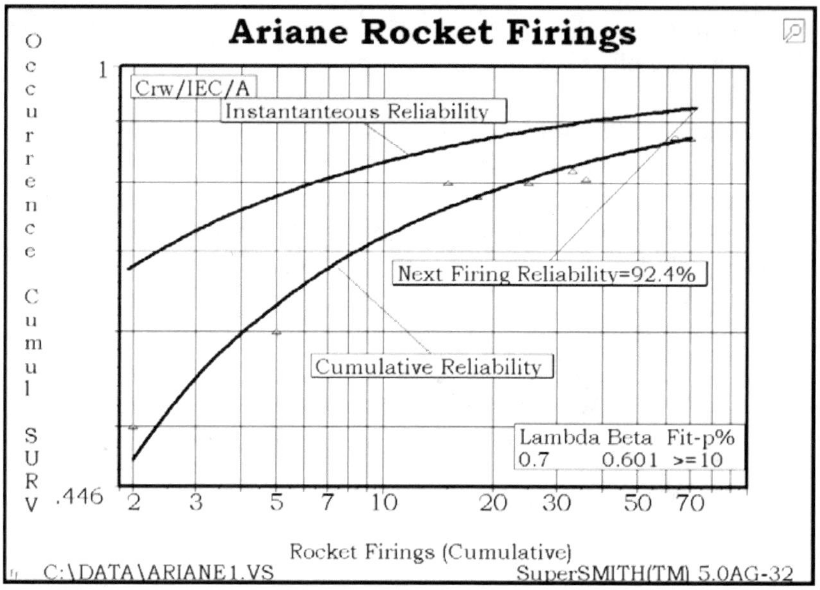

Figure 9-8A. CA Solution using Success-Failure Data

9.3 Comparisons of the IEC and Regression CA Methods

The CA *regression solution* is still available in SSV and is most accurate for n<5. Paul Barringer employed Monte Carlo simulation to estimate the critical 90% lower values of r squared for the Crow-AMSAA (CA) plot line for data where the exact times of failure are known. It works exactly the same way as the CCC^2 for Weibull plots. These Monte Carlo results are plotted in Figure 9-9. Note that these results only apply to regression solutions where the exact times to failure are known, *not to interval or grouped data.* Barringer also developed Monte Carlo sampling of CA data to compare the three methods (IEC Unbiased, IEC MLE, and regression) for estimating beta and lambda. See Figures 9-10 and 9-11. The results indicate from most accurate to least accurate the order is unbiased, MLE and regression except for small samples, n<5. In this region regression is best. These simulations have been verified independently by Geoff Coles of Rolls Royce Bristol.

There are two interesting CA case studies in Chapter 11: 11.12 & 11.14.

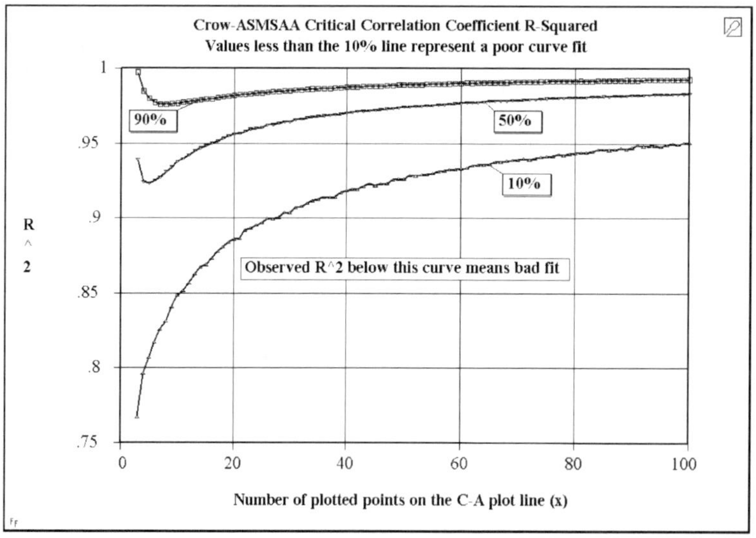

Figure 9-9. Crow-AMSAA Critical Correlation Coefficient R-Squared 10% Percentile

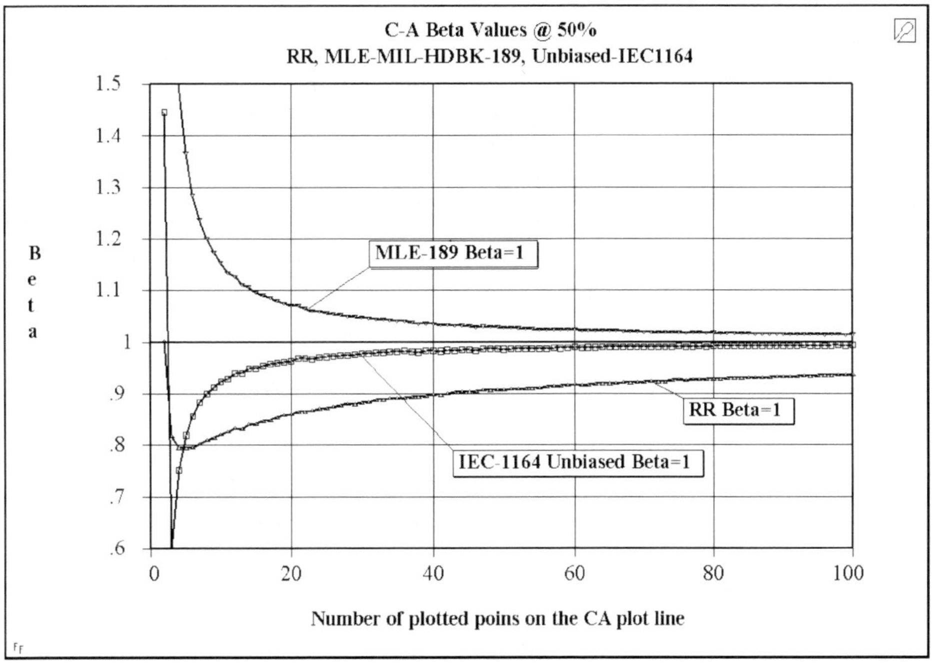

Figure 9-10. CA Beta

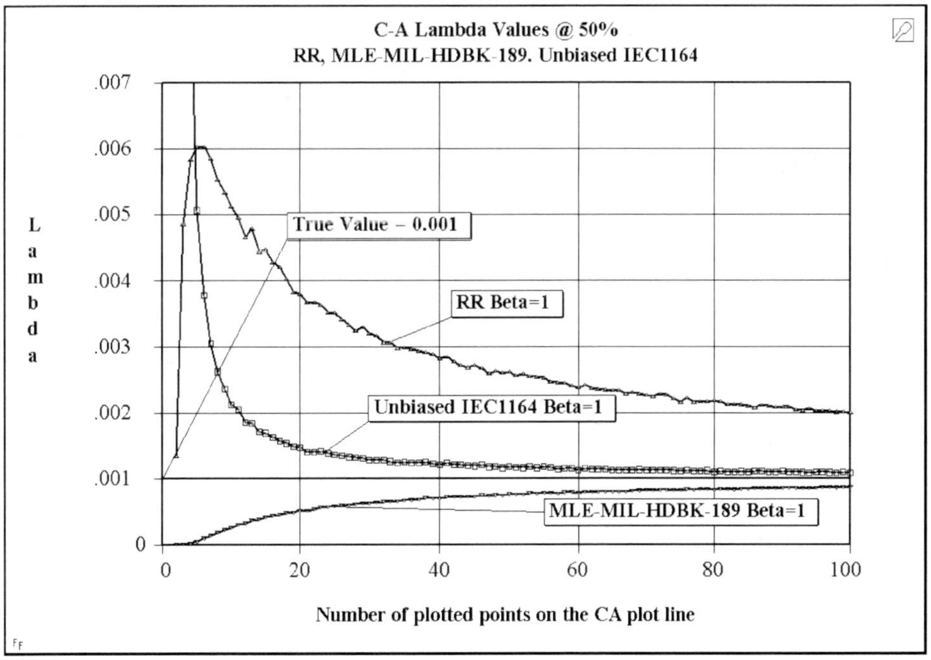

Figure 9-11. CA Lambda

9.4 CA Input May Be Confusing

Data input from in-service units may be confusing. For example, assume 100 systems are produced in each of three months and there is one failure or warranty claim in each of those months plus the following three months. The data may be input to SSV as *cumulative, the default*, or interval.

Table 9-4. Interval Versus Cumulative Data

Months	Produced	Operating (Interval)	Cumulative Operating	Failures (Interval)	Cumulative Failures
1	100	100	100	1	1
2	100	200	300	1	2
3	100	300	600	1	3
4	0	300	900	1	4
5	0	300	1200	1	5
6	0	300	1500	1	6

Input: Interval 100x1, 200x1, 300x1,..., 300x1 or
Cumulative: 100x1, 300x2, 600x3,...,1500x6

Also note that if cumulative failures *do not increase,* we do not plot a point. For example, assume there was no failure in the second month above. The data input would be:

Input: Interval 100x1, 300x1,..., 300x1 or
Cumulative: 100x1, 600x2,...,1500x5

9.5 Missing Early Data with CA

Many analysts ignore the missing data and start anywhere. However, it is more accurate to estimate the missing data. Here is one method: Make the CA plot with the data you have and obtain an estimate of cumulative failure rate at the first data point. Make the best guess possible for the missing cumulative time, say ΔX. Multiple ΔX by the cumulative failure rate to estimate the missing cumulative failures say, ΔY. Add ΔX to the X values and ΔY to the Y values, respectively, and redo the CA analysis and plot. This should provide an improvement in accuracy when early data is missing.

9.6 First Time Failure Mode Analysis, Reliability and Availability

The author has reviewed data on aircraft jet engines in service for as much as 100 times their original design life, hundreds of millions of hours of operation for each model. The data was cumulative fleet time when the **first failure of a new failure mode**, Figure 9-12, was observed. Data from two different engine manufacturers showed good CA plots, all with slopes close to 0.91. Derivative engines had lower lambdas but the same slope as all new designs. The surprising inference from these plots is that there are always failure modes beyond the ones we know about! All failure modes are never found. The consistency of beta at 0.91 is curious but the author has no explanation. First failure CA plots can be used to estimate how many new failure modes will be uncovered with additional testing. A good application is software reliability testing where every "bug" is fixed. How many "bugs" will be found with 1000 more hours of testing?

The author has shown that with Binomial Success-Failure data, if the number of trials, N, is large and the probability of the event, p, is small, Crow-AMSAA may be used to obtain a much more precise estimate of **reliability growth** compared to the Binomial. An example would be satellite launch reliability. Also, if the mean-time-to-repair is available for Crow-AMSAA failures, **availability** may be plotted. Both plots are available in the SSV software.

Thank You We are indebted to Paul Barringer, David Weber, Wes Fulton and Larry Crow for all their efforts to improve Crow-AMSAA modeling and provide an understanding of this most valuable tool.

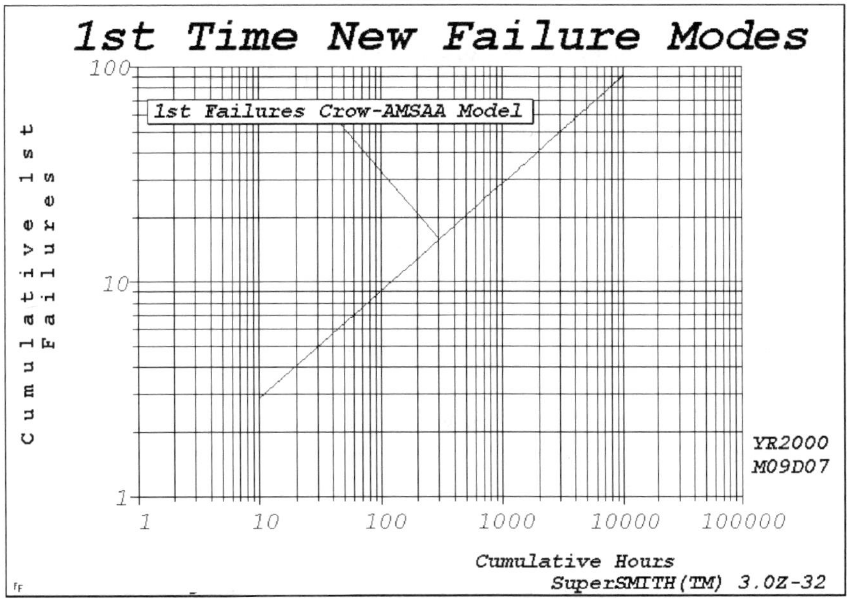

Figure 9-12. 1st Time New Failure Modes

9.7 Warranty Claims Analysis

Contributed by Joe Wauben – Honda Corporation and Wes Fulton-Fulton Findings™

Many businesses back their products for some initial service time with warranties. A common warranty period for automotive products, for example, is 5 years or 50,000 miles, whichever occurs first. Electronics firms may provide a 1-year or 2-year warranty period or even longer. Often customers can buy extensions to the warranty period at the time of product purchase. Covered products are refurbished or replaced during the warranty period if there is a verifiable problem reported by the customer. There has been a recent increase in warranty guarantees and extensions of warranty periods to appeal to more customers. Some customers are demanding these improvements.

There is a need to predict the quantity of expected claims *either by (1) equipment age or (2) by calendar time.* In Chapter 8 the Kaplan-Meier model was applied to the problem of predicting warranty claims by *equipment age*. The Inspection Option was also recommended for the same purpose. In Example 9.4 above the CA model was recommended as best practice for predicting warranty claims by *calendar time*. There is a third objective for warranty analysis: *(3) to improve the product based on failure analysis* of the warranty data. Techniques presented in this section will address all three applications.

There is a wealth of information within warranty and service data waiting to be unlocked. Service data is any data coming from actual usage of equipment (or services) by customers and includes warranty claims. Both trouble-free usage records and reports of product complaints and warranty claims are important. The main benefit of Weibull and Crow-AMSAA analysis of service data is the feedback gained from:

1) Starting and maintaining a continuous improvement program
2) Preliminary reliability models
3) Responding to quality defects and customer concerns more quickly
4) Determining whether all products are affected or whether it is only a smaller portion (a batch issue)
5) Projecting the quantity of claims expected in the next month, the next year, etc.
6) Projecting the quantity of claims within the warranty period.
7) Evaluating effects of alternatives within maintenance policy
8) Right-sizing of contingency monies for customer satisfaction concerning product complaints
9) Identifying and targeting high cost root causes for product returns
10) Predicting future potential cost/customer impact liabilities
11) Cost justification for quality improvements

Continuous improvement should be a goal of any long-term enterprise. New design requires innovative response avoiding the past failure modes. A Weibull Library (the collection of all previous Weibull and CA analyses) becomes a data source of lessons learned. See Chapter 6. Service data is a valuable starting point along with laboratory test results for developing these resources.

There has been an increase in warranties lately to make the products more attractive to customers. As a result there is a need to forecast warranty claims by the age of the unit and by calendar months. Customers may buy extended warranties. Cost estimates for basic and extended warranties depend heavily on forecasts of the return quantity expected. Claims forecasting starts with the CA and the two parameter Weibull plots. Distribution analysis may lead to the lognormal or three parameter Weibull distributions.

Caution: Simply "plugging" data into the software will not always produce a reliable forecast. Different problems have different forecast "personalities." Understanding how, why, and when to apply the tools along with underlying assumptions is important for accurate forecasting. This comment actually applies to every method in this *Handbook*. The Logic Diagram in Chapter 10 may be helpful.

Service data analysis is not restricted to warranty data. Even without a warranty, service data should be collected and analyzed. To pursue this gold mine of information top management must commit to good data collection practice and a reasonable level of up-front investment. The warranty data matrix format described below is one way to organize warranty data for analysis.

9.8 Warranty Data Matrix

MOP & Month Put Into Service

Quantity Put into Service Each Month

	1	2	3	4	5	6	7	8	9	10	11	12	13
	815	765	808	787	372	67	42	34	21	18	14	5	2
1	3	1	6	4	1	0	0	0	0	1	0	0	0
2	2	0	0	1	0	0	0	0	0	0	0	0	
3	2	0	4	3	0	1	0	0	0	0	0		
4	2	4	3	3	0	1	0	0	0	0			
5	3	2	0	1	0	0	0	0	0				
6	3	2	2	1	1	0	0	0					
7	3	0	2	1	1	0	0						
8	0	1	2	2	0	0							
9	2	1	0	0	1								
10	4	1	2	0									
11	1	0	1										
12	0	2											
13	1												

Claims Reported Each Month

MIS Months In-Service

Figure 9-13. Warranty Data - Typical Format

A popular warranty data collection format conforms to Months Of Production (MOP) as columns versus Months In Service (MIS) as rows. This MOP-MIS format allows both Weibull analysis and Crow-AMSAA modeling. Another name for this format is "Layer Cake" data because it displays a step-like triangular appearance. Figure 9-13 shows the data format for this type of data using months as the time interval. Months are the most commonly used time interval, but some manufacturers use weeks. Note: This layer cake matrix can be Copied/Pasted directly into the SSW for Weibull plots and the SSV Software for Crow-AMSAA plots.

In Figure 9-13 numbers in the first column and those in the first row indicate MIS and MOP. The remaining data (the main body of data) is referred to as the *analysis data*. In Figure 9-13, the analysis data begins in the upper left with the number 815 and includes all the numbers to the right and below. For MOP-MIS format columns represent a calendar period with the first row in each column indicating the units put into service from that month's production and numbers below representing the quantity of units returned under warranty by months in service.

At this point it may be useful to revisit the definition of suspensions for Weibull analysis. *For failure analysis*, suspensions are equipments that have not failed by the failure mechanism being analyzed as stated earlier in this handbook. Suspensions are successful equipments or they can be equipments that have failed for a different root cause. Suspensions are extremely important to proper analysis but are sometimes difficult to obtain from actual in-service data collection. Most warranty data analysis has to overcome the difficulty with getting good data. Some references call suspensions "censored" equipments instead. *For warranty claims predictions*, a unit that is beyond the warranty limits "disappears," it can no longer generate warranty claims. It is not technically a suspension. It is considered to be a suspension only for model solution but it is not considered to be a suspension for forecasting. However, a unit within the warranty limits that has not failed is a suspension.

Suspensions directly impact the end results of total number of failed units projected to fail within the warranty period. Careful attention should ensure that the suspension data is accurate. Spending time to truly understand the impact of including suspensions is recommended. Understating or overstating your company's warranty liability can easily occur if the incorrect suspensions are used.

In building the Warranty Data Matrix it is very important to understand what goals are desired. If the expected goal is to forecast total failures among all units produced, then the first row of the analysis data should be the *production number of the units*. If, however, the desired goal is to forecast how many will fail in the service in the next few years month by month, use only those *units that are in service* as the first row of the analysis data. If production units are produced but not put into service and this data is used as the quantity in the first row of the analysis data, a near-term forecast can be overstated. Obviously, those units sitting on the shelf will not fail from usage until they are actually put into use.

9.9 Warranty Data Rules

Simple Rules to follow:

Rule 1: Forecast One Problem at a Time (Separate Failure Modes)

- 1.) Example:
    - a.) Weld Failure
        - 1.a.) Insufficient Weld Depth
        - 1.b.) Bad Weld Material
        - 1.c.) Overstress

Rule 2: Must have a Claim History (How long was each unit in service)

- 1.) Measure of Time or Distance)
    - a.) Service Seconds/ Minutes/Days/Months/etc...
    - b.) Inches/Cm's/Miles/Kilometers/etc...(useful for automotive and construction)

Rule 3: Suspensions Must Be Included

- 1.) Ages of Units Produced that have not submitted a warranty claim to date.
    - a.) The measure of time or distance must match the units that have already failed. (Miles to Miles, Days to Days, etc... Apples to Apples Comparison)
    - b.) Suspensions with shelf time but no in-service time should not be used
    - c.) Shelf life failures should be considered separately

Rule 4: Accurate and Appropriate Data Is Required. It may be that the data is not available and a forecast cannot be completed. In this case the priority should be on acquiring the correct data. Warranty data can be dirty for several reasons such as:

1) Claim reporting and payment are not strictly the result of product malfunction, false claims
2) Problems with warranty data collection and transmission to the data collection point
3) Warranty term length complications (e.g. limited in terms of months and also limited in miles)
4) Opportune warranty return spikes (just before end of warranty)
5) Factory recalls and campaigns
6) Duplicate data issues
7) Mixed failure Modes (Difficult to determine separate failure modes due to similar failure symptoms)
8) Missing warranty data (various reasons)
9) Unknown "Batch"

9.10 Warranty Data Matrix Conversion & Analysis

The warranty data format may be converted into *Weibull* input in terms of age or *Crow/AMSAA* type input in terms of cumulative operating time. Conversion of the layer cake format depends on which of these two different inputs is desired. The SSW and the SSV software automatically converts the matrix format into data import format according to the following guidelines.

Weibull input is age data. The rows of the analysis data correspond to age and so the quantity of occurrences (claims) at each age is given by accumulating quantities in each row. Suspensions are calculated by subtracting the quantity of occurrences in each period from that period's quantity of production. All of the suspensions from the same period have the same age, the length of time since the period's production was put into service.

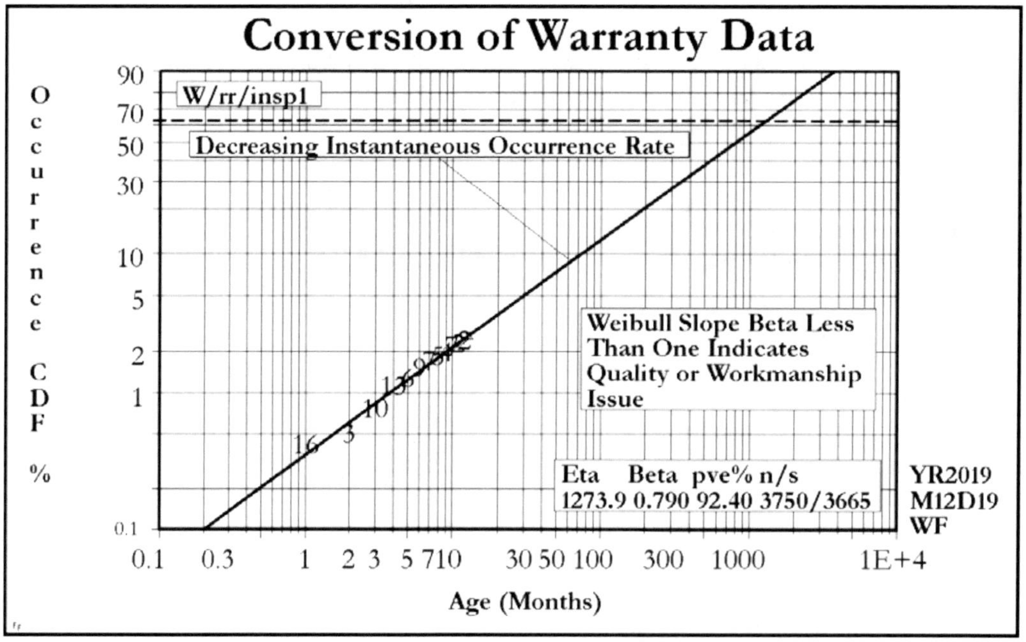

Figure 9-14. Weibull analysis of Data in Figure 9-13

Each *diagonal* (lower left to upper right) in the analysis data represents a calendar month and the sum of occurrences is the cumulative events for *Crow/AMSAA*. Operating age accumulates from the upper left and proceeds diagonal by diagonal toward the lower right. Suspensions are accumulated in the same sequence.

When the analysis data is imported into the SSW for the Inspection Option or Kaplan-Meier methods or into the SSV software for Crow-AMSAA there is an option for specifying the imported data as warranty format data. When imported into SSW the end result will be a Weibull-type probability analysis. When imported into the SSV program the result will be a Crow/AMSAA reliability growth analysis.

There is an option for whether renewal of the equipment should be modeled. If you are a manufacturer or a supplier, then the renewal option is not needed. This is because renewal units are ordered from you and are already included in the production quantities going into service. However, this is not so if you are a customer or end-user acquiring parts from manufacturers or suppliers and installing them into larger assemblies. Here replaced parts are not included in the production quantities and the renewal option should be selected for the model.

If there is a delay between production and the beginning of service operation, only those data representing units in-service should be selected for analysis. From the data shown in Figure 9-13, all data should be selected if there is no delay or if the data represents equipments already in service. Alternatively, if the data represents total production and there is a delay of say 2 months, then selecting only the first 11 columns from the left would be appropriate for analysis.

Failure or claims forecasting uses either "Abernethy Risk Analysis" for failure forecasting based on the Weibull Inspection Option or straight line extrapolation on the standard Crow/AMSAA display.

The data represented in Figure 9-13 is automatically converted to the Weibull plot as shown in Figure 9-14. All of the data was used for the analysis without renewal. This would be appropriate for a manufacturer or supplier making the unit under investigation and when there is not a significant delay between production and deployment. Figure 9-15 is a Crow/AMSAA analysis of the same data to allow tracking analysis of reliability performance and changes with calendar time.

Using Abernethy Failure Forecasting one of the most valuable formats of output would include "Total Forecasted to Fail vs. In-Warranty Forecasted to Fail".

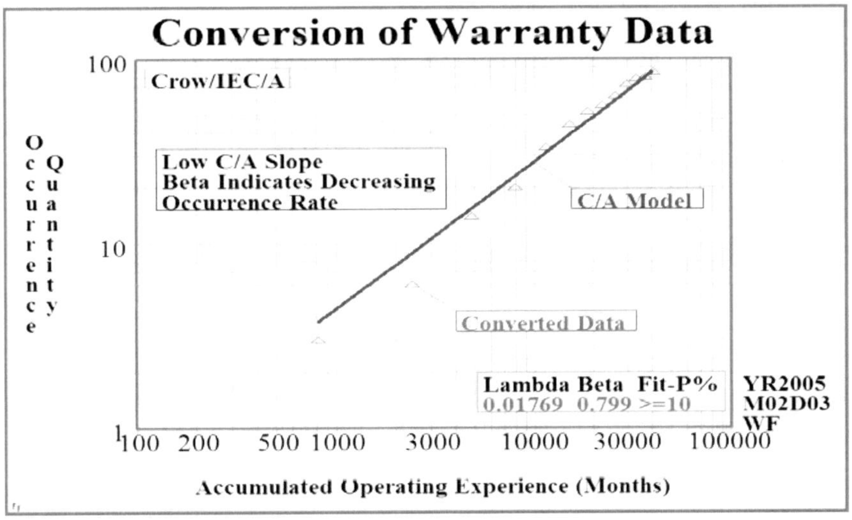

Figure 9-15. C/A Analysis of Data in Figure 9-13

9.11 Warranty Analysis Methods Within Weibull analysis there are **four methods** that work well with warranty data. The first three are used for predicting claims by the age of the unit and for failure analysis. Data from one failure mode at a time is needed. Crow-AMSAA is used for predicting claims by calendar month and may handle either mixtures of failure modes or one at a time. Each method will be explained below.

1) **Inspection Option #1**
2) **Kaplan-Meier**
3) **MLE Interval**
4) **Crow-AMSAA**

9.11.1 Inspection Option #1

For periodic data collection, say monthly or weekly, inspect option #1 is a method that produces good results. It is also used for inspection data where all units have the same inspection schedule. Multiple claims per month at the same time are common. Graphically this produces columns of points on probability scaling, like Weibull scaling. Inspect Option #1 is a graphical solution using only the top point in each column of identical data values for regression analysis. This simple method generally agrees better with higher-valued data points and preserves the suspension count for Abernethy Risk forecasting. See Chapter 5 for more details.

9.11.2 Kaplan-Meier

Kaplan-Meier (KM) is a non-parametric analysis technique for discrete distribution-free probability estimates. These KM estimates can also be used for graphical plotting positions in order to produce a continuous model on any probability scaling. Kaplan-Meier generally agrees well with other methods especially for large data sets. Care must be taken if the suspension count is altered with the actuarial correction for Kaplan-Meier grouped data. When the actuarial correction is used to establish the model, removed suspensions must be added back for failure forecasting. See Chapter 8 herein for a description of the Kaplan-Meier method.

9.11.3 MLE Interval

The MLE-Interval technique works on any type of grouped data, even when inspection intervals are not regular, for example when the inspection sequence varies from unit to unit. There are four types of data including failures, suspensions, discoveries, and interval data. Higher resolution may be obtained that is not possible with standard failures plus suspensions only analysis. Recent studies [Marquart] indicate that MLE interval method produces slightly more accurate results than the methods above. See Chapter 5 for more details.

9.11.4 Crow AMSAA

The CA method is best practice for predicting warranty claims by calendar time. How many warranty claims will we receive this month, this year? It is widely used for this purpose. It is also the most robust method in the sense that it will handle mixtures of failure modes, missing data and changing conditions. The early part of this chapter is devoted to the CA method.

9.12 Case Studies

Please review Case Study 11.14 in Chapter 11 as an excellent example of using in-service warranty with CA to do failure forecasts with batch problems in the data.

9.13 Tracking Your Results

Tracking the results of your efforts, comparing forecast claims versus actual claims, is an excellent idea. Much effort is employed toward producing an accurate forecast and should be followed by similar effort to measure the forecast accuracy. Benefits of doing this comparison include:

1) Building management's confidence in forecasting methods
2) Checking Data accuracy
3) Learning which methods work best with your data

Understanding the market situation when interpreting your results is also extremely important. There are many external influences that affect the way warranty claims occur. One is the consumer's reaction to failure; the consumer is easily swayed to react in different ways. Other influences may be changes in our political process, a new president, economic depression, war, interest rates.

9.14 Warranty Conclusions and Recommendations

Warranty costs can erode company profit. Proficiency in forecasting warranty claims gives any company unique insight into future liabilities and enables a company to more easily absorb related losses, maintain customer satisfaction, and meet federally mandated requirements for warranty programs. Management must commit resources for establishing good service data gathering, data storage, and analysis capability. This will provide the basis for many substantial returns on investment and enable company planners, design engineers, and upper management to better accommodate the future.

9.15 Life Cycle Cost
Contributed by Paul Barringer, CEO, Barringer and Associates

Life cycle costs (LCC) are all costs associated with the acquisition and ownership of a system over its full life. At the top level, **LCC = Acquisition Costs + Sustaining Costs**. The devil lies in the cost details which must be found for each year of the project life. Acquisition costs often have quotations. Sustaining costs must be inferred by calculations.

About 85-95% of component lives are adequately described by Weibull distributions. Thus, Weibull distributions are helpful for calculating failure costs in each time interval of the project. About 85-95% of most repair times are adequately described by lognormal distributions. Thus, lognormal distributions are useful for describing out of service conditions which in turn generates costs associated with repair times. So, Weibull analysis and lognormal analysis help get to the money issues for sustaining costs.

9.16 Net Present Value (NPV)

The usual figure of merit for life cycle cost is **net present value (NPV)**. NPV is a single value found by adding the present value of expected future cash flows plus the cost of the initial investment. Flows of money can be both positive and negative. In short, find the costs incurred each year of the project, apply the time value of money for each year to get the present values, and then sum all the present values to get the net present values. Present value is the now-value of money expected to be received or spent in the future where the flows of cash are altered by discount rates to adjust for the investor's time-value of money. NPV, a financial term fits the concept of "give me a single number for the figure of merit".

9.17 Discount Rate

The **discount rate** is the interest rate that makes an investor indifferent between cash amounts received or paid at different points in time. A crucial factor for NPV financial calculations is the discount rate. Typical discount rates for industry are ~12% whereas governments avoid bankers and use financing based on long-term Treasury bond rates described in USA Government publication NISTIR 85-3273-19 which shows 4.0% real discount rate or 4.8% nominal discount rate including general price inflation. There are two discount rate approaches in dealing with inflation which yield the same present values: 1) exclude general price inflation by specifying all costs in constant dollars and use the discount rate excluding inflation, and 2) include general price inflation by specifying all costs in current dollars and use a nominal discount rate including inflation. Typical industrial discount rates have these general elements: 6% prime interest rate + 3% banking charges + 3% internal organization handling charges = 12%--of course if industry is cash rich then banking charges are not required and the discount rate would be ~9%. Government discount rates are set at T-bill rates of 4% plus inflation rates of 0.8% for 4.8%.

Money **interest rates** go up/down on short intervals, while discount rates take a longer view and change more slowly. Each organization may have different opportunities for acquiring cash and in the same market conditions a rate of 12% may exist for one group but a less fortunate organization unable to borrow money at low rates may carry a burden of a 20% discount rate. Thus the "right" discount rate decision for each organization can vary considerably for acquiring and sustaining ownership costs. Discount rates are established as high level financial decisions; they are not set at the engineering level. Discount rates change with time. Project life span decisions change with business condition and they may involve engineering decisions for selection of the study period.

Smart technical decisions without money considerations are usually viewed by owners/managers as bad engineering. Thus, engineers must be able to speak in financial terms of NPV and LCC.

9.18 Life Cycle Cost and NPV

LCC is a financial description of the birth/death cycle. LCC is the total cost of ownership described in financial terms. LCC involves the time value of money (the financial aspects) for expenditures to arrive at the single figure of merit for NPV. NPV calculations require annual time buckets of positive/negative cash flows. Cash flows are discounted for future values at a specified discount rate. All annual discounted flows are summed, including the initial investment. NPV can be a positive or negative number.

When **NPV is positive**, the most desirable choice is usually the **largest positive NPV**. When **NPV is negative** (this occurs when equipment is selected without knowledge of "profits" contributed by the individual equipment), then the most desirable selections are based on the **least negative NPV**. LCC is about alternatives, and the first alternative is finding the cost of doing nothing.

Business wants to know about both positive and negative NPV. For the big picture, they will have both gains and losses. Furthermore, financial decisions will be made without knowledge of how much profit an individual component or machine will contribute to the overall positive NPV for the project. This means that many components will be selected based on the least negative NPV.

Government projects don't have profits, they don't pay taxes, and discount rates are smaller than for industry. Their calculations (which when viewed by industry will be negative) but will not be presented with inclusion of the negative presentations. Watch for the difference in calculation details when they appear in the Government literature.

Businesses also pay different tax rates depending upon their costs and investment strategies. A typical business tax rate is 38%. Governments pay no taxes and their rates = 0%.

9.19 LCC Calculations

In LCC calculations you must separate capital cost. Capital costs are long lasting assets of a business that remain after payment of debts and other liabilities. Expense costs are values of resources consumed during the current period that can be charged against revenues for the period. These two costs are handled differently in the NPV calculations.

What goes into the LCC calculations? — everything, don't include the trivial costs or ignore the obvious costs but be practical as LCC is more about the $1000's than about the $0.01's. Figure 9-16 shows a memory jogger for the details for both acquisition costs and sustaining costs.

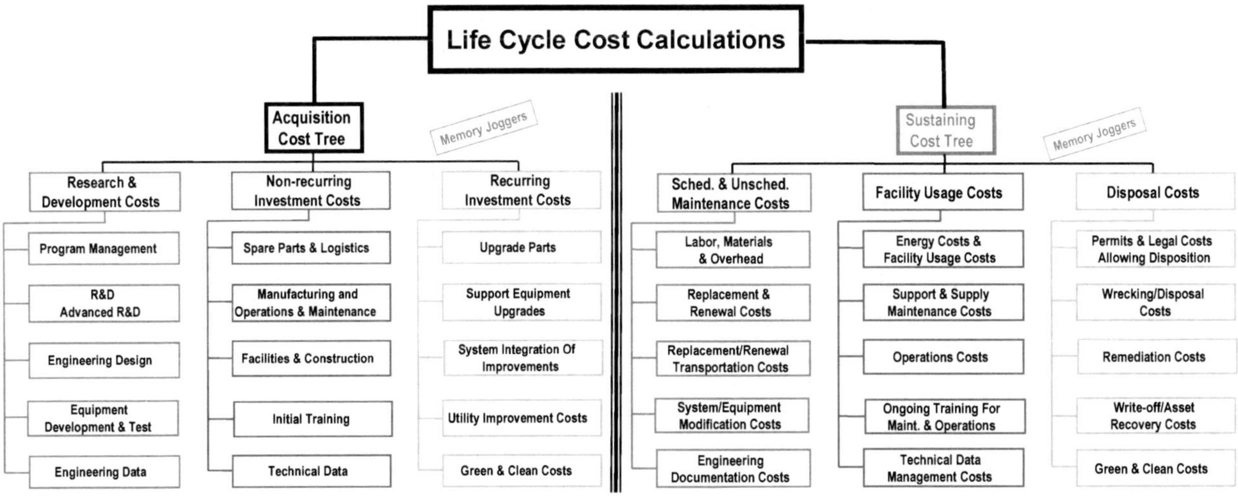

Figure 9-16. LCC Cost Elements—Memory Joggers

Furthermore, most of the items are also capital acquisition costs which mainly occur in year 0 of the project. Money spent in year zero for acquisition usually has some finality and decisiveness. Details on the right-hand side of Figure 9-16 are more difficult to acquire and they are cost items which must be inferred. Finding cost data often requires the use of Weibull analysis for when the parts will fail, and it also requires lognormal analysis for how long repairs will take with all data reduce to currency measures in the year of occurrence. We often lack visibility of how much will be spent and in which year the expenditure will occur—so the Weibull and lognormal analysis is valuable for producing the money details for the time buckets to arrive at NPV.

Figure 9-17 provides consideration for important items along with feedback loops for the question: Is this item necessary and does it represent the best of the alternatives?

Note the word alternatives in step 2 of Figure 9-17 is plural! Usually the first alternative considered by Accounting (who has their grip on the purse strings) is simply: What is the cost of doing nothing? Of course, the do-nothing case is hated by Engineering, and Engineering often fails to establish the do-nothing datum against which to measure their "new and improved" alternatives. Engineering must develop costs for the do-nothing alternative recognizing that the Accounting department, who provides the cash for Engineering projects and endorses/rejects projects, highly values the do-nothing case. Management values a process for LCC procedures as shown in Figure 9-17:

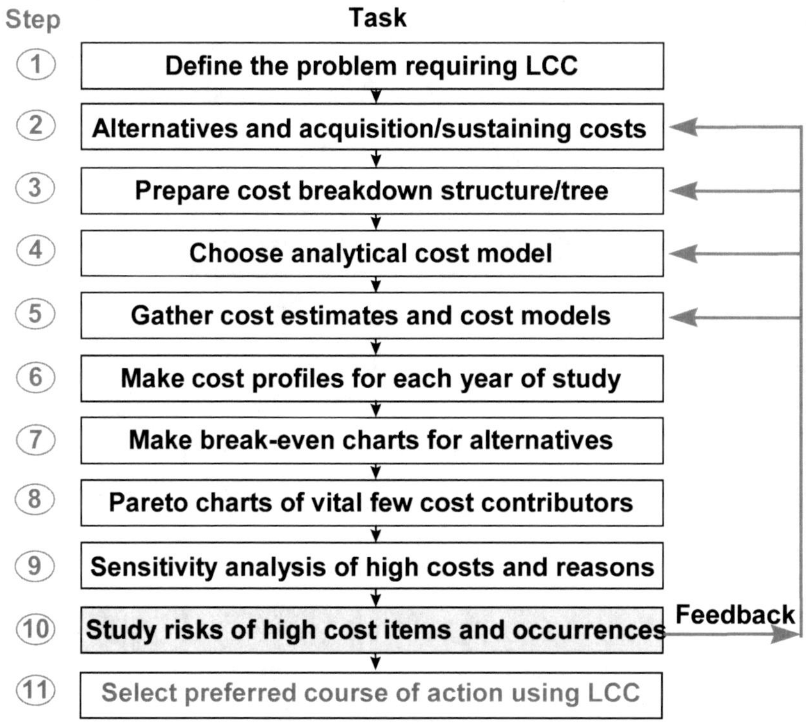

Figure 9-17. The Process For LCC Calculations

9.20 Case Studies

Consider the following example for the do-nothing case versus the new idea alternative. See how NPV's are determined using a NPV spreadsheet which is Paul Barringer's website (see References).

Use a project period of 10 years. Take the discount rate at 10%. Take the tax rate at 38%. For the do-nothing case, the capital is a sunk cost (no further capital is required) and the annual costs are $50,000. The new idea requires capital of $50,000 and the annual costs are $30,000. Which is the best alternative considering straight line depreciation? First solve the individual problems and combine them for a net result. The do-nothing case has NPV = -$175,157.

The new idea alternative (excluding savings) has a NPV = -$144,359.Now these two must be combined to see the net savings with an NPV Δ = -$144,359 – (-$175,157) = +$30,789 (which is a gain for the new idea alternative and thus most desirable between the two alternatives).

A second way is to combine both the do-nothing case and the new idea case into a single spreadsheet calculation.

Solve the problem by spending the $50,000 in year 0 and taking the net saving of $20,000/year. The answer is NPV = +$30,798 (the same practical answer) along with an internal rate of return (IRR) = 25.69%. The IRR is an obvious benefit not as easily obtained from the independent calculations. IRR is defined as the interest rate received for an investment consisting of payments (negative values) and income (positive values) that occur at regular periods.

If capital funds are available, reject the do-nothing case, and accept the improvement for better financial results. Remember also that capital funds may not be available even when cost reductions are obvious because the demand for cash elsewhere in the company may have a higher alternative.

Complicated NPV cases require Monte Carlo simulations to find how much cost should be applied to each year of the project as equipment lives and dies. Also SuperSMITH Weibull can be very helpful in making the risk prediction required to drive the financial calculations. Consider the following example: Three vendors offer to provide us with the same fabricated component for our product. The mean time to failure for each component is quoted as 3,000 hours.

We ask for the vendor's Weibull's: vendor A quoted $\beta = 0.7$ and $\eta = 2370$ hours with an item cost of \$700, vendor B quoted $\beta = 1.4$ and $\eta = 3290$ hours with an item cost of \$700, and vendor C quoted $\beta = 3.0$ and $\eta = 3360$ hours with an item cost of \$700. Use a project period of 10 years. Take the discount rate at 12%. Take the tax rate at 38%.

Find the annual cost using a replace-when-broken strategy. There is an Excel simulation in the internet material at Paul Barringer's website. The title is **LCC-Failures-Fix-When-Broken.zip**. That simulation was employed before to find costs for each year to be used in the NPV calculation. With such a simulation it is best to use at least 50,000 iterations (more will give smoother numbers). Results of a previous such Weibull simulation are shown below in Figure 9-18. NOTE: Paul Barringer's website contains this information when it is available (see References).

Data For Use In NPV Calculations			
Vendor	A	B	C
MTTF	3000	3000	3000
Beta	0.7	1.4	3
Eta	2370	3290	3360
Cost/failure	**\$700**	**\$700**	**\$700**
Year	Cost For Each Year		
1	\$2,412	\$1,882	\$1,740
2	\$2,079	\$2,048	\$2,050
3	\$2,057	\$2,053	\$2,030
4	\$2,020	\$2,045	\$2,043
5	\$2,049	\$2,056	\$2,045
6	\$2,037	\$2,038	\$2,048
7	\$2,025	\$2,046	\$2,046
8	\$2,053	\$2,035	\$2,042
9	\$2,071	\$2,053	\$2,037
10	\$2,061	\$2,037	\$2,057
Project Life	10	10	10
Discount	12%	12%	12%
Tax Rate	38%	38%	38%
NPV	-\$7,384	-\$7,078	**-\$6,991**

Figure 9-18. Facts For NPV Decisions

Vendor C has the least negative value and is the winner as shown in Figure 9-18. While the absolute cost differences are small, the calculation concept is important. Also note the variability in the numbers after year 2 results from noise in the Monte Carlo process from using only 50,000 iterations which takes less than one minute to complete on a fast computer.

Note that each of the vendors LCC annual cost numbers approaches a fairly constant rate after the first few rounds of replacement. Thus, failures occur earlier in the years with maximum impact of present value are very significant. A stronger wear out failure mode trend will postpone expenditures until a later date which is always important for LCC decisions.

LCC methodology provides facts and figures for making decisions. Thus intuition and emotion are minimized in most cases. Decisions are based on the facts rather than opinions.

This page is left intentionally blank.

CHAPTER 10. SUMMARY

10.1 The Beginning of the End

The scope and hope of this final chapter is to "Put it in a box, tie it up with ribbon…," borrowing the title of a song from long ago. Much has been said. Many methods, distributions, dirty data sets, have been described. It may be helpful to try to summarize all of that, and try to "put it in a box..."

The rest of that song was, "… ♫ ♫ and throw it in the deep blue sea. ♫ ♫ …" Please don't do that.

10.2 Which Method? What Kind of Data?

To summarize the special methods it may be helpful to describe the methods as a function of the type of deficiency in the data. [Nelson 1990] contains a good discussion and some excellent graphics on this subject. Our graphics in Figures 10-1 through 10-4 are similar:

Standard Point-By-Point Data: See Figure 10-1. Both failure and suspension times are known.

- With small samples, 20 or less failures, ***Weibayes is best practice and*** should be used ***if beta is known***. See Chapter 6.
- ***If beta is unknown, two parameter, X on Y, median rank regression Weibull*** is *the* ***best practice*** for small samples. This is the standard method for engineers. See Chapters 2 and 5.
- With larger data sets, more than twenty failures, a ***distribution analysis*** may be useful to help select the most appropriate distribution from a statistical view. Both the ***rank regression and MLE-RBA methods*** should be employed. Usually they will compare well and provide some assurance that the Weibull, normal, or lognormal is appropriate. The rank regression and MLE-RBA are more accurate for small and medium size data sets say up to 100-500 failures. Above 500 MLE is slightly more accurate. In spite of all you have read herein about dirty data, probably 95% of your data sets will produce useful Weibulls.

Grouped, or Interval Data: This category includes several types with different formats. There are techniques for each type providing the most accurate analysis of this dirty data. It is dirty data because the exact failure and suspension times are unknown. See Chapter 5.

- Inspection data involves a benign failure mode that is only found when the system is shut down and inspected. See Figure 10-2. If the inspection times are the same for all the units and the failure is obvious when inspected ***Inspection Option One*** is recommended. See Section 5.8.1.
- Coarse data is generated when the collection intervals are too long, say months instead of days or hours. ***Inspection Option One*** may improve the fit for coarse data.
- If nondestructive evaluation (NDE) methods are used to find small cracks or flaws, the probability of finding small cracks or flaws will be less than one. The ***Probit Method*** is recommended if the same number of pieces is inspected at each inspection and the inspection intervals are the same for all units. See Section 5.8.2 & 5.8.3. The ***Kaplan-Meier Method*** with the actuarial correction may be used if the number inspected at each interval is not the same. However, Kaplan-Meier is not recommended for failure forecasting. See Section 8.5.
- If destructive inspection is employed at intervals of age, the ***Probit Method*** will be the best choice. In this case each unit is only inspected once. [Nelson 1990] calls these data quantal response. See Figure 10-3.
- If the inspection intervals are not the same with all the units as shown in Figure 10-2, the ***MLE interval method*** should be used. MLE interval is slightly more accurate than the inspection option or Probit based on [Marquart's] study. The MLE interval will handle all of the above types of data.
- If there is a mixture of failure modes or mechanisms that cannot be unraveled or if there is large portions of data missing, or a batch problem, the ***Crow-AMSAA model*** is **best practice** for failure forecasts as a function of additional test time or calendar time. This method is also excellent for

tracking critical parameter rates such as in-flight shutdowns, outages, fires, uncontrolled accelerations, accidents, etc. See Chapter 9.

- Warranty claims predictions are usually based on interval or grouped data. The ***Crow-AMSAA model*** is best predicting claims by calendar time, and the ***Interval MLE*** is the choice for predicting claims by the age of the systems although the inspection option is more often used.

No Failures: ***Weibayes*** is **best practice** if beta is known or estimated from past experience. Prior knowledge of β may vary from excellent to none, and the corresponding Weibayes uncertainties from small to large. In preliminary design, Weibayes with large uncertainties are expected. The alternative, no analysis, has even larger uncertainties. See Chapter 6.

One Failure: If there are no late suspensions, use ***Weibayes***. With late suspensions, the choice is ***MLE or Weibayes***. The choice will be based on the quality of the prior knowledge of β. If the prior knowledge is good, Weibayes will be more accurate; if bad, MLE may be the choice.

Few Failures: In this case there are three alternatives, ***Weibayes, MLE-RBA, or median rank regression***. If β is known or can be estimated, use Weibayes. Otherwise, a comparison of MLE-RBA and rank regression is a good idea. See Chapter 5, Appendix D, or use Monte Carlo simulation to choose between MLE-RBA and RR. Employing both may be the best choice.

Failure Unidentified: Failure and suspension times are known as well as the number of failures, but the failures and suspensions are not identified. This happens with overhaul depot data. ***Weibayes*** is the only method.

Suspension Times Unknown: Assume a histogram of suspensions if possible with some accuracy. Otherwise, use the ***Dauser Shift*** as a last resort. See Chapter 5.

Which Aging Parameter? If your physical understanding fails to tell you, plot each parameter, and use the one that gives the ***best fit***. See Chapter 5.

Accelerated Test Data: The methods in Chapter 6 may be appropriate. [Nelson 1990] is the hallmark textbook. SuperSMITH Weibull has a special menu for accelerated test data that will allow a fixed or variable beta with eta a function of the load that may be temperature, stress, pull, vibration, etc. There may be more than one type of load

Success-Failure, Go-No Go, Counted Data: This is not "life" data. If the events are random the binomial or the Poisson described in Chapter 8 should apply. If the number of events is large and the probability of an event is small, Crow-AMSAA may be the best choice. See Chapter 9.

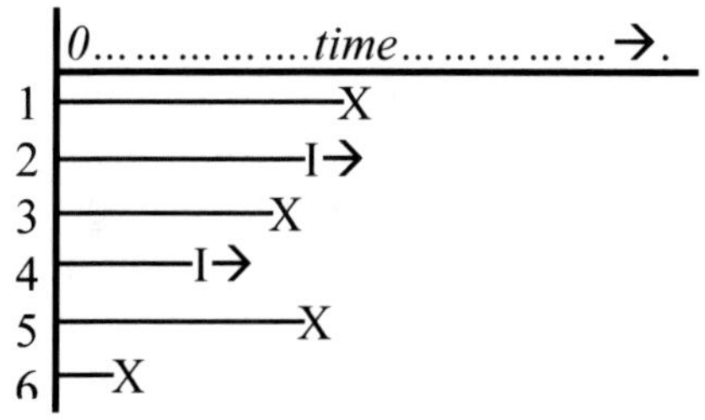

Figure 10-1. Standard Data, Legend: X - Failure Time, I→ Censored or Suspended

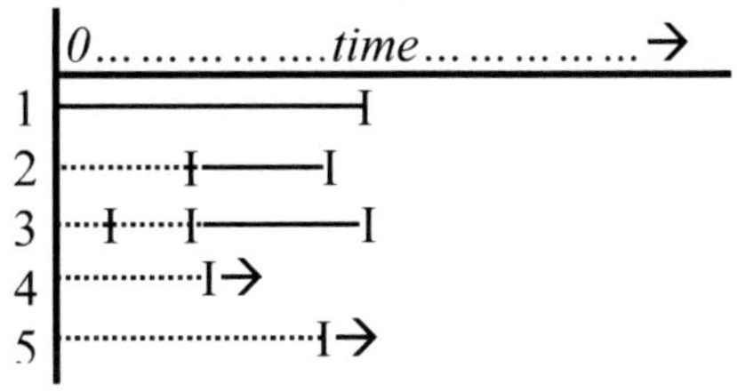

Figure 10-2. Grouped or Interval Data, ----- no failure found, —— I Failure Found

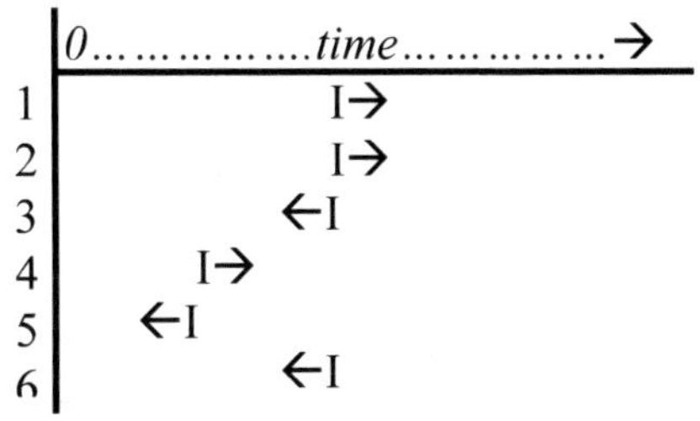

Figure 10-3. Destructive Inspection, Quantal Response Data

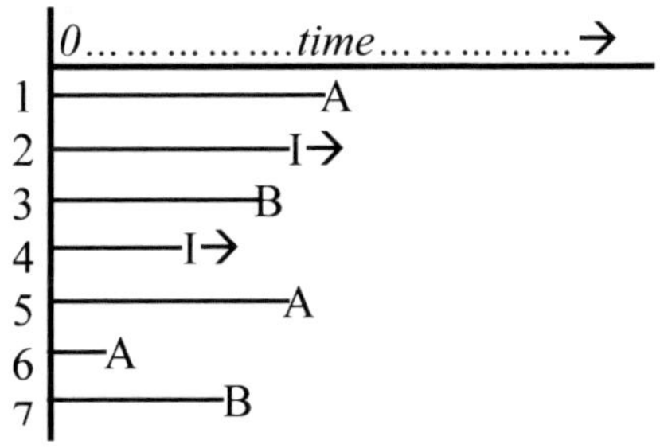

Figure 10-4. Competitive Failure Modes A and B

10.3 Looking at the Plot, What Do You See?

Plot has Cusps and Corners: Suspect a mixture of two or three failure modes. Figures 10-5A and B.

Otherwise, use the likelihood ratio test to find the best split. The pattern will often be the ***Classic BiWeibull*** with a shallow β, followed by steep β. See Figure 10-5A. A gentle curve upward may indicate a negative t-zero correction is needed. A batch problem may be involved.

Figure 10-5A

Figure 10-5B. If the plot shows a steep slope followed by a shallow slope look for a ***batch problem***. (Sometimes this plot will look more like a curve than a cusp.) There may be multiple modes. See Chapter 8, and Appendixes F and J.

Figure 10-5B

Plot is Curved Down: The ***lognormal*** or the positive t_0 ***three parameter Weibull*** are contenders. See Figure 10-5B. However, at least 21 failures are needed to have minimal confidence in the distribution analysis. See the more extensive discussion below. Past experience and physical understanding are extremely important and hopefully, may be consistent with or supported by the statistical results. Figure 3-4 should be employed to provide the statistical comparison with the SSW distribution analysis. Curves and cusps and corners may be confused on some plots. Engineering analysis will be the only way to unravel these mysteries. The classic BiWeibull may also appear as a curve.

Distribution Arguments: If there are serious criticism of your distribution assumptions, the ***Kaplan-Meier Survivor*** analysis may provide answers because no distribution is assumed. Further, the Kaplan-Meier estimates may be plotted on probability papers to verify the distribution assumption. See Chapter 8.

Success-Failure and Go-No Go Data: The binomial and Poisson apply to counted random events that are not related to age. The distinction is whether they occur in the binomial's n trials or the Poisson's continuum. If all we know is the average number of random events, the Poisson will be used. If all we know is the average rate of failures, and we think the rate is stable, the **exponential** will be used. If the failure modes are irrevocably mixed, the **Crow-AMSAA** model would be better even if the average is unstable. Here the best choice is only partially a statistical issue; deficient data may force a selection, right or wrong. See Chapter 8 for the binomial, Poisson, exponential and Chapter 9 for the Crow-AMSAA.

Failure Rates and Averages: If there is no other data available and the process is believed stable, the ***Poisson*** may be useful. If the process is unstable, try the Crow-AMSAA. See Chapters 8 and 9.

10.4 Which Distribution is Best?

Only a few of the many statistical distributions are described herein. On the other hand, there may be too many for the novice reader. (And not enough for the experts...) Here are some recommendations for distribution analysis:

- Prior experience with the same data source should be weighed heavily.
- Without prior experience at least 21 failures are required to do a distribution analysis; if less than 20 use the Weibull $[Liu].
- Knowledge of the physics of failure is extremely important. Nothing in statistics should ever be considered a substitute for engineering analysis. Statistics is just another tool for engineers, an addition, but not a substitute.
- You may lack experience with a new problem; you may not have a clear picture, no prior experience, no understanding. A Weibull plot is the best place to start. It will often be the best choice, but if not, the plot provides clues as to what distribution or method might be best. All other things being equal pick the simplest distribution. More information is needed to support complex distributions such as prior information and larger sample sizes. Without prior information, a t_0 three parameter Weibull or a lognormal needs at least 21 failures to support your choice.
- Consider the statistical evidence, the correlation coefficient squared, for each distribution compared to the critical correlation coefficient. Better yet, use SSW pve or prr, estimates of the percentile of the observed correlation coefficient squared. The highest percentile is the best statistical choice. Weigh this with your engineering analysis and your prior experience.
- Use MLE-RBA and the likelihood ratio to do a distribution analysis by a different method than pve. If the results agree with the pve method you will feel confident that you have selected the best distribution.

- The probability of selecting the correct distribution decreases with the number of distributions considered. For life data analysis use the two and three parameter Weibulls and the lognormal. For non-life data use the normal and the Gumbel distributions. See Appendix D.

Try to **relate whatever you know about the failure** to selecting the distribution. This is important. Hopefully the data plot will support and verify your engineering knowledge. [Tobias] is recommended. The **Weibull distribution** is the "weakest-link-in-the-chain" model. This concept follows from considering the Weibull as the Type III smallest extreme value distribution. It applies to situations where there are multiple opportunities to fail and our interest is the first failure. It also applies to system deterioration, which is linear rather than accelerating. For example, loss of turbine efficiency from blade tip erosion is often a linear function of operating time, producing a Weibull. On the other hand, if the deterioration is non-linear, a function of the current deterioration, the **lognormal** will apply. Crack growth to rupture is an example. If the crack region is highly stressed, as the crack grows, the uncracked material will have rapidly increasing stress, accelerating the crack growth. The lognormal will be a good choice. If the crack is in a low stress region the growth-to-rupture may well be linear and Weibull. In other words progressive deterioration relates to the lognormal, linear deterioration relates to the Weibull.

Little has been said about the **normal distribution** as it is not of major importance in life analysis. (This is due in part to the fact that normal data sets make good Weibulls, so most analysts do not check for a normal fit.) However, in the world of measurements, dimensions, performance, etc., the normal is of overwhelming importance. The Central Limit Theorem tells us that the sums or averages of samples from any distribution tend to be normal. The strength and wide application of the theorem is truly amazing. For example, sums as small as seven from rectangular random distributions are usually quite normal. If the physics of failure involves some multiplication, the Central Limit Theorem provides the lognormal. Materials Engineers relate laboratory tests of specimens to in-service failures with a model like this:

$$\mathbf{Time_{Service} = K_1 * K_2 * K_3 \ldots K_n * Time_{lab}}$$

The time to fail in the laboratory is often **Weibull.** If the K factors say for stress, vibration, corrosion, and temperature, are close to one, the time to fail in-service will also be **Weibull.** However, if the K factors are significantly different from one, the service times will be **lognormal.** Taking the logarithms of the above equation and applying the Central Limit Theorem shows this. The service time is the sum of the logarithms, i.e., lognormally distributed.

For Weibull plots that curve down, analysts have argued the merits of the **three parameter Weibull with** t_0 **versus the lognormal** for years. Usually the Weibull lower tail is more conservative. The distribution option in SSW, used with Figure 3-4, is a quick and easy statistical method for selecting the best distribution for life data. The lognormal is a simpler distribution, two parameters instead of three. However, for the same data, the three parameter Weibull lower tail, say B.1 life, will usually be less than the lognormal B.1 life. This makes the three parameter Weibull slightly more conservative in the lower tail than the lognormal. However, the three parameter Weibull requires more data to obtain a good fit than the two parameter lognormal. [Meeker] says a minimum of one hundred failures are needed to fit the three parameter Weibull.

Greater importance should be placed on **Weibull libraries** that allow Weibayes applications instead of the less accurate Weibull, particularly for small sample analysis and sudden death tests.

The **Crow-AMSAA model** is good for tracking "non-homogenous Poisson" processes (those processes that are changing). It handles some kinds of the dirty data very well; e.g., if the failure modes are irrevocably mixed together or major portions of data are missing. It is **best practice** for tracking (1) reliability growth for development of a component, and (2) significant events for management, and lastly for tracking (3) fleets of repairable systems as a process. It is also useful for tracking warranty claims by calendar month. It is easy to do, and is recommended, in addition to the more accurate Weibull analysis.

The **Kaplan-Meier** survival model is distribution free, so it provides a way of avoiding distribution arguments. It is also excellent for handling large data sets with large numbers of suspensions. It may be useful for tracking warranty data by age and for analyzing inspection data. It is the **best practice** for snapshot data. (Treated in Section 8.5.) It may illuminate problems to identify the chronological test order and part serial numbers on the Weibull plot. The numbers should appear to be in random order.

When your organization has established a library of Weibull, lognormals, and Crow-AMSAA studies, it will be time to consider the usefulness of a **Monte Carlo simulation** of your product. This handbook touches this subject lightly in Chapter 4, but in aerospace these models are the foundation for all of RMS engineering including risk analysis, supportability, spare parts forecasting, support costs, warranties and guarantees. [ASAC, Bucci, Crosby, Law, Pritsker, RAPTOR, Redman, Saunders, Tocher, Trimble]

10.5 Substantiation and Accelerated Testing

Weibayes test planning is **best practice**. Extending the required test length along with allowing more failures can reduce costs. Sudden Death should also be considered with Weibayes. See Chapter 6.

10.6 Confidence Intervals

Monte Carlo pivotal bounds are **best practice** but you may use other methods for large samples to save computer time. If MLE-RBA is used, likelihood ratio confidence bounds are recommended. See Chapter 7.

10.7 Presentations and Reports

No matter how complex your analysis becomes, simplify the presentation as much as possible. Your objective is to communicate with decision-makers. Remove labels, legends, and confidence intervals from your plot unless they are absolutely required. Managers are not dumb, but they are often rushed. They reward those that provide accurate, quantitative recommendations in clear, concise language. Avoid mathematical verbiage. Of course, there are exceptions. You may have a manager with a strong statistical background, but they are rare.

10.8 Logic Diagram - Flowchart

The author tried several times to construct a logic diagram for the material in this chapter without success. However, Bob Rock of PACCAR has succeeded and presented his solution at the SAE 2000 Weibull User Conference. His flowchart is reproduced on the pages that follow. We are indebted to Bob for this valuable contribution and hope you will find it useful.

10.9 The End

The author congratulates you for persisting to the end. If you have questions or comments about the material in this Handbook, please feel free to contact Wes Fulton or Carl Tarum. Our contact information is listed on the beginning pages.

10.1 Best Practice Flow Chart

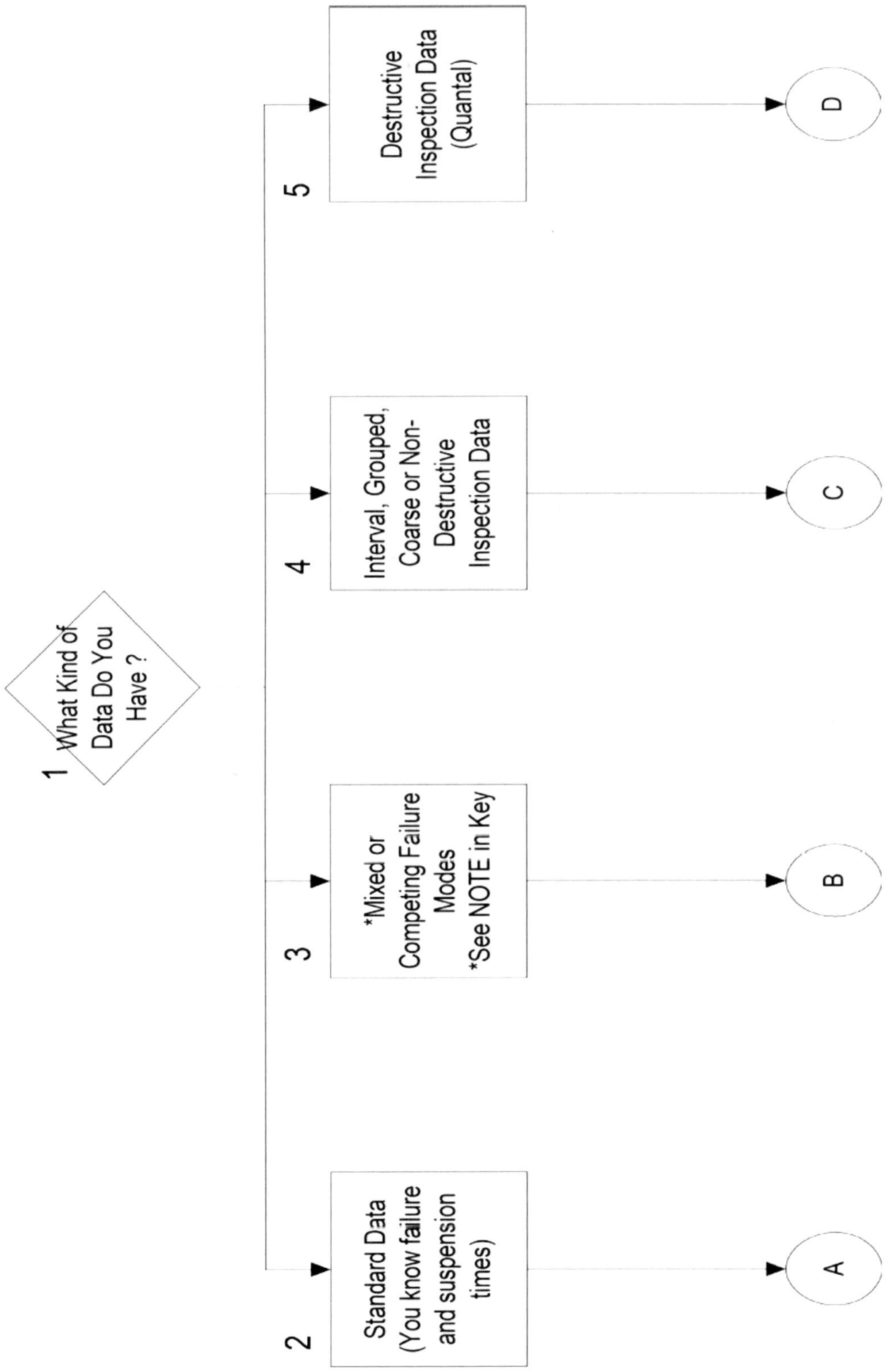

Figure 10-5. Flowchart for Selecting "**Best Practice**" for Life Data Analysis

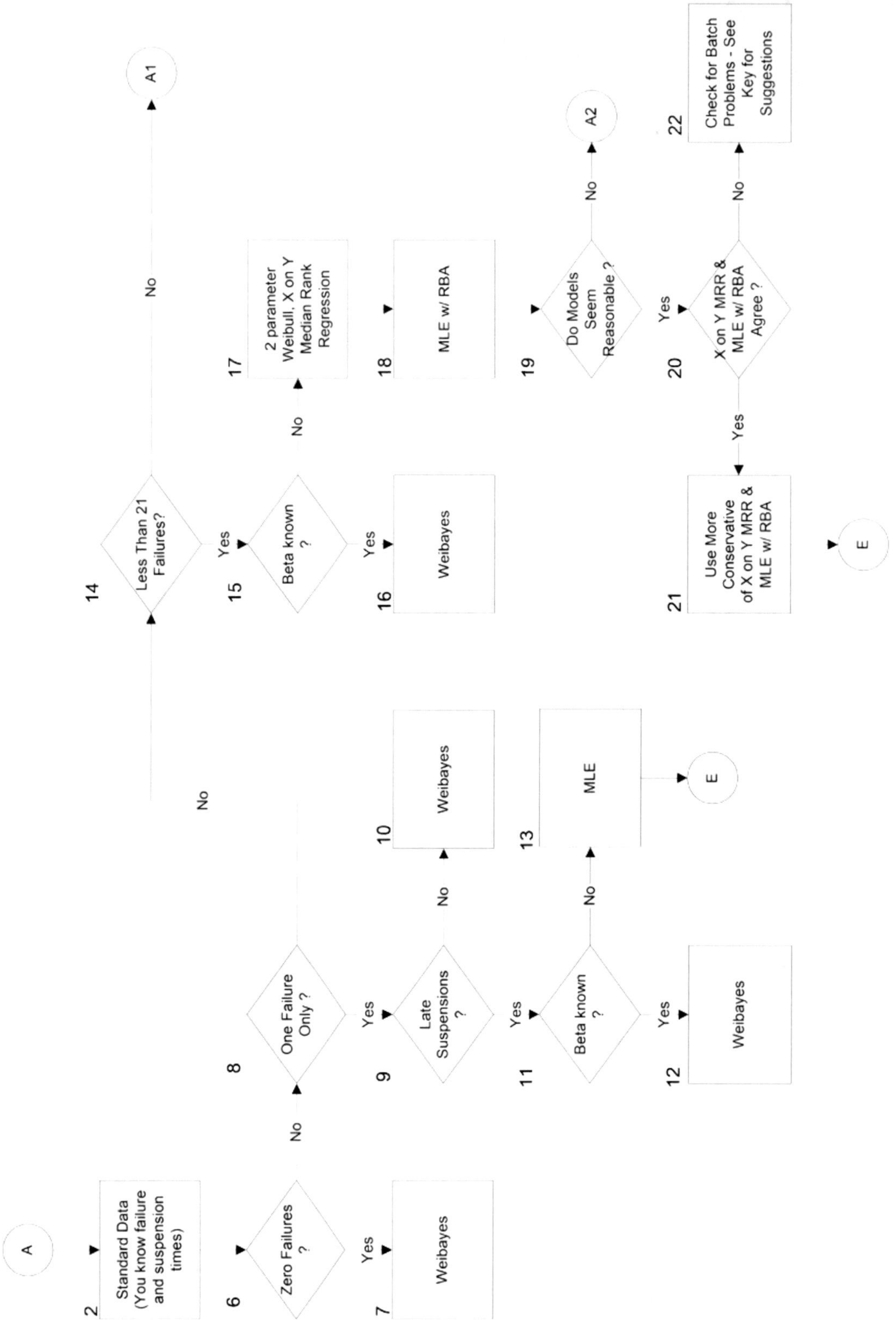

Figure 10-5. Flowchart for Selecting "**Best Practice**" for Life data Analysis (Cont'd)

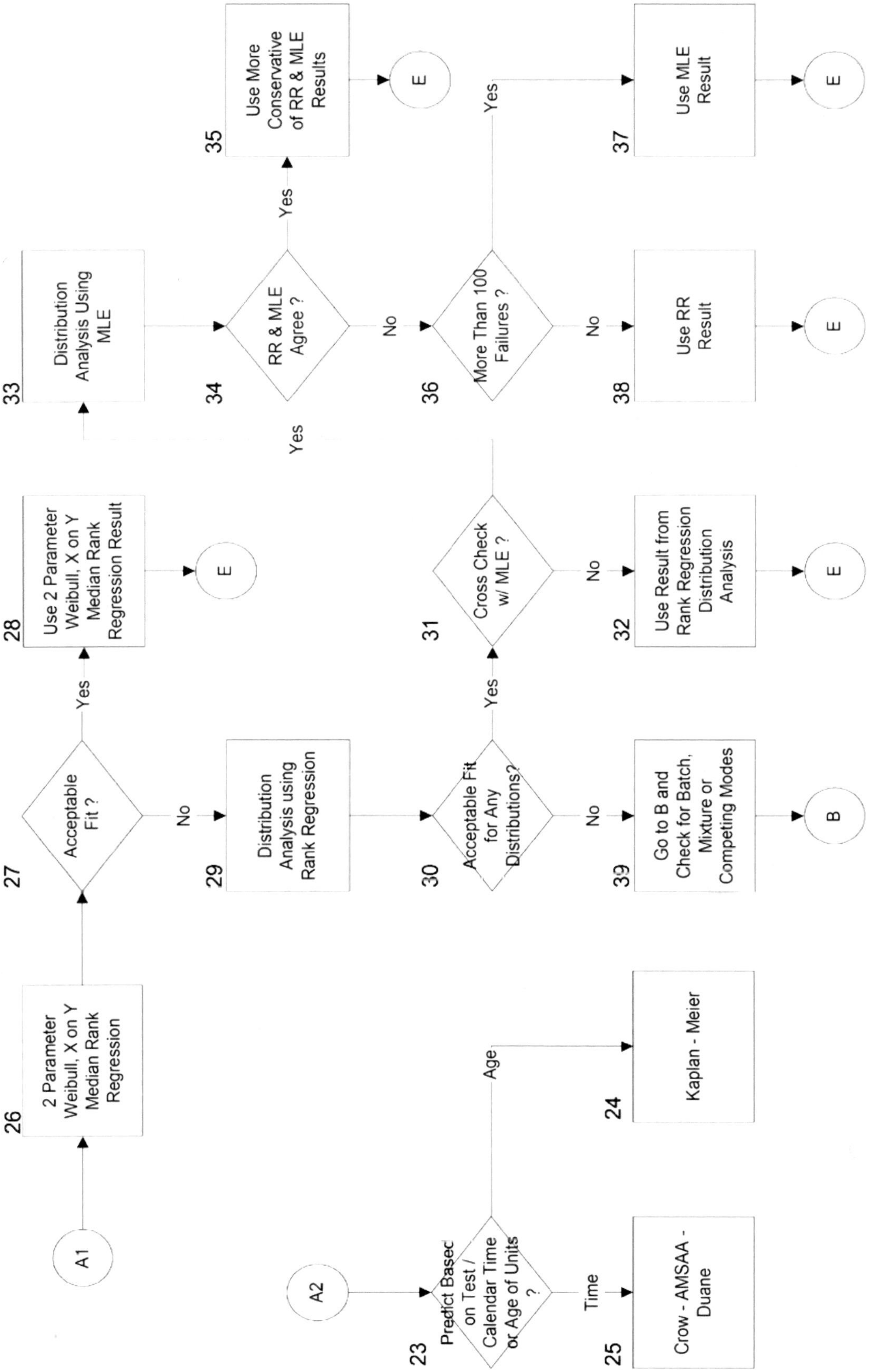

Figure 10-5. Flowchart for Selecting **"Best Practice"** for Life Data Analysis (Cont'd)

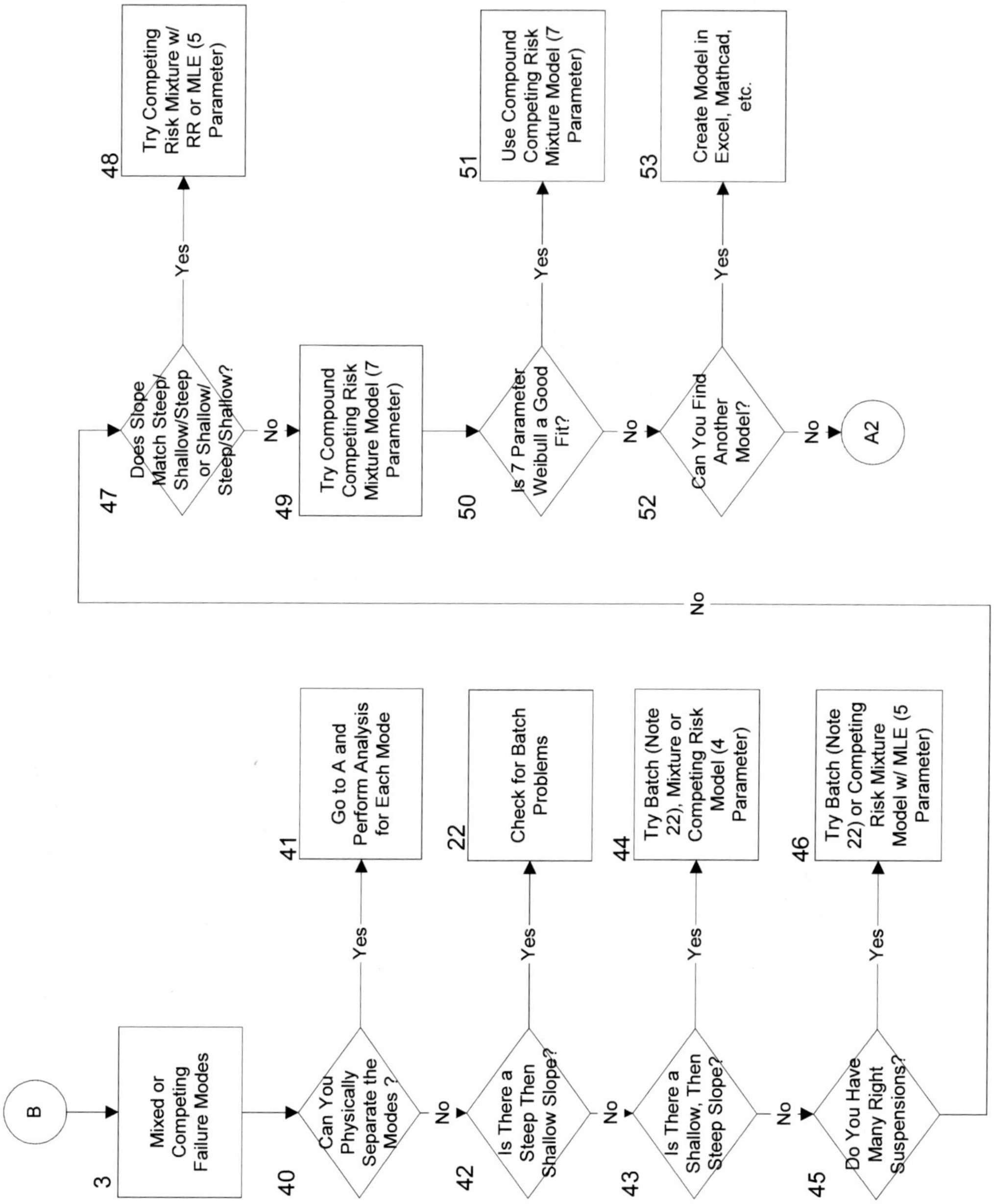

Figure 10-5: Flowchart for Selecting "**Best Practice**" for Life Data Analysis (Cont'd)

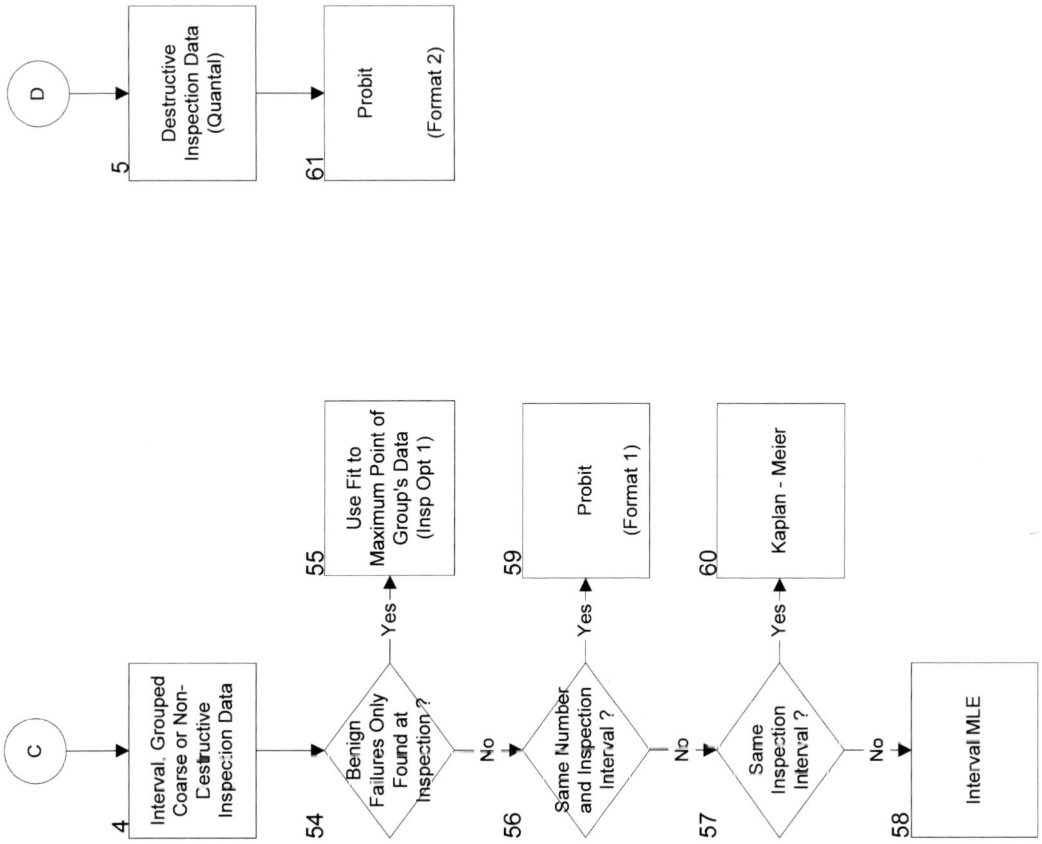

Figure 10-5: Flowchart for Selecting "**Best Practice**" for Life Data Analysis (Cont'd)

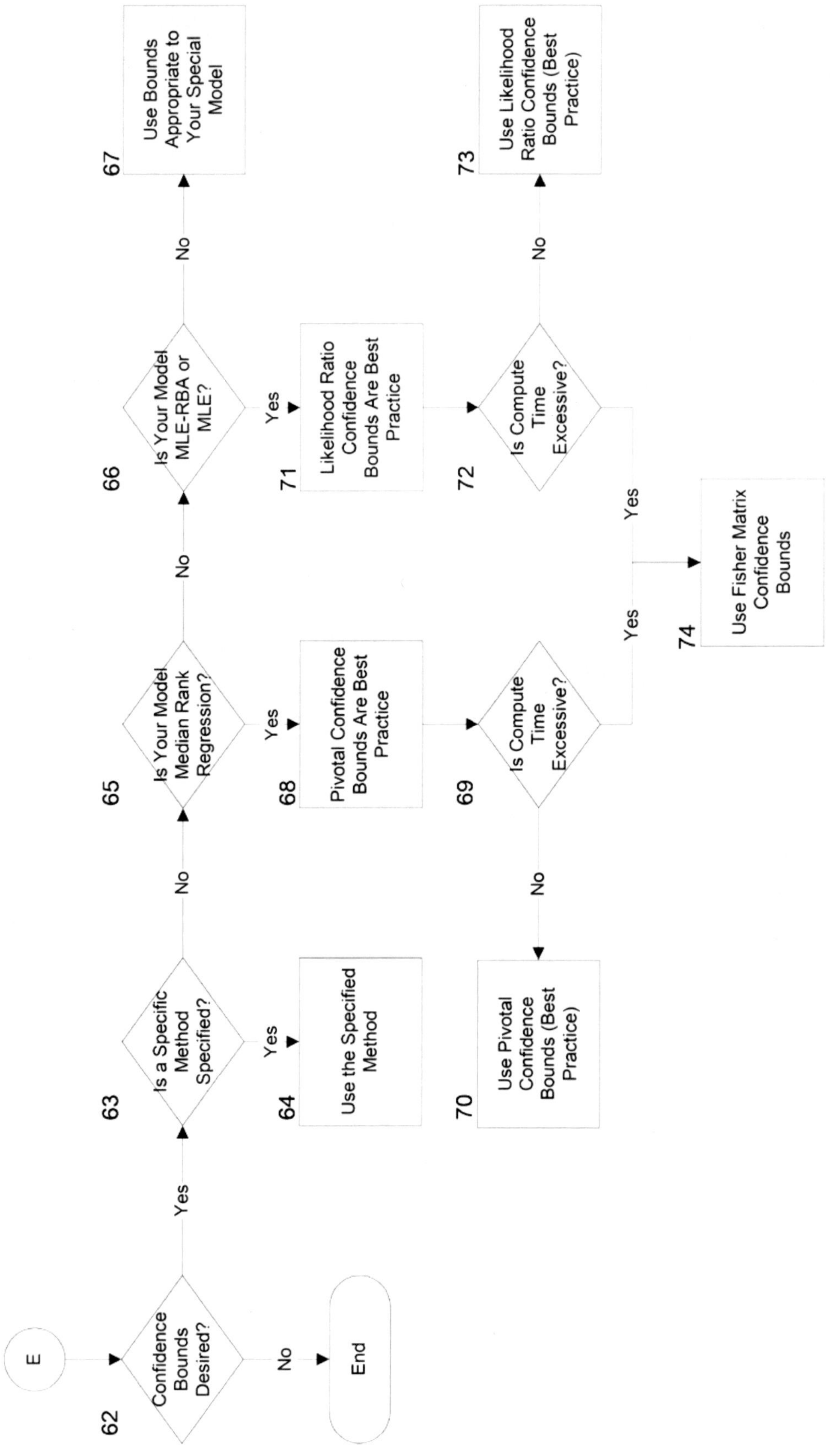

Figure 10-5: Flowchart for Selecting "**Best Practice**" for Life Data Analysis (Cont'd)

Key to Flowchart for Selecting "Best Practice" for Life Data Analysis

NOTE: This key does not represent a sequential flow of steps. You must use the flowchart to follow the logical flow and branching through the choice of analysis techniques. Use the flowchart as the primary document. Refer to this key when you want more detailed discussion of one of the numbered flowchart items.

Main Branch Starts Here

1. "What Kind of Data Do You Have?" – Understanding the data that you have, or are able to get, is the most important first step in finding the best practice for your analysis. Your data most likely can be classified into one of the four data types described in 2, 3, 4 and 5 below. (For discussion & diagrams of these data types see Ref 1 & Ref 2.)

2. "Standard Data" – You have standard data if you know all of your failure times and all of your suspension times. Most controlled test data is of this type because you know how long each sample has run and whether it has failed or survived. Warranty data may also fall into this group if you know all failure times and are able to make a good estimate of your suspension times. (Go to Branch A.)

3. "Batch, Mixed or Competing Failure Modes" – (* You usually do not know you have this type of data until you see evidence of it from plots and preliminary analysis. Therefore start with the analysis for "Standard Data" and allow the flowchart to guide you to this section as necessary.) This is data where you may have multiple failure modes included. You know failure and suspension times but you do not know if all the failures are the same mode. Warranty data may fall into this group if your claims show that failures have occurred but modes of failure are not known. (Go to Branch B.)

4. "Interval, Grouped, Coarse or Non-Destructive Inspection Data" – This is data where inspections are made at multiple times and the failures are found only at these inspections. The exact failure time is not known – only the interval in which the failure occurred is known. Warranty data can fall into this category if it is handled in monthly counts of failures without exact failure and suspension times. (Go to Branch C.)

5. "Quantal or Destructive Inspection Data" – This is data where each piece is inspected only once. You know only a component failed when inspected or that failure would have occurred after the inspection. Each observation is either censored or failed. Testing of bombs and missiles produces this type of data. When tested the missile either works or does not work. (Go to Branch D.)

Branch A Starts Here

6. Do you have zero failures in your data?

7. When you have no failures the only analysis you can do is Weibayes. You must know the number of units you have and how long they have run. You must make a slope assumption to use Weibayes. You can choose a slope based on previous experience with expected failure modes, data from your Weibull library or from knowledge based on physics of failure. This approach is often used for zero failure testing to see if a known mode of failure has been eliminated.

8. Do you have exactly one failure?

9. Do you have late suspensions? Late suspensions are units that have not failed that have already run longer than the failed units.

10. If you have one failure and no late suspensions use Weibayes. You must know the number of units you have and how long they have run. You must make a slope assumption to use Weibayes. You can choose a slope based on previous experience with expected failure modes, data from your Weibull library or from knowledge based on physics of failure.

11. If you have late suspensions do you have good knowledge of slope?

12. If you feel confident in your knowledge of slope use the Weibayes as described in item 10 above. You must make a slope assumption to use Weibayes. You can choose a slope based on previous experience with expected failure modes, data from your Weibull library or from knowledge based on physics of failure.
13. If you do not feel confident in your knowledge of slope use MLE.
14. Do you have more than one failure but less than 21 failures?
15. Do you have good knowledge of slope? Good knowledge of the slope could be knowledge based on previous experience with expected failure modes, data from your Weibull library or from physics of failure.
16. If you are satisfied with your knowledge of slope based on previous experience with expected failure modes, data from your Weibull library or from knowledge based on physics of failure you can use Weibayes.
17. If you are not satisfied with your knowledge of slope fit a 2 parameter Weibull with X on Y median rank regression.
18. Fit a Maximum Likelihood Estimate (MLE) model using the Reduced Bias Adjustment (RBA).
19. Do the models fit in 17 and 18 seem reasonable based on your knowledge of the failures?
20. Are the models in 17 and 18 in reasonable agreement?
21. If the models in 17 and 18 are in reasonable agreement use the more conservative of the two models.
22. If the models in 17 and 18 are not in reasonable agreement consider the possibility of batch problems. There are several indicators to look for: 1) Is the lower confidence bound for present risk above the number of observed failures? 2) Is the MLE slope less than the Median rank Regression slope? 3) Does the Aggregate Cumulative Hazard indicate a batch effect?

BRANCH A2 STARTS HERE

23. Do you want your life prediction to be based on test / calendar time or age of the units?

NOTE: Methods 24 & 25 are particularly useful for warranty data since warranty data often suffer from mixtures.

24. If you want to predict life based on the age of the units use Kaplan-Meier.
25. If you want predict life based on Test or calendar time use Crow-AMSAA.

BRANCH A1 STARTS HERE

26. Fit a 2 parameter Weibull with X on Y median rank regression.
27. Is the fit acceptable for the 2 parameter Weibull with X on Y median rank regression? Consider the Critical Correlation Coefficient, Critical Coefficient of Determination and pve%. (See section 3.2.1 in [Ref 1])
28. Use the results from the 2 parameter Weibull with X on Y median rank regression.
29. You have more than twenty failures. Perform a distribution analysis to check for an appropriate distribution.
30. Do you have an acceptable fit for any distribution? Consider the Critical Correlation Coefficient, Critical Coefficient of Determination and pve%. (See section 3.2.1 in [Ref 1])
31. Do you want to cross check your rank regression result with MLE?
32. Use rank regression result.
33. Perform MLE distribution analysis to check for an appropriate distribution.
34. Do the rank regression and MLE results agree?
35. If the rank regression and MLE results agree use the more conservative of the two models.
36. If the rank regression and MLE results do not agree do you have more than 100 failures?

37. If you have more than 100 failures use the MLE results.

38. If you have less than 100 failures use the rank regression results.

39. If the fit is not acceptable check for mixture of failure modes.

BRANCH B STARTS HERE

Before starting this branch it is helpful to review Appendix J in The New Weibull Handbook [Ref 1] for analysis of mixtures. The SS software HELP file has some tips for estimating various models. It may also be helpful to review Appendix F in [Ref 1] for using the Aggregate Cumulative Hazard (ACH) model.

40. Can you physically separate the failure modes in the mixture? This requires failure analysis to be performed on the failed parts. Does failure analysis show that two modes of failure are present? For example, does one group fail in shear and another in fatigue?

41. If you are able to physically separate the failure modes go to 2 and perform the analysis for each of the modes of failure. Treat failures of other modes as suspensions in the analysis. For example, when analyzing failure mode 1 treat failure mode 2 as suspensions, and vice versa.

42. Do you have a Steep/Shallow slope pattern? If so, go to note 22 and check for batch problems.

43. Do you have a Shallow/Steep slope pattern. The first part of the data looks like a good Weibull fit, but then the slope increases. This is not a curve like a 3 parameter, but a distinct slope change.

44. This may be a batch problem (see note 22) or competing risk mixture. Try the methods in note 22, the mixture icon in SSW software or use SSY and a 4-parameter Weibull Competing Risk model. The parts may have 2 failure modes on a homogeneous population like a tire puncture and wear out.

45. Do you have many Right-Tail suspensions, far beyond the last failure mode?

46. This is often a batch problem so try the methods in note 22. Try using a Weibull Competing Risk Mixture (5 parameter) model in YBath. Use MLE model with suspensions.

47. Do you have a Steep/Shallow/Steep or Shallow/Steep/Shallow slope pattern? Depending on the relationship of the 2 sub-populations, a mixture may have one of these shapes. The two steep or two shallow slopes may not be similar slopes. Also be aware that you may not see all of the slope changes due to suspensions or sample size.

48. Try using a Weibull Competing Risk Mixture (5 parameter) using rank regression or MLE.

49. Try using a Weibull Compound Competing Risk Mixture model (7 Parameter). This typically has a shallow/steep/shallow/steep slope. This fits a population that has infant mortality, chance, and wear out failures.

50. Is the Weibull Compound Competing Risk Mixture model (7 Parameter) a good fit?

51. Use the Weibull Compound Competing Risk Mixture model (7 Parameter) results.

52. Can you find some other model that might fit the data? If the model is more complex, you may need to set up a model in Excel or another numerical analysis software.

53. Most mixtures can be solved for Y on X analysis in Excel, using the Solver Add-In. MLE can also be done using derivatives. X on Y rank regression is more difficult; requiring a macro to solve for each X so the error can be calculated.

BRANCH C STARTS HERE

54. Do you have benign failures that are only found at inspections?

55. When you have benign failures found only at inspection fit a model to the maximum point of the interval's data. (See Inspection Option 1 in [Ref 1].)

56. Do you use the same inspection interval and inspect the same number of units for all inspections?

57. If you do not inspect the same number at each inspection do you use the same inspection interval for all inspections?

58. Use the interval MLE method.

59. Use Probit (Format 1).

60. Use Kaplan-Meier.

BRANCH D STARTS HERE

61. Use Probit (Format 2).

BRANCH E STARTS HERE

62. Do you want confidence bounds on your model?

63. Has a specific type of confidence bound been specified by your customer or management?

64. Use the method specified by your customer or management.

65. Is your model a rank regression model?

66. Is your model a MLE model?

67. Use bounds appropriate to your special model.

68. Pivotal confidence bounds are the best practice for your model.

69. If you have a large set of data, compute time may become excessive and you may be willing to use a somewhat less than "Best Practice" for your confidence bounds.

70. Use pivotal confidence bounds as best practice.

71. Likelihood ratio confidence bounds are the best practice for your model.

72. If you have a large set of data, compute time may become excessive and you may be willing to use a somewhat less than "Best Practice" for your confidence bounds.

73. Use likelihood ratio confidence bounds as best practice.

74. Use Fisher matrix confidence bounds.

REFERENCES

1. Abernethy, Robert B., ***"The New Weibull Handbook,"*** 4th edition, 2000. (*Actually the 5th edition if the "**USAF Weibull Analysis Handbook"** is considered the original.)*

2. Nelson, W., "Applied Life Data Analysis", Wiley & Sons, 1982.

CHAPTER 11 - CASE STUDIES AND NEW APPLICATIONS

11.1 Foreword

The objective of this chapter is to illustrate case studies and new applications. The author is indebted to his students and clients for their fine contributions to this chapter. Each new Edition has additional new case studies. You will find examples from many industries and applications. The reader may want to "hop, skip and jump." A list of the case studies may be helpful for this selection:

- **11.2 Stress Corrosion Failure Forecasting**. Key words: risk, failure forecast, renewal, replacement, maintainability rhythm, spare parts
- **11.3 Optimal Component Replacement - Voltage Regulators**. Key Words: optimal replacement, block replacement, voltage regulators, unplanned and planned costs, machine tools, nuclear refueling
- **11.4 Locomotive Power Units Overhaul Life**. Key Words: Block replacement, classic Bi-Weibull, failure forecasting
- **11.5 Cost Effective Calibration Intervals**. Key Words: Calibration, Weibull, Cost Effective
- **11.6 FPL Turbogenerator Failure**. Key Words: One failure Weibull, MLE, failure forecasting
- **11.7 TVA Bull Run Fossil Plant - Controller Cards**. Key Words: Truncated data, t_0, risk analysis, electronic failures
- **11.8 Repairable Systems Reliability Growth Assessment**. Key Words: Failure forecast, repairable systems, Crow-AMSAA
- **11.9 Front Jounce Bumpers**. Key Words: Three parameter Weibull, tracking development progress, "Covered" Weibulls
- **11.10 Transfer Case Seal**. Key Words: Kaplan-Meier, new design versus old design
- **11.11 Dental Acrylic Adhesive Fatigue**. Key Words: Distribution Analysis, Comparison of Data Sets, Outliers
- **11.12 Crow-AMSAA Reliability Modeling**. Key Words: Crow-AMSAA, Cost Analysis, Cost Savings, Process Industry
- **11.13 Weibull Analysis of Boiler Tube Failures**. Key Words: Electric Utility, fatigue, cracks, failure forecast, unplanned shutdowns, corrective action
- **11.14 Gas Turbine Seal Failures - A Batch Problem**. Key Words: Batch, Failure Forecast, Now Risk, Aggregate Cumulative Hazard, Crow-AMSAA
- **11.15 NASA Challenger Space Shuttle Weibull**. Key Words: Temperature Weibull, O-Ring Failures

11.2 Stress Corrosion Failure Forecasting

In case the reader would like to study and analyze this case further, here is the data. The histogram of suspensions is approximately normal with a mean of 1000. Data set notation: "23x4" means 4 failures at time 23, ">16x4" means 4 suspensions at time 16. If SSW is used the data set listed by Figure 11-1 may be entered directly.

Stress corrosion cracks develop in many components. Boiler-steam generator tubes, gears, and turbine blades are examples. As described in Chapter 4, the first step in failure forecasting is to develop the Weibull plot as shown in Figure 11-1. The aging parameter is operating time in hours. This data is simulated, but typical of stress corrosion data. The SSW failure forecast is shown in Table 11-1, based on a usage rate of 200 hours per month, without renewals. All of the 1025 parts are forecast to fail within about 30 months. This forecast is plotted in the lower left hand corner, Figure 11-2. Note the replacement or renewal forecast is also shown. Renewal implies a constant fleet size. The difference between the two risk projections is the predicted failure of the replacement parts.

>475x1	>1075x121
>575x1	>1125x99
>625x7	>1175x59
>675x16	>1225x51
>725x23	>1275x34
>775x50	>1325x11
>825x50	>1375x11
>875x91	>1425x2
>925x115	4200x1
>975x144	4500x1
>1025x139	6000x1

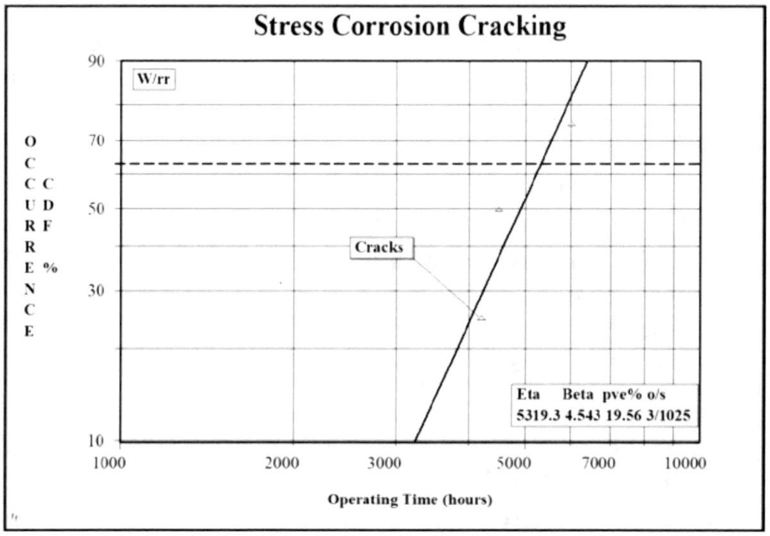

Figure 11-1. Stress Corrosion Cracking

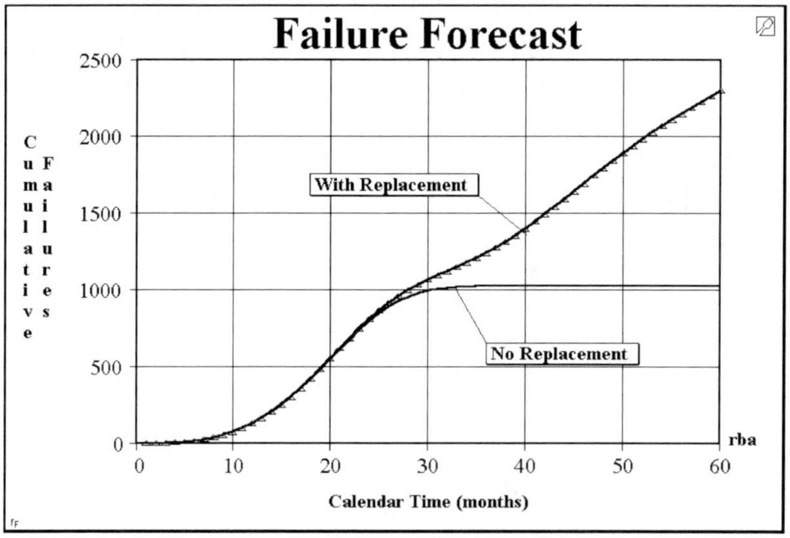

Figure 11-2. Failure Forecast

Table 11-1. Failure Forecast Without Replacement

Abernethy Risk (Standard Forecast) Quantity Expected Weibull
Stress Corrosion Cracking, Set#1 Cracks D04-M12-YR2024
Present (Now) Risk=3.599472 Occurrence /Suspensions = 3/1025
Eta=5319.299 Beta=4.54263
Usage [Additional X-Value Age (hours) Each Month / Item] = 200
No Renewal
Expected Additional Occur.(Cumulative.): Next Event [Day]=+25

Month	Risk	Month	Risk	Month	Risk	Month	Risk
1	1.235	16	299.9	31	1005	46	1025
2	2.525	17	356.1	32	1013	47	1025
3	4.624	18	416.6	33	1019	48	1025
4	7.845	19	480.1	34	1022	49	1025
5	12.56	20	545.4	35	1024	50	1025
6	19.19	21	610.9	36	1025	51	1025
7	28.23	22	674.9	37	1025	52	1025
8	40.19	23	735.9	38	1025	53	1025
9	55.63	24	792.3	39	1025	54	1025
10	75.1	25	842.8	40	1025	55	1025
11	99.13	26	886.6	41	1025	56	1025
12	128.2	27	923.3	42	1025	57	1025
13	162.7	28	952.9	43	1025	58	1025
14	202.8	29	975.9	44	1025	59	1025
15	248.6	30	992.9	45	1025	60	1025

Careful examination of the upper curve will show a slight wave which crests twice. A better method for depicting this wave is to plot the failures per month, the failure rate, as shown in Figure 11-4. The crests are the peak failure rate of the original and the first replacement parts. If the forecast is extended further out in time, the wave action will dampen and the failure rate will approach $1/\eta$. See [Denson 1991]. The time for the failure rate to reach the asymptotic value is a function of β, the steeper the slope, the longer the wave persists. For maintainability and logistics planning, spare parts forecasts, and labor requirements, knowledge of this maintainability rhythm is required. It is always a good idea to plot the failure rate in addition to the Weibull and the cumulative risk.

Note that the first wave should approximate the probability density function (PDF). Compare the first wave with the PDF shown on the previous plot, Figure 11-3. This plot is valuable to show the responsible engineer how close his fleet is to the Weibull. This plot indicates how much lead time he has for a recall or retrofit program if the failure consequences are severe. The intersection of the suspension fleet and the PDF produce failures. All the plots were done with the SuperSMITH software. In the past this more complete analysis required gross hand calculations and hand plotting, so it was rarely done. Today, with the new software, all the effort is obtaining and validating the data.

11.3 Optimal Component Replacement - Voltage Regulators

(Key Words: optimal replacement, block replacement, voltage regulators, unplanned and planned costs, machine tools, nuclear refueling)

Data Set: 17x2 18x4 20x3 21x5 22x6 23x3 24x2 25x5 26x2 27x15 28x2 29x1-10x100 -11x100 -12x100 -13x100 -14x100 -15x100 -16x100 -17x100 -18x100 -19x100 -20x100.

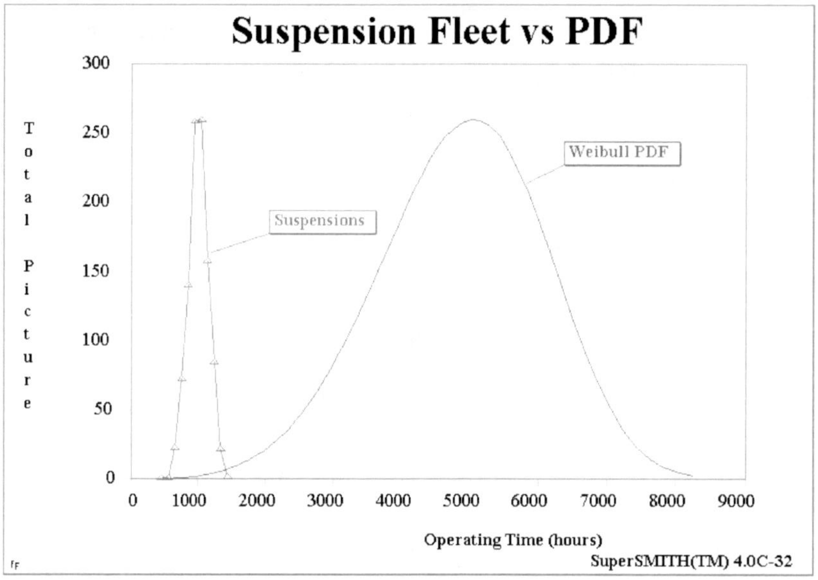

Figure 11-3. Suspension Fleet vs. PDF

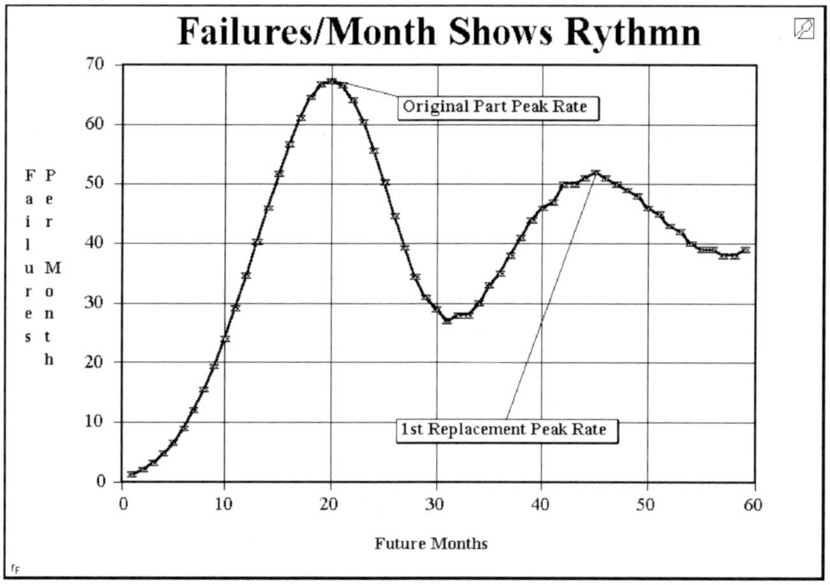

Figure 11-4. Failures/Month Shows Rhythm

There is an optimal replacement interval for parts that have a Weibull wear out failure mode and unscheduled failure costs higher than planned replacement costs. This is a common situation in many industries.

Electric utilities use voltage regulators by the thousands. Figure 11-5 depicts electricity transmission cable towers. Unplanned failures of voltage regulators are often costly. The failure data in Figure 11-6 is real, but the histogram of suspensions was assumed, 1100 units uniformly distributed from 10 to 20 years. As this is obviously interval inspection data from the vertical columns of points, the inspection option gave a better fit. The replacement failure forecast is plotted in Figure 11-7.

Figure 11-5. Electricity Transmission Cable Towers

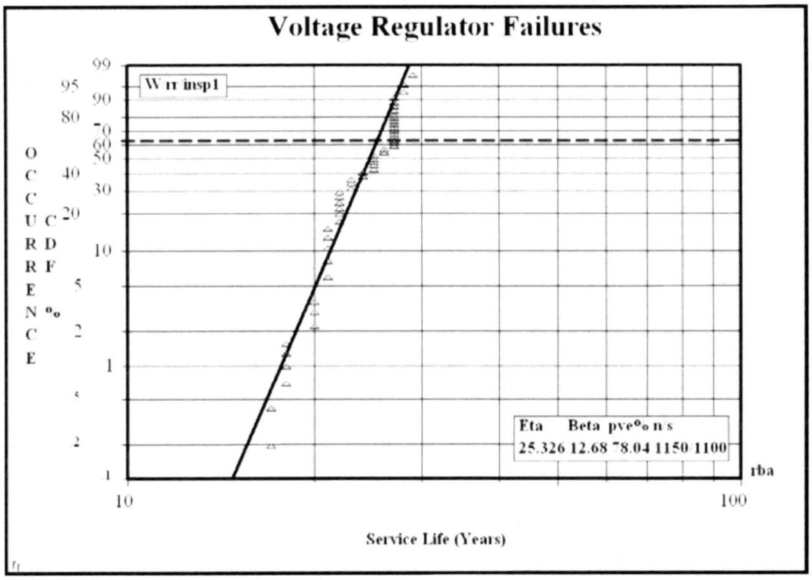

Figure 11-6. Without the Inspection Option pve% = 10.3

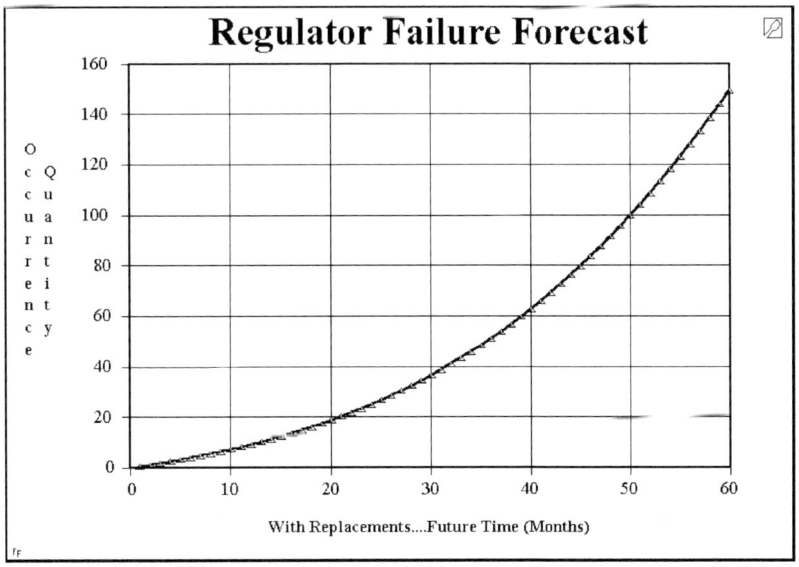

Figure 11-7. Regulator Failure Forecast

Table 11-2. Optimal Replacement for Minimum Cost

Weibull Optimum Replacement Interval 06-09-1996
Voltage Regulator Failures

Eta = 25.322647 Beta = 12.67963
Cost Each ($) Planned = 10 Cost Each ($) Not Planned = 200

Cost/Time [$/(Years)]:

(Years)........Cost		(Years)........Cost		(Years)........Cost		(Years)........Cost	
.4167	24	6.667	1.5	12.92	.7771	19.17	.809
.8333	12	7.083	1.412	13.33	.7542	19.58	.8785
1.25	8	7.5	1.333	13.75	.7333	20	.9682
1.667	6	7.917	1.263	14.17	.7144	20.42	1.082
2.083	4.8	8.333	1.2	14.58	.6977	20.83	1.224
2.5	4	8.75	1.143	15	.6833	21.25	1.399
2.917	3.429	9.167	1.091	15.42	.6715	21.67	1.611
3.333	3	9.583	1.044	15.83	.6628	22.08	1.867
3.75	2.667	10	1	16.25	.6576	22.5	2.17
4.167	2.4	10.42	.9602	16.67	.6567	22.92	2.524
4.583	2.182	10.83	.9234	17.08	.661	23.33	2.932
5	2	11.25	.8895	17.5	.6715	23.75	3.392
5.417	1.846	11.67	.858	17.92	.6897	24.17	3.901
5.833	1.714	12.08	.8289	18.33	.7173	24.58	4.449
6.25	1.6	12.5	.882	18.75	.7562	25	5.025

The optimal replacement interval is about 16.5 years as shown by the total cost curve in Figure 11-8 and the Table 11-2. The unplanned and planned cost curves were also generated with SSW by using a zero cost for one, and then the other cost variable, and plotting the results with SuperSMITH Visual.

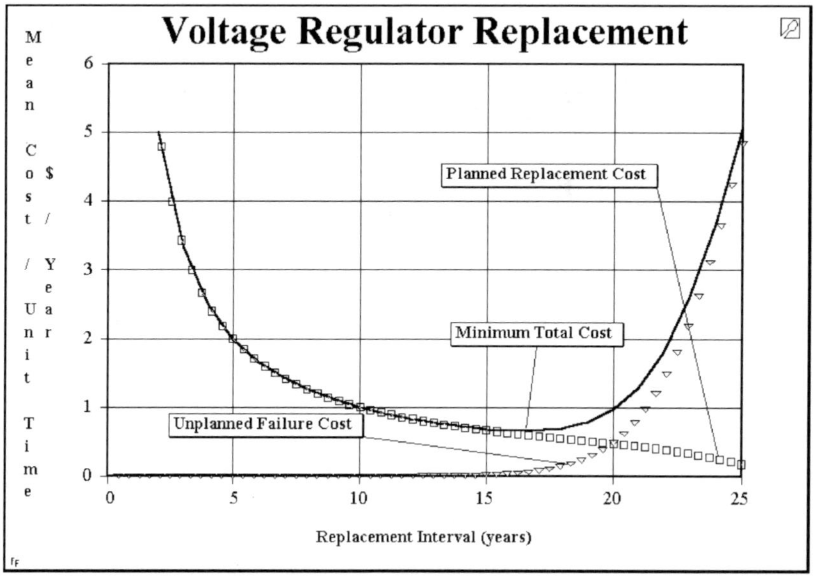

Figure 11-8. Voltage Regulator Replacement

If a part is replaced at fixed intervals regardless of its age, it is called block replacement. It is appropriate where records are not kept or maintenance can only be done at certain times. The cost per unit time is higher but there are no record keeping costs. The methods are presented in Chapter 4. SSW provides both age and block replacement analysis. See [Billatos] for machine tool applications and [Stanton] for nuclear refueling block replacement. A Southern Pacific railroad example with block replacement will be illustrated next.

11.4 Locomotive Power Units Overhaul Life

Contributed by Mr. Robert J. Scanlon, Director Quality and Reliability Engineering, Southern Pacific Railroad

Key Words: Mixture of modes, block replacement, classic Bi-Weibull, Abernethy risk analysis. Data Set: 1x1 3x2 4x1 9x2 11x1 17x2 22x1 23x3 31x1 35x1 36x1 37x4 39x3 40x2 44x2 45x1 46x3 47x8 48x2 49x3 50x6 51x2 52x14 -9.5x15 -12.5x16 -52.5x76 -53.5x131.

Southern Pacific's high horsepower diesel electric locomotives had 16 power assemblies each. These power assemblies were one of the largest sources of unreliability. Failure of one of the units reduces available horsepower and eventually forces return of the locomotive to the shop. This event produces transportation, labor, locomotive rental, and spare parts costs. These are estimated at $10,000 per failure. All sixteen units are made good-as-new at overhaul which costs $56,000, $3500 per power assembly. This is essentially block replacement. The problem is to determine the most cost effective block replacement (overhaul) period based on total costs.

The Weibull is a classic Bi-Weibull. See Figure 11-9. Note that the β and η were estimated by Southern Pacific by simply splitting the data set as detailed failure reports were not available. (An improved failure data collection system was put in operation to solve and eliminate the infant mortality produced by the overhaul process.) The SSW Mixture capability did not exist when the original analysis was done. For comparison of the SSW results with the earlier results see Figure 11-10.

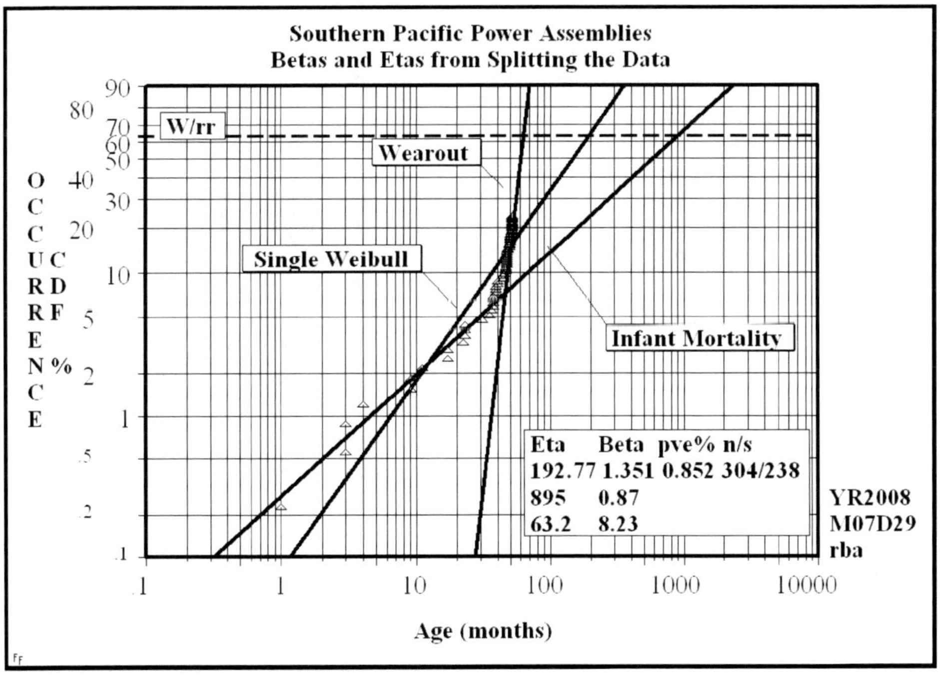

Figure 11-9. Southern Pacific Power Units BiWeibulls From Splitting the Data at the corner.

Update 2005: With improvements to SSW, it is now possible to do a risk forecast using the two Weibulls from the mixture analysis with and without replacement. Based on the 238 power assemblies operating (suspensions) in the data set, the Abernethy Risk for 12 months with renewal is 118 failures. This is reduced to 2.7 failures by overhaul-replacement at 46 months. Southern Pacific Railroad is now part of Union Pacific.

The calculation of the block overhaul period is complicated by the mixture of two modes. For example, if a unit fails by the first mode and is replaced, it is not exposed to the second mode. If the wear out mode fails a unit, the replacement may fail from the first mode, infant mortality. In other words, there is an interaction between the two failure modes. They are not independent. However, with the new capability in SSW accurate risk calculations are possible with the results of the BiWeibull mixture analysis.

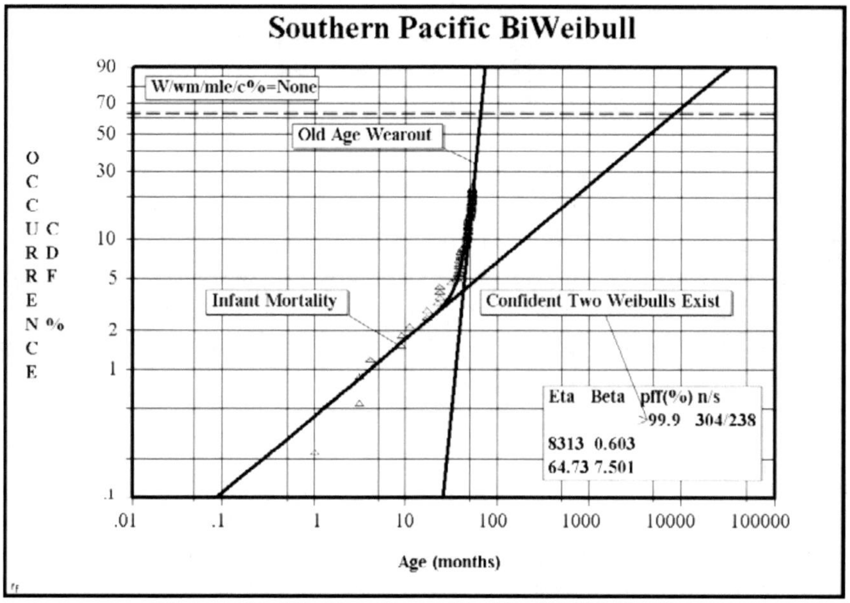

Figure 11-10. Southern Pacific Bi-Weibull

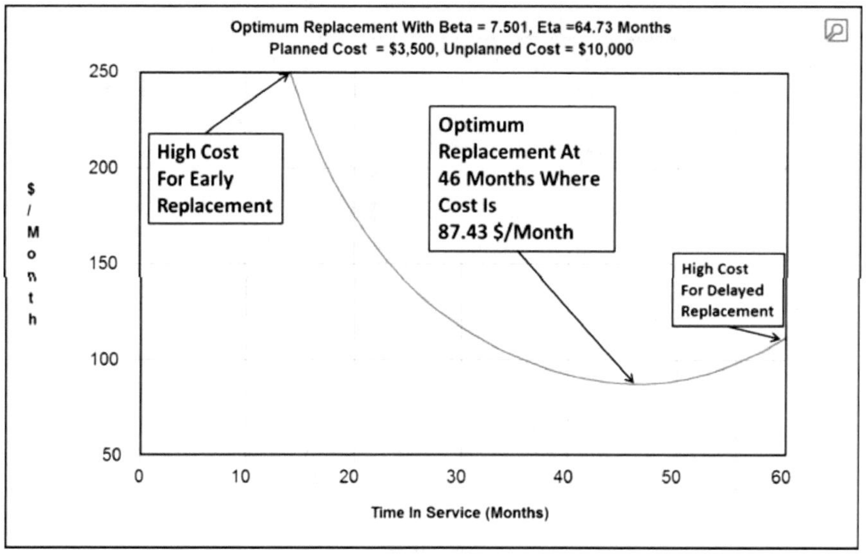

Figure 11-11. Locomotive Drive Motor Optimum Replacement For Minimum Cost

The SSW block replacement results are shown in Figure 11-11. If the infant mortality problem is solved by improved quality in the overhaul depot, the optimal overhaul period is 46 months for minimum costs.

11.5 Cost Effective Calibration Intervals

Key Words: Calibration, Weibull, Cost Effective

In the past decades, most organizations calibrated all instrumentation at fixed intervals, perhaps monthly or quarterly. In the mid-seventies some types of instrumentation were found to be much more stable than others [Greb 1979]. Some instrumentation should be recalibrated daily and other types yearly or longer.

If instrumentation that is out-of-calibration is identified when returned to the laboratory, a Weibull plot is ideal for determining optimal calibration intervals. We define failure as out-of-calibration and suspensions as in-calibration.

There are large costs associated with calibration. A large organization may have 30,000 instruments which may cost $50 each, on average to calibrate. If the calibration intervals are too short there will be large nonproductive charges, instrumentation availability will be lowered, maintenance labor will be increased. On the other hand, if the intervals are too long, defective products may be produced, scrap and rejects are increased. What is needed is just-in-time calibration. That requires an accurate method for forecasting out-of-calibration for each type of instrument. Weibull is ideal for this purpose. The objective is to have high probability that all operating instrumentation is in calibration at minimum cost.

For critical instrumentation, the B1 life might be employed as the calibration interval. This assures that 99% of the instrumentation in service is in-calibration. For less important instrumentation, B5 (95% in service in-calibration) or even B10 (90% in service in-calibration) is used. The Weibull in Figure 11-14 is based on ten failures (out-of-cal units) and 90 suspensions (in-cal units). The calibration intervals are read from the Weibull plot: (B1 = 21 months, B5 = 33, B10 = 40).

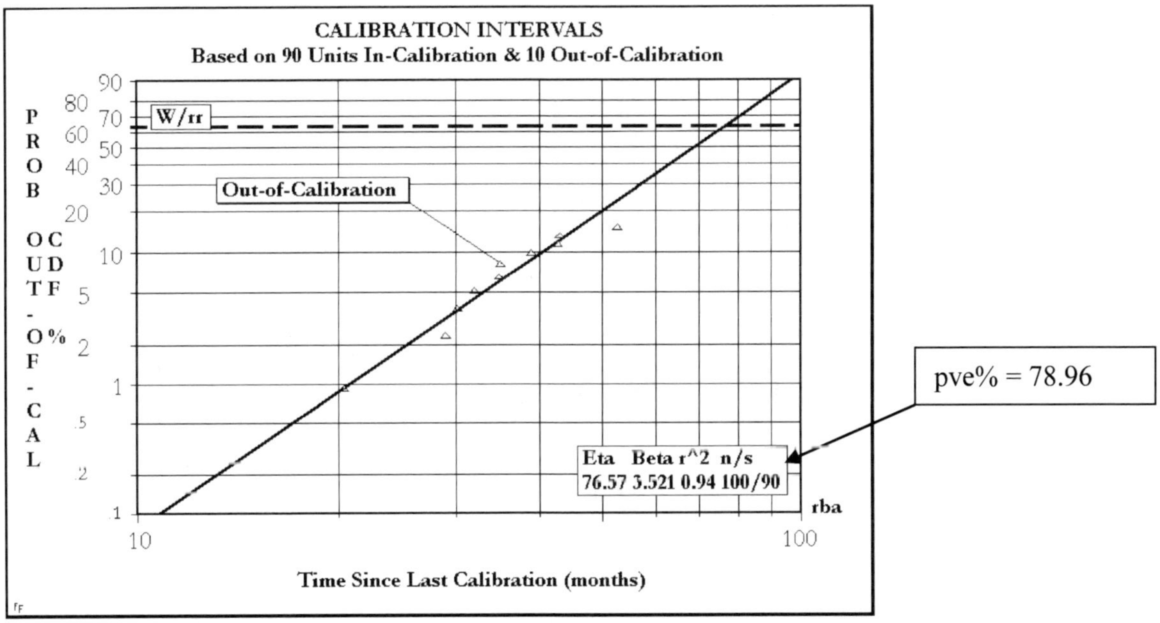

Figure 11-14. Calibration Intervals Based on 90 Units In-Calibration & 10 Out-of-Calibration

If the instrumentation is returned for recalibration at fixed intervals, the SSW inspection option should be employed because the data will pile-up at discrete intervals as mentioned above. (See Chapter 5). If beta is greater than one, a wear out mode, and costs can be estimated, there will be an optimal replacement interval. (See Chapter 4.)

All of the usual Weibull slope interpretations apply for calibration data. For example, shallow slopes show infant mortality, perhaps poor quality in the calibration process or installation problems. Slopes of one indicate random events are causing the out-of-cal failures. Random events may relate to abusive practices such as over pressure, over voltage, dropping the instrument, contamination, etc. Slopes greater than one imply wear out modes. One of the newest calibration interval Weibull methods employs the absolute differences between redundant instruments. The first few differences after the pair of instruments has been recalibrated are used to plot the Weibull. (Zero differences are omitted.) This is the Weibull for in-calibration instruments. If either instrument shifts out, the difference will increase significantly. The B95 difference, 95% cumulative failure, is recommended as a signal to recalibrate again. The 95% in-calibration would be responsive to both domestic and international standards such as ISO 10012-1-1992 and ISO DIS

5168. This procedure is easily computerized as part of the data reduction to automatically produce out-of-calibration signals when the B95 difference is exceeded. Figure 11-15 illustrates the graphical Weibull calibration technique for showing data trends plus the zone of gage inaccuracy and gage unreliability. Differences larger than allowed (110 microinches) fall into the zone of gage inaccuracy on the X-axis while values above a cumulative occurrence fall into a zone of gage unreliability on the Y-axis. In this case, a 10:1 gage ratio, a measurement tolerance of □0.0015 inches, and a requirement that 95% of all gages returned must be within tolerance at the end of the calibration period are combined with variables data spanning a six year period. The results justified extending micrometer standard bar recalibration periods far beyond the current one year calibration period which provided substantial cost reductions.

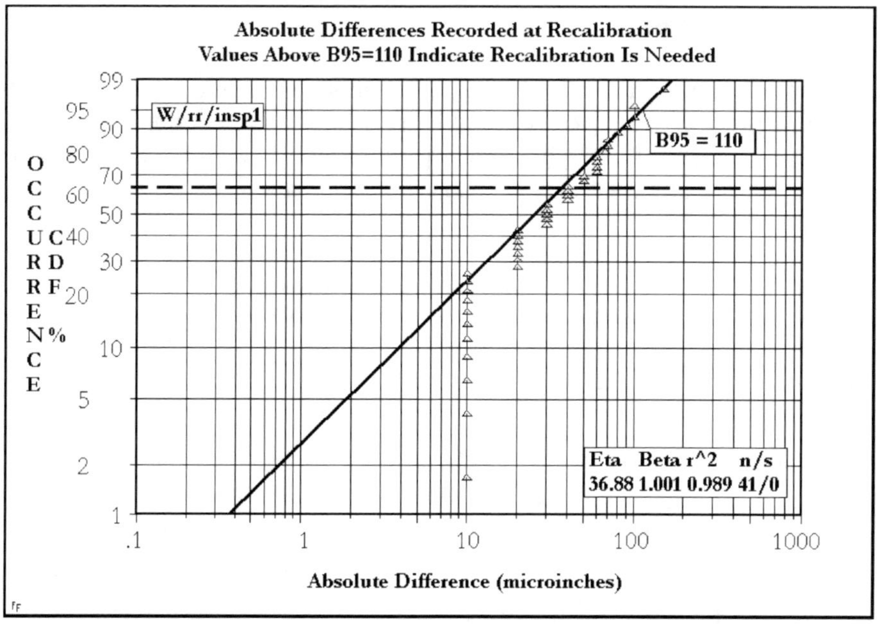

Figure 11-15. Absolute Differences Recorded at Recalibration Values Above B95 = 110 Indicate Recalibration Is Needed pve% = 99.9

11.6 FPL Turbogenerator Failure

Contributed by Jim Keener, Mel Thomas and Don Paxson of Florida Power & Light.

Key Words: One Failure, MLE, Risk Analysis

At the Florida Power & Light (FPL) Indiantown, Florida facility, there were two electric power generating units in 1988, Martin 1 and Martin 2. The Martin 2 turbogenerator failed which put the unit out of commission for over a year. The cost of the turbogenerator was approximately fifty million dollars but the replacement, including the outage cost, was about three times that amount. The manufacturer of the turbogenerator had built six units of this design and FPL owned four of them, Martin 1, Martin 2, Manatee 1 and Manatee 2. The other two units were at a utility in upstate New York. Management was concerned about the risk of failure at the other three installations. The question they faced was whether they should order replacement turbogenerators or risk failure and long outages.

Jim Keener led the failure analysis team that produced two possible failure theories. One theory would have produced random failures by age, and the other was a wear out mode with on-off cycles as the aging parameter. They did a one failure MLE analysis, Figure 11-21. The MLE β = 1.8, favored the wear out failure analysis. A risk analysis indicated that the next unit might fail in two years and would probably be Martin 1 as it had the highest cycles. The FPL Board of Directors reviewed this one point Weibull and decided to replace Martin 1. The original turbogenerator removed from Martin 1 was refurbished and is now available for quick replacement of Manatee 1 or 2 if they should fail.

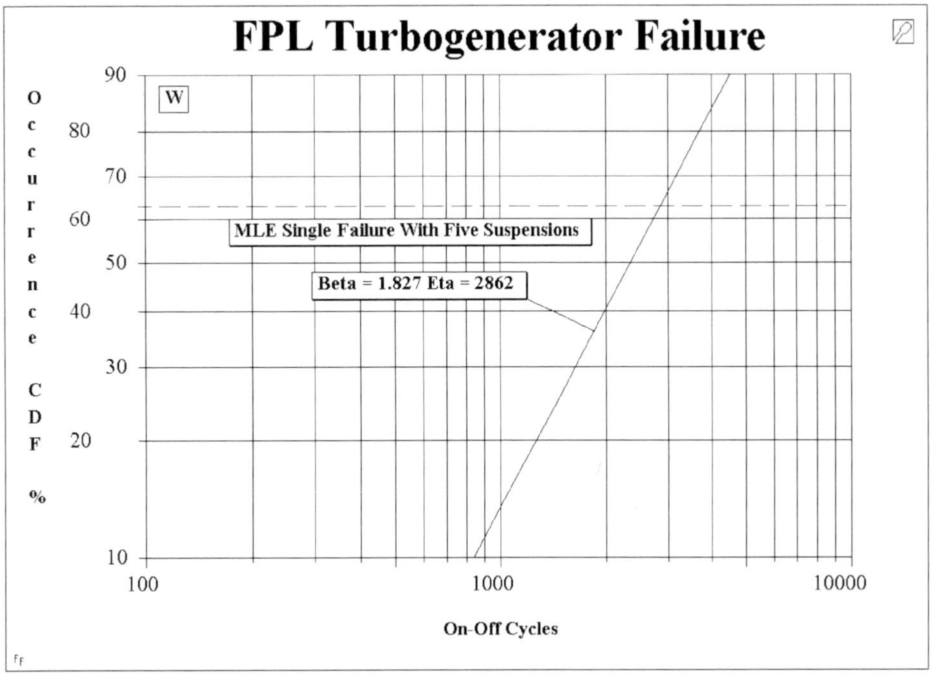

Figure 11-21. FPL Turbogenerator Failure

Author's Comment: Large samples of failure data are always desired but sometimes impossible to obtain. It is unnerving to base an important and costly decision on a one point MLE Weibull, but there was no alternative. This case study illustrates two important points; the ability to do an analysis with a single point plus some right suspensions, an advantage of MLE, and the Weibull capability of producing meaningful results with extremely small samples. Incidentally, the Weibull plot was made with the software listed in the original USAF handbook as SSW was not available at that time. FPL is one of the leaders in the electric power industry as recognized by the Deming Award. In this case, FPL management took decisive action based on the Weibull analysis.

11.7 TVA Bull Run Fossil Plant - Controller Cards

Contributed by Bill Pritchard and Charles H. Goodard, Tennessee Valley Authority.

Key Words: Truncated data, t_0, risk analysis, electronic failures

The Combustion Control System at Bull Run controls critical parameters such as feed water flow, coal flow, air flow, throttle pressure. Automatic control of these parameters is necessary to achieve the maximum reliability, efficiency, and safety from the unit.

The printed circuit cards in the Control System have required frequent removal for adjustment or replacement. Impacts include unit trips and generation control problems (affecting start up reliability). Weibull analysis was performed to gain some insight about the failures on the system. The data on the number of cards was gathered from plant maintenance records on PAT (Position adjusting type) and CAT (Current adjusting type) controllers. Records on failures were not kept for the first 156 months of operations, so any failures that may have occurred during that time cannot be included in the analysis. Additionally, there is no way of determining if a failure was of an original board or one repaired previously. A Weibull plot of all controller failures effecting manual control of equipment was constructed. See Figure 11-22.

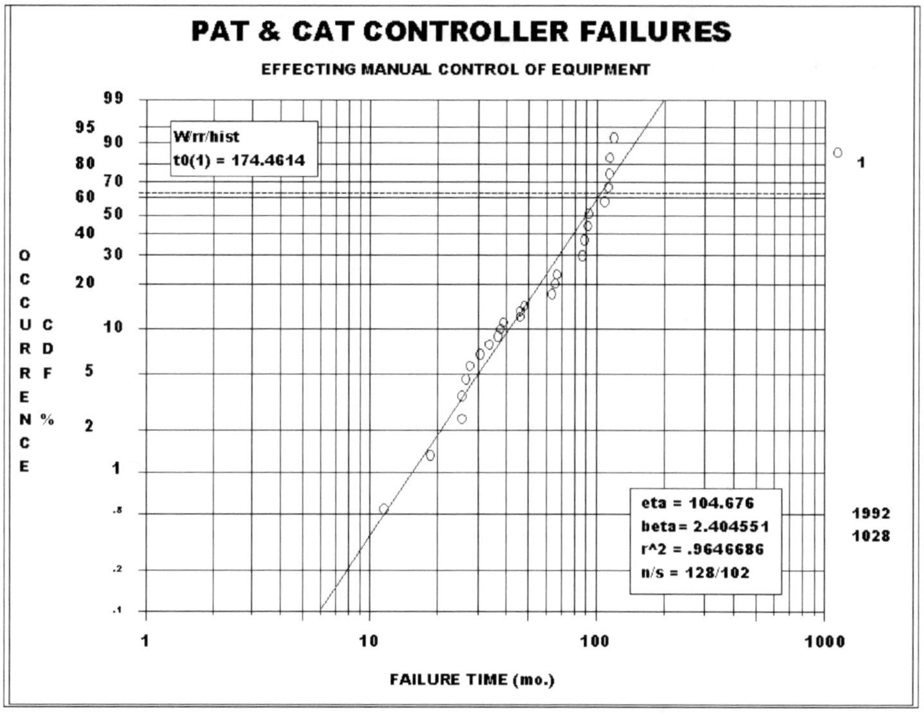

Figure 11-22. Pat & Cat Controller Failures

As shown above, a t_0 correction was applied after viewing the initial plot. The value of 174 months is consistent with no data available for the first 156 months and poor data for the following 24. Although the value for r^2 is adequate, the plot does appear to contain a mixture of failure modes. A risk analysis was performed to estimate future failures. The Weibull and risk analysis was performed with data ending December, 1991 [Figure 11-23]. There have been 11 board failures since that time; the analysis predicted 9 failures. Some of the discrepancy could be explained if the repaired boards have failed as well.

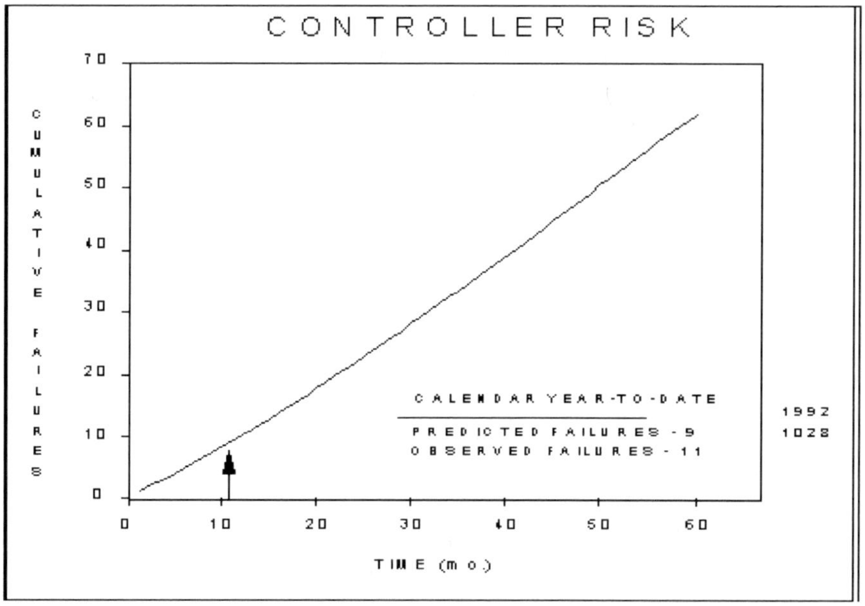

Figure 11-23. Controller Risk

The Combustion Control system is composed of 9 other major components. An analysis was also performed on transmitter boards with similar results. The risk analysis in that case correctly estimated 5 failures occurring since the start of the year. The remaining components of the system are currently undergoing analysis. This stratification improves both the validity of the analysis and will allow the risk of the critical components to be separated from the more benign. See Figure 11-23.

Author's Comment: Almost another year later, a TVA executive asked Bill Pritchard how good were the risk forecasts on the PAT and CAT units. Bill responded that one prediction was one failure too high and the other was one failure too low.

11.8 Repairable Systems Reliability Growth Assessment

Contributed by Richard J. Rudy, Chrysler Corporation Minivan Quality Executive.
KeyWords: Failure Forecast, Repairable Systems, Crow-AMSAA

The objective was to determine reliability growth from the prototype design configuration through the production release of a system consisting of the engine, transaxle (front wheel drive transmission and axle half shafts), battery, alternator, starter, and engine wiring. The requirement was to have fewer than five incidents per vehicle at 100,000 miles with none of the incidents being vehicle-disabling.Two different level prototypes, an engineering release, and the final production release were tested throughout the development cycle in sequential test phases to 100,000 equivalent customer miles. Fixes were incorporated into each subsequent design level to correct deficiencies found in the previous test phases. Data were analyzed using the Crow-AMSAA model for repairable systems. The results are as shown below. Reliability growth is seen from the prototype through the production release.

Author's Comments: Dick Rudy taught the author something new here. If you have a Crow-AMSAA plot in SuperSMITHVisual and you select "ordinary scaling" the result is a failure forecast plot against cumulative time or mileage, Figure 11-24.

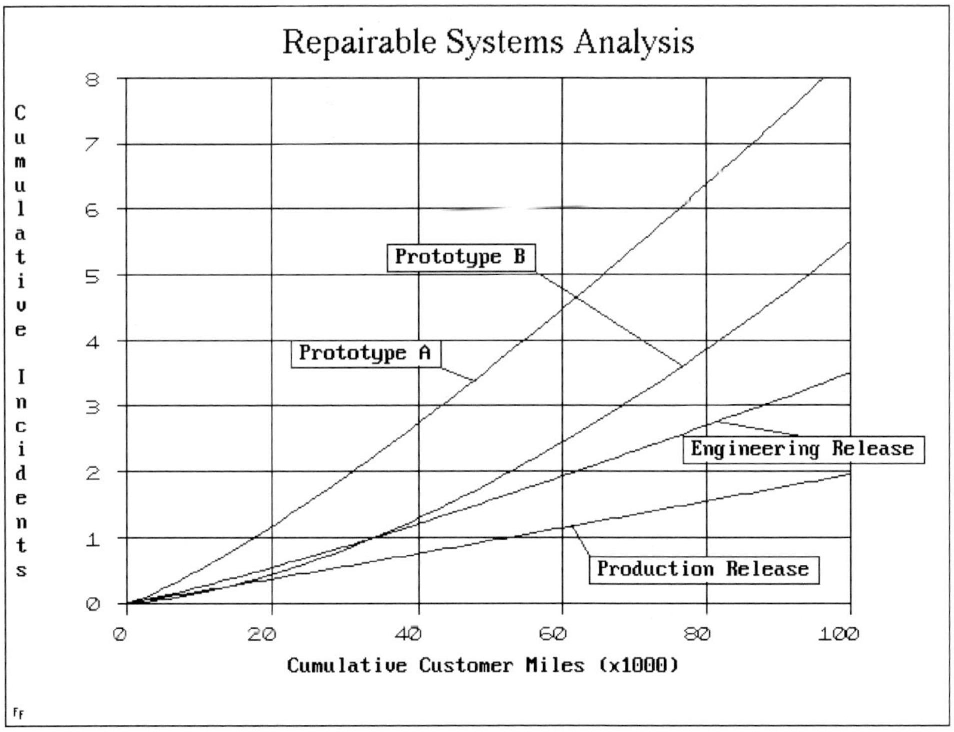

Figure 11-24. Crow Plot with Linear Scales

11.9 Front Jounce Bumpers

Contributed by Richard J. Rudy, Chrysler Corporation Minivan Quality Executive.

Key Words: Three parameter Weibull, tracking development progress, "Covered" Weibulls.

This case study involved the design of the front rubber jounce bumpers for a Chrysler carline. The requirement was a B5 life at 600,000 cycles. Testing consisted of three different designs: An Engineering Design (ED) test phase, which is basically development testing before design release; a Design Verification (DV) test phase, which is supposed to be the design that will be released for production; and the released design, in this case, the design that was finally released.

ED testing revealed a failure mode that was totally unanticipated. A redesign of the bumpers to eliminate that failure mode was undertaken. This was expected to be the released design. However, during the DV testing of this design, another failure mode was discovered. It had been masked by the early failure mode of the previous design. An FMEA analysis of the DV design had indicated that this new failure mode would not occur as frequently as it did. A subsequent material change was incorporated and that design tested (the released design shown). This design exceeded reliability requirements. See Figure 11-25.

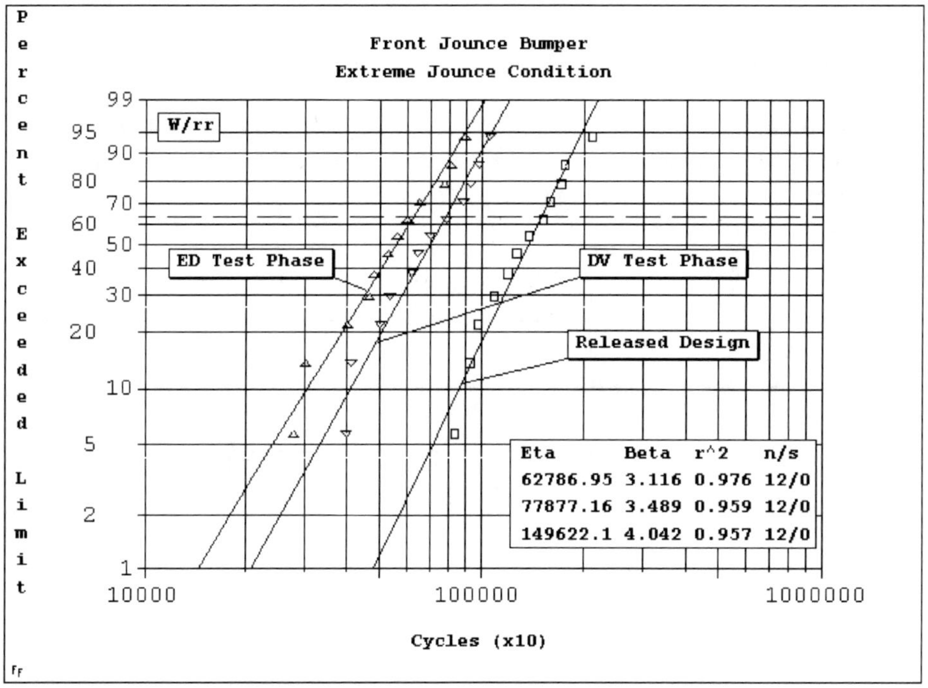

Figure 11-25. Two Parameter Weibull

Data was presented using the three-parameter Weibull shown. See Figure 11-26. A discussion ensued during the presentation of the results of the tests as to whether failures would occur before the 600,000 cycle requirement. For service planning purposes, and because the sample size is low for a three-parameter Weibull study, it was decided to accept the two-parameter analysis. Service planning was based on the lower 90% confidence limits of the two-parameter Weibull line.

Author's Comments: Mr. Rudy used a Beta test copy of SSW to make Figure 11-26, the first application of this new option. We hope this capability to plot t-zero plots in real time will reduce errors from misinterpreting the t-zero scale. I also like the concept of choosing the simpler two parameter model that gives more conservative estimates of B5 life.

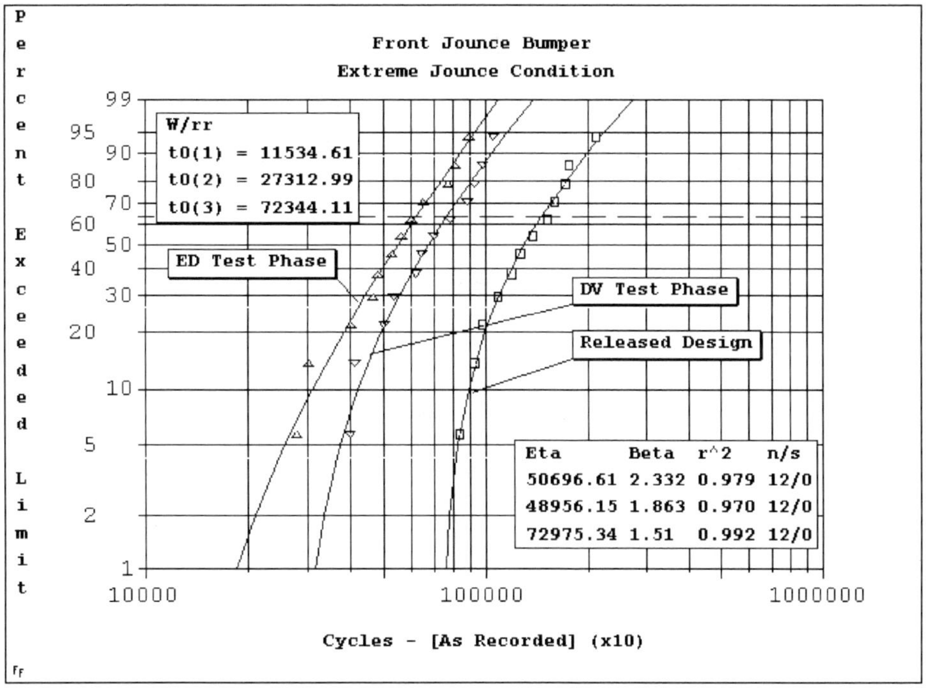

Figure 11-26. SSW 3 Parameter Weibull Plot with Real Time Scales

11.10 Transfer Case Seal

Contributed by Richard J. Rudy, Chrysler Corporation Minivan Quality Executive.

Key Words: Kaplan-Meier, New Design versus Old Design.

This case study involved the transfer case design for one of our four-wheel drive vehicles. A study of warranty records for the previous design revealed an unacceptably high defect rate at 200,000 cycles. The data analysis was conducted by our Warranty Analysis department using the Kaplan-Meier (K-M) method. At the time of the study, no vehicle had compiled more that 35,000 cycles. The K-M estimate could not project out to 200,000 cycles. We came up with the idea of plotting the data on various probability papers to examine the fits. The Weibull paper provided a good fit and allowed us to extrapolate to the 200,000 requirement. See Figure 11-27.

Prior to the study, it had been determined to use this carryover transfer case in the new vehicle design. Failure analysis of the field units was undertaken and a leak issue uncovered. The subsequent redesign was thought to be significantly better. A customer usage test was designed to simulate the environment and usage that would allow us to directly compare test results with past warranty data. The test could be completed within eight months. The results are shown as the new design on the graph.

Author's Comments: Dick Rudy has given us an excellent Kaplan-Meyer case study using warranty data.

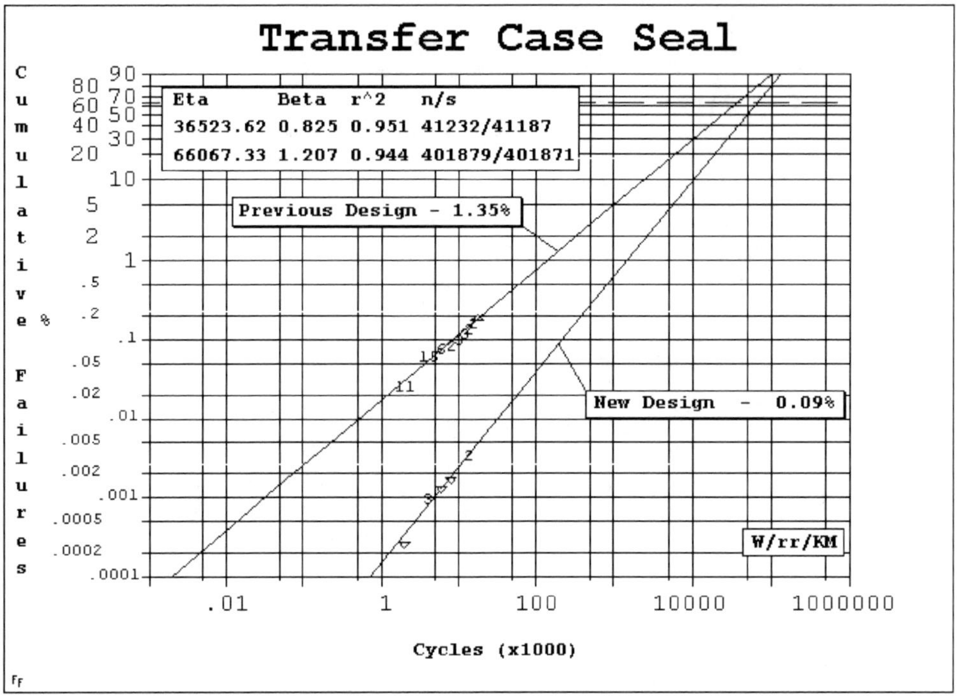

Figure 11-27. Kaplan-Meier Analysis of Leakage Failures

11.11 Dental Acrylic Adhesive Fatigue

Data and Reference Paper Contributed by Drs. Daniel L. Menis and Gene Lautenschlager of Northwestern University.

Key Words: Distribution Analysis, Comparison of Data Sets, Outliers.

Author's Comments: The reference paper [Menis et al] describes a designed experiment to compare the fatigue lives of different dental formulations of acrylic bone cement using different mixing methods and temperatures. The paper was published in 1991 using a very early version (circa 1988) of WeibullSMITH. There were three different materials and three different mixing procedures tested. The paper identified the best material and mixing method. Using Beta-Binomial confidence set at 98.5%, the dentists concluded that the three sets were significantly different at 95%. The paper states, "Most groups displayed a wide range of fatigue lives, so much so that modeling the conventional Gaussian distribution to the data implied the existence of significant numbers of specimens breaking at <0 cycles. A more realistic modeling of the data was found when the Weibull distribution was used." The authors, dentists, are commended for this application of Weibull to dental life data analysis, an excellent study.

However, it is interesting to note that there was a strong need here for several methods that have been developed in the intervening years, but were not available at the time of the study. Just a few moments with our latest software indicates:

- The plot, Figure 11-28, shows a classic Bi-Weibull for the Formatray set. SSW and YBath confirm that there are two modes, with betas of about 0.95 and 24. Examination of the fracture surfaces might have been useful to separate the data into two sets.
- The first point in the Simplex-P set appears to be an outlier. The Critical Correlation Coefficient (SSW distribution option) confirms that it is an outlier; the fit is unacceptable with the outlier, but acceptable without the outlier.

- The Weibull fit is good but distribution analysis shows the lognormal distribution is much better for the Simplex-P and Jet sets.
- The likelihood ratio test applied to the three sets confirms that the sets are significantly different with a p-value approaching 100%.
- The sample size is really too small to do an accurate distribution or mixture analysis. 20 or more failures are recommended as minimum sample size.

Although these comments would probably not have changed the conclusions of the study, it is encouraging that our new methods provide more insight and accuracy with very little effort.

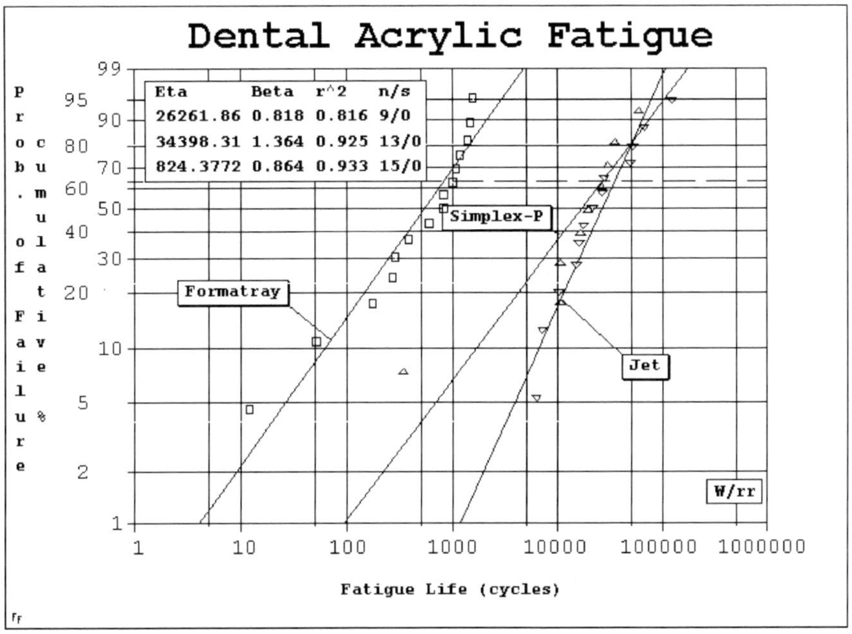

Figure 11-28. Acrylics Mixed at Room Temperature

11.12 Crow-AMSAA Reliability Modeling

Contributed by: David Williamson, Reliability Specialist, Dow Chemical - Canada

Key Words: Crow-AMSAA, Cost Analysis, Cost Savings, Process Industry

Problem Introduction

The data for this case is age to failure data from sets of 48 heat exchanger tubes arranged in 6 identical systems. All the failures were due to tube plugging. After a tube plugged it could only be cleared after it had been removed from the exchanger at significant maintenance cost and lost production. The problem was difficult to understand due to the number of variables and as the failures were analyzed, we found they were due to several different causes. The tubes in the exchangers are in a service that is expected to foul and they are cleaned periodically using an on-line cleaning process. The following is a Crow-AMSAA plot of the complete data set. Note how the slope of the line changes with time indicating changes in reliability. As the slope increases the reliability decreases and conversely as the slope decreases the reliability increases. These changes in the slope of the line reflect real changes in the plant reliability as changes were made to the maintenance and operating procedures to solve the reliability problems. These system changes and the different failure modes are the reasons Weibull analysis did not work for this case. The areas of interest for further analysis are the cusps between segments of the line as shown in Figure 11-29.

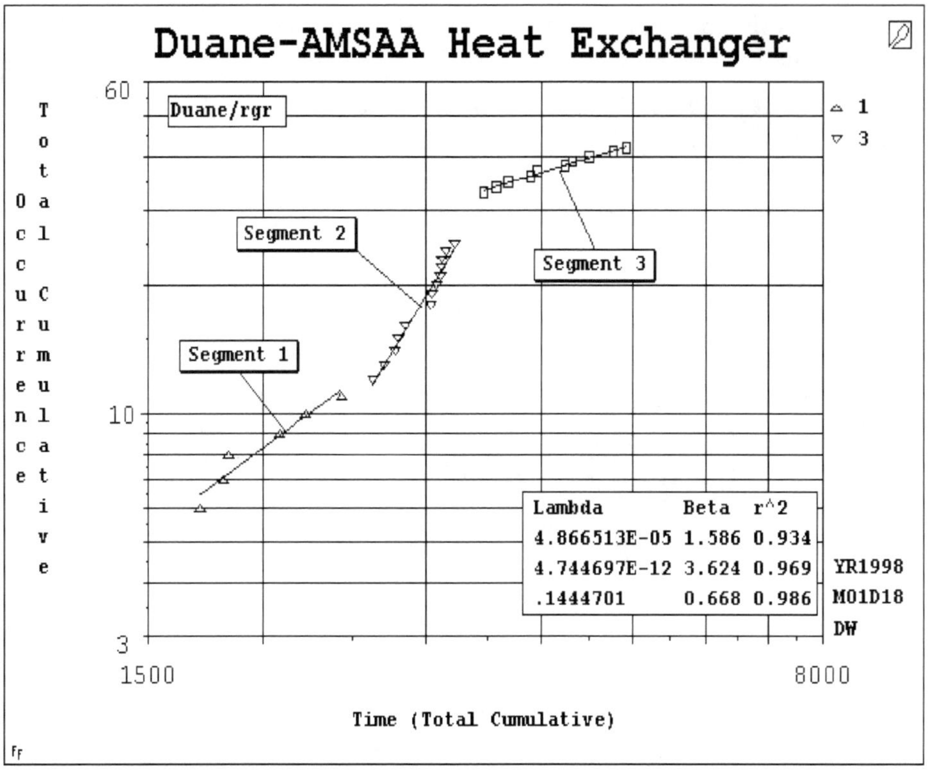

Figure 11-29. Evidence Showing Three Failure Phases

Background

The tube plugging problem was recognized during the segment 1 time period. Analysis work was started to find and solve the problem. There were several theories for the cause of the tube plugging. The one that was initially thought to be the most probable was the tube cleaning process was not removing all the fouling material and over time the tubes were plugging off. Changes to the cleaning process were tried, cleaning times and temperatures were increased but the failure rate increased. The analysis work continued, including discussions with the manufacturer and other industry experts. In the segment 2 time period, the failure rate and cost of the failures was very high, this was a large and expensive problem that really had the attention of upper management. The analysis work started to provide some answers, the design of the exchangers included measuring the flow in only 4 of the 48 tubes in each unit and the fouling rate in each tube was very variable, there was no way to know if a particular tube was fouled and going to plug. The other problem was as the tubes aged and as the exchangers were taken off line to replace plugged tubes the fouling rate and variation between the tubes increased, this was why the failure rate was so bad during the segment 2 time. The time between tube cleaning was reduced and the production rate was reduced to slow the fouling rate. Also the number of times the exchangers were taken off-line was reduced by solving other plant reliability problems and the failure rate improved as shown in segment 3 of the line. New tubes were installed in the exchangers with flow instruments on each tube and the failures were eliminated.

Statistical Analysis

The lines on this plot look like they fit the points well. The time to the next failure can be calculated by extrapolating the line with the following formula. **N = lambda (T0)^beta.** N is the number of failures calculated. Note: if the line fit is good this will be close to the actual number of failures, use this as a check. T0 is the time of the last failure and lambda and beta are functions of the line. The time T1 of the next failure N+1 is found by solving:

T1 = the root beta of [(N+1)/lambda]

In this case there are six systems in the process, so the time to the next failure "T" is:

$$T = (T1 - T0)/6$$

For this case it works out to 29 days or about 12 failures per year for the last segment of the line.

This prediction is a valuable management tool when the failure prediction is converted to a cost of unreliability. For example if each failure costs $10,000 the cost of unreliability is $120,000 per year. To calculate the savings from the reliability improvements that were shown by the cusp or change in slope of the line on Crow-AMSAA plot (Figure 11-30). We follow the same calculation as before but on the second segment of the line, failures between twelve and twenty-five. The prediction is that the next failure will be in 6 days, or that there will be about 60 failures per year and that the cost of unreliability is $600,000 per year.

This allows us to speak to management in terms that they can relate to, dollars saved per year. In this case we can show that changes to operating procedures made between failure 25 and failure 33 saved $480,000 per year.

Further reliability work was done to determine the root cause of the failures. Design and operational improvements were made to improve the reliability and since that time there have been no more failures. Figure 11-30 was created as a communication tool for management to show the results of these changes. The last point on the plot was entered by assuming that a failure occurred today. This point is located far to the right of the original line indicating that the system changes have substantially improved the system reliability with about a six hundred thousand dollars a year savings.

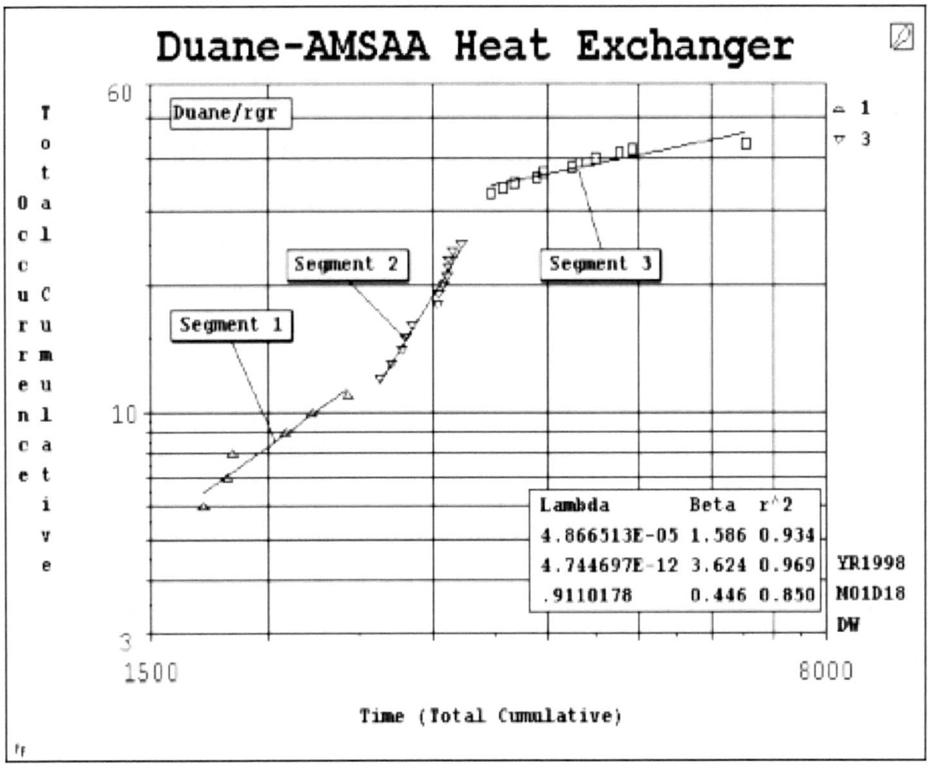

Figure 11-30. Improvement Hypothesis: Next Failure Occurs Tomorrow

Author's Comments: An excellent example of using the C-A model to evaluate cost improvements. Note that the time to the next failure, two failures, 5 failures, etc., is now available on the report tab in SSV for all Crow-AMSAA plots. The cost per event may by added from the C-A menu.

11.13 Weibull Analysis of Boiler Tube Failures

Contributed by David B. Smith, Detroit Edison.

Key Words: Fatigue, cracks, failure forecast, unplanned shutdowns, corrective action.

The majority of electric power in the United States is generated using large coal-fueled boilers supplying steam to turbine generators. A typical power boiler consists of hundreds of thousands of feet of 1-1/2" to 4" diameter alloy steel tubing. This tubing contains water and steam at high pressure, typically 2500 pounds per square inch (PSI), and temperatures up to 1000°F. Boiler tube failures are the major cause of unplanned shutdowns of generating units. The following is a case study of Weibull analysis of boiler tube weld failures and its use in the economic analysis of options and decision making.

The Belle River Power Plant consists of two coal fired electric generating units that were placed into service in 1984 and 1985. Each boiler has 82 superheater tube elements which heat the steam to 1000° F before supplying it to the steam turbine. The superheater is located high in the boiler, at the exit of the furnace where combustion of coal takes place. The superheater tubes are exposed to the radiant heat of the coal flame and combustion gasses at 2400° F and are subject to accumulation of soft and sticky ash, the non-combustible portion of the coal. Due to the nature of the ash from the coal burned at this plant (Powder River Basin coal from Wyoming), the regular ash removal systems were not fully effective. Consequently hand-held high-pressure water lances were utilized to remove the ash accumulations while the boiler was operating. This process permitted extending the operating time between boiler shutdowns for off-line furnace cleanings to remove the accumulated ash.

One of the boilers began experiencing failures of superheater tube welds. Metallurgical examination of the failed welds determined the failure mechanism was thermal fatigue. It was concluded that the failures were caused by quenching of the tubes associated with the high-pressure water lance operation combined with stress concentration in the welds. Boiler #1 experienced failures of three welds over an 18-month interval, with failures occurring at 13.81 years, 15.01 years, and 15.44 years. Each weld failure resulted in the shutdown of the boiler and turbine generator unit for approximately five days. This required the utility to generate replacement power at a higher cost generator or purchase power on the wholesale market, at an average cost of $1,000,000 per event. The third failure occurred in January, a time when this northern utility experiences more moderate power demands. After returning the boiler to service following the third failure, a Weibull plot was made indicating eta of 20.59 years and beta of 11.99. From the Weibull plot it was concluded that weld failures could be expected about every three months.

A shutdown of the #2 boiler approximately one month later provided the opportunity to inspect and repair the superheater welds. After determining that #2 boiler superheater welds were also cracking, plant management decided to replace all 82 welds utilizing additional crews, requiring the shutdown to be extended four days.

Almost right on schedule, # 1 boiler experienced the fourth superheater weld failure 86 days later at 15.68 years. Although the Weibull failure forecast had been developed previously, the concurrent shutdown of a large generator at another plant dictated returning #1 boiler to service in the shortest possible time, so only the failed weld was replaced during a four day shutdown. At this point concern was raised regarding the reliability of the boiler during the upcoming summer. The utility experiences highest electricity demand during the summer months of June, July and August due to air conditioning loads. The cost of replacement power in the summer months can be 10 or more times higher due to increased demand and shortage of reserves.

Earlier than expected, another #1 boiler superheater tube weld failed in early June after 48 days of operation, at 15.85 years. Fortunately, all other generators were available and the weather was cool, so the decision was made to replace all of the remaining original welds at this time. A 10-day shutdown was required to replace the remaining cracked welds. The Weibull plot, see Figure 11-31, showing only the first three failures and all five failures. It is evident that the first three failures provide an accurate forecast.

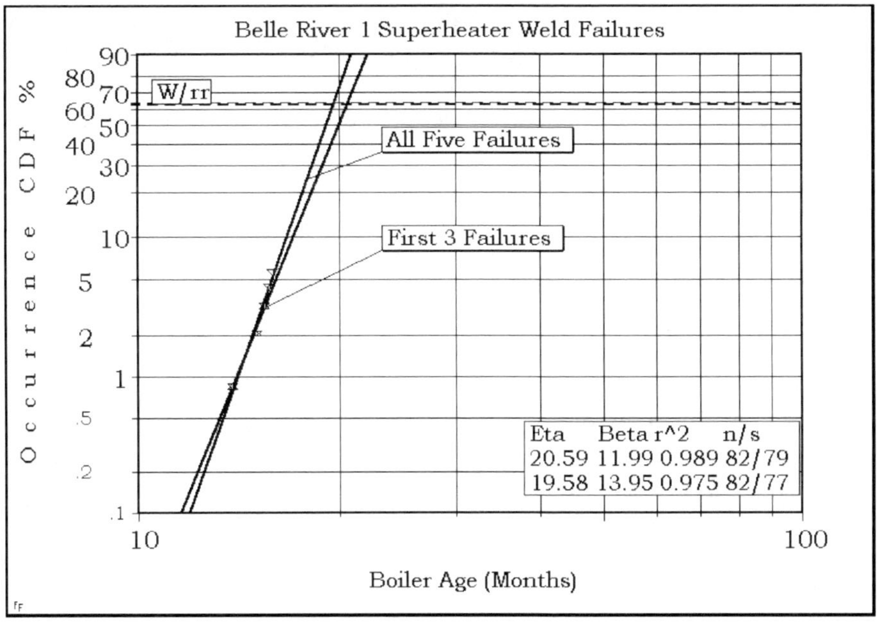

Figure 11-31. Weibull Plot of Boiler Tube Weld Failures Set 1 pve%=79.57, Set 2 pve%=93.75

At this point a forecast and analysis of future failures was conducted. It was desired that the boiler be operated for the next three months through the critical summer period without a failure. The failure forecast for the end of summer was an additional 1%, or 0.9 tubes, approximately one. The instantaneous Mean Time Between Failures was calculated to be 102 days. The Reliability or probability of no failures occurring over the near-term time interval was calculated as shown in the following graph:

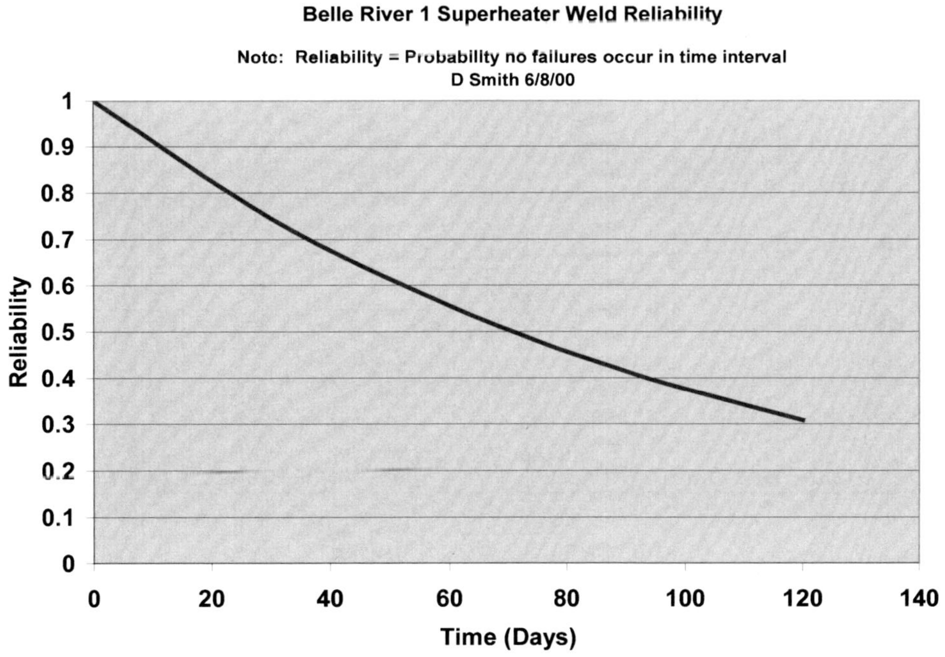

The probability of zero to four weld failures was calculated from a Poisson distribution for the next four-month period.

The process of replacing the welds provided the opportunity to destructively examine the original welds as they were replaced and measure the depth of the cracks. Fifteen welds had depths less than 0.020" which were not recorded and 15 welds had previously been replaced. Measured crack depths are shown on the following chart; the 15 welds under 0.020" depth are assumed to be 0.015" and the 15 previously replaced welds were shown as suspensions greater than 0.250".

The extent of measured cracks confirms the extent of the problem, providing support to the decision to replace the welds and avoid future failures and expensive downtime. See Figure 11-32. Note that the inspection option was employed to significantly improve the fit because of the 15 welds assumed to be 0.015". With the inspection option on, distribution analysis indicates the Weibull is best, but without the inspection option the lognormal is the choice.

Utilities normally schedule periodic shutdowns of generating units to conduct preventive maintenance at one to four year intervals. This unit has a preventive maintenance shutdown scheduled in two years, at which time these weld replacements would normally occur. Weibull failure forecasting indicates eight additional failures are likely to occur before the scheduled shutdown. Consequently extending the current shutdown by five days to replace all of the welds is expected to save approximately 40 days of shutdowns over the next two-year period. The Weibull failure forecast, supported by actual crack measurement data, indicates replacement of the remaining original welds at this time is a sound business decision.

Author's Comments: This is an outstanding example of life data analysis. The accuracy of the failure forecast is worthy of note as we are frequently questioned on this issue. See also Case Studies 11.6, & 11.7.

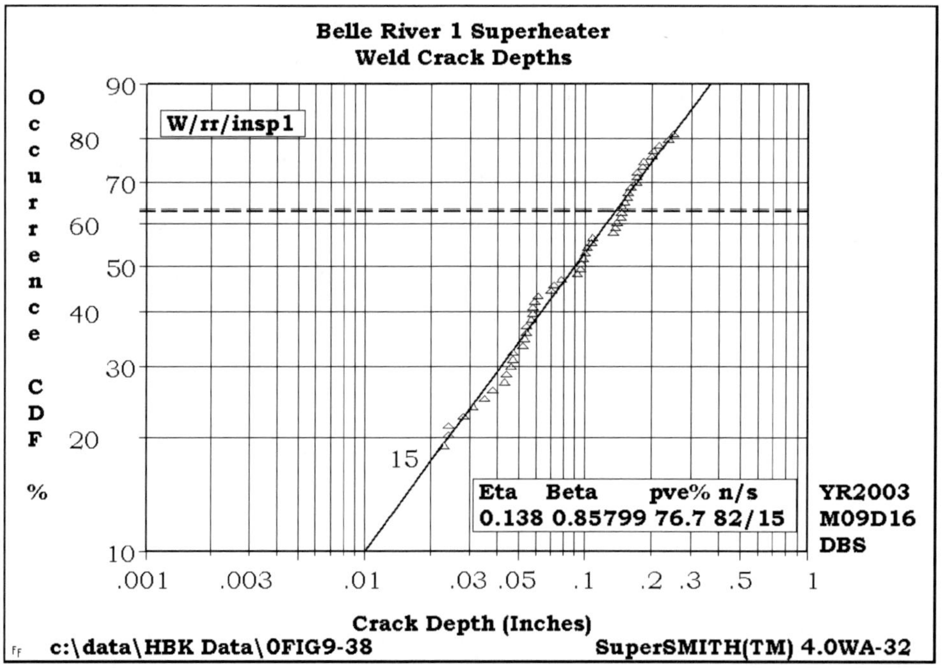

Figure 11-32. Crack Depths

11.14 Gas Turbine Seal Failures - A Batch Problem

Data and Analysis Contributed by David Langanke, Honeywell Engines.

Key Words: Batch, Failure Forecast, Now Risk, Aggregate Cumulative Hazard, Crow-AMSAA Failure Forecast. *(Clues that this is a batch problem are numbered.)*

In-service failures occurred on the newer seals of two engine models including seals from two different vendors using a variety of lubricants. Only one set of the data is shown here for simplicity. There are five failures in a fleet of 1355 engines. See Figure 11-33.

1. The fact that there are many late or right suspensions indicated a possible batch problem.
2. The Now Risk is 41, much greater than the 5 observed failures, indicating a batch problem.
3. The lower 90% bound on the Now Risk is 21, confirming a batch problem.
4. The batch inference implies that some of the seals may never fail which is consistent with the plot pattern, steep followed by shallow curve, perpetual survivors. (Note however, with only five failures making inferences about the plot pattern may be fantasy.)
5. The MLE beta, 1.86, is less than the MRR beta which is another clue that a batch problem exists.
6. An Aggregate Cumulative Hazard plot is shown as Figure 11-34. It indicates the batch size is approximately 10% of the fleet (5/48). The Now Risk to observed failures would estimate more like 12%, (5/41).

With the wrong data the failure forecast will over predict from batch data. In this case, the 12 month forecast is 68 failures.

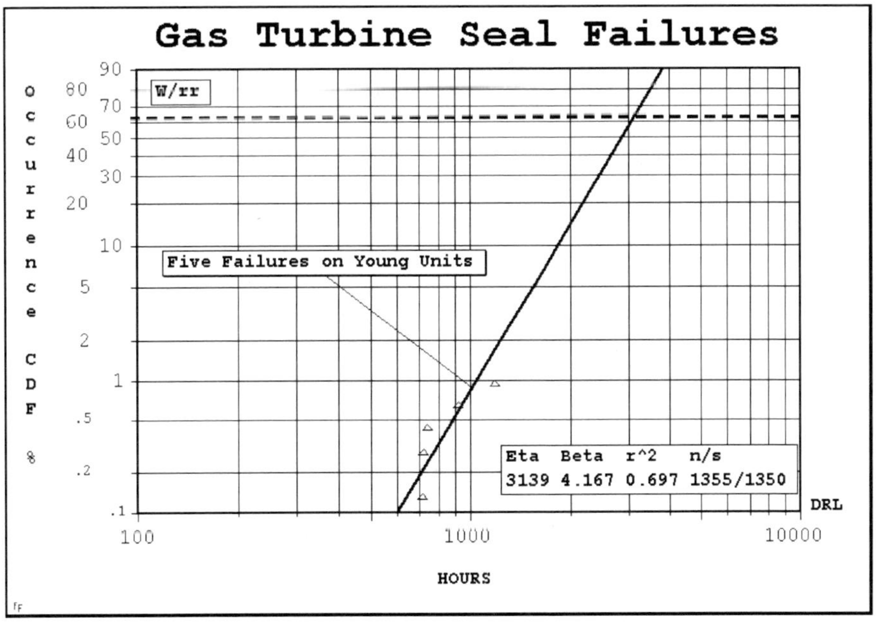

Figure 11-33. Gas Turbine Seal Failures pve% = 2.06

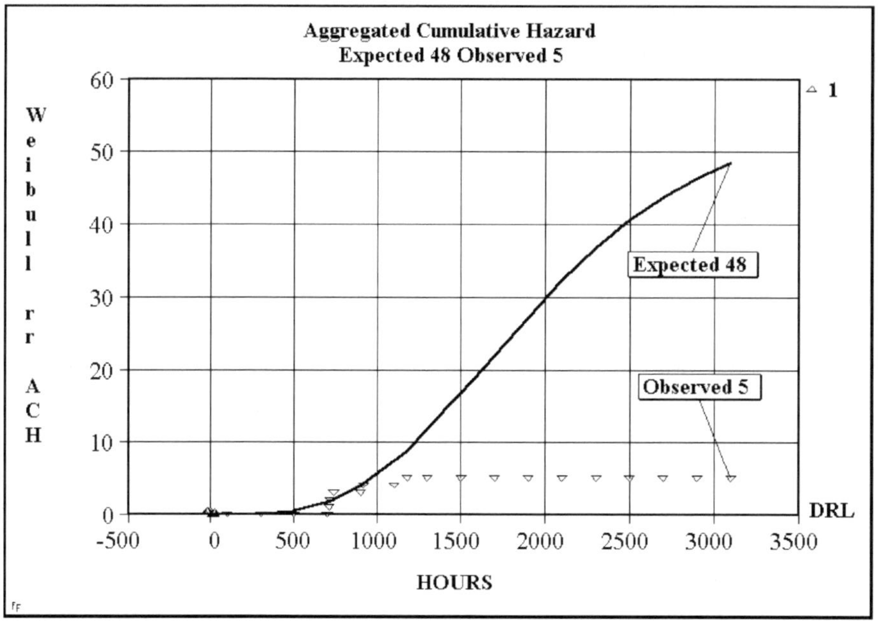

Figure 11-34. Aggregate Cumulative Hazard Expected 48 Observed 5

The number one priority task for engineering is to identify the units in the batch in order to do a correct failure forecast. Batch effects are most often produced by a change in production but in this case as both vendors seals have failed this is unlikely, i.e., the probability that both vendors would produce the same batch failure mode simultaneously is small. The batch is more likely caused by improper installation, different usage, different engine conditions, but probably not production.

With batch data, a Crow-AMSAA failure forecast should be more accurate than a Weibull failure forecast. In Figure 11-35 we see a cusp, indicating a shallower slope. Using the last three points to estimate the current process, the 12 month forecast would be 3 failures. Note the use of regression in the Crow-AMSAA analysis, the new **best practice**, instead of MLE.

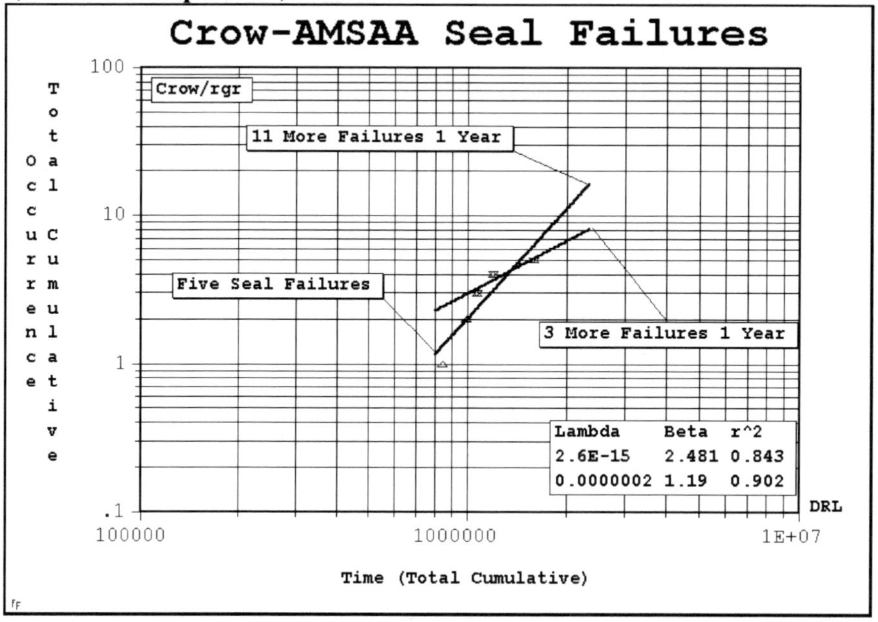

Figure 11-35. Crow-AMSAA Seal Failures

This data illustrates many of the batch clues and analysis described in earlier chapters and summarized in Section 8.9.

11.15 Challenger Space Shuttle Weibull

By Dr. Robert B. Abernethy

Key Words: Temperature Weibull, O-Ring Failures

Recently I was honored by being invited to do a Weibull Workshop at the NASA Goddard Space Flight Center. One of the NASA engineers asked me if he could do a Weibull plot with the data on O-ring damage from the Challenger-Space Shuttle history. He provided the data and linear plot shown below (Figure 11-36) which are available on the Internet [Dunar]. The curve fit on the linear plot is obviously not very good. More complex regression modeling is attempted in [Wujek] but it also has poor goodness of fit. These models are better than nothing but inadequate.

My Weibull experience with temperature is based on increased temperatures accelerate failures but here the reverse is true. What to do? I thought perhaps if we took the reciprocal of absolute temperature as the accelerating parameter that might work. Figure 11-37 below shows my results. Beta is 83 indicating extremely rapidly increasing O-ring damage with colder temperatures.

The fit is excellent with a p-value for the coefficient of determination of 74.5%. (Any value above 10% is considered an acceptable fit.) The probability of damage at the Challenger's 31 degrees F is 100%. It is easy to be smart retrospectively and I do not mean to criticize NASA for whom I have great respect. However, it certainly would have been enlightening if they had made this Weibull plot.

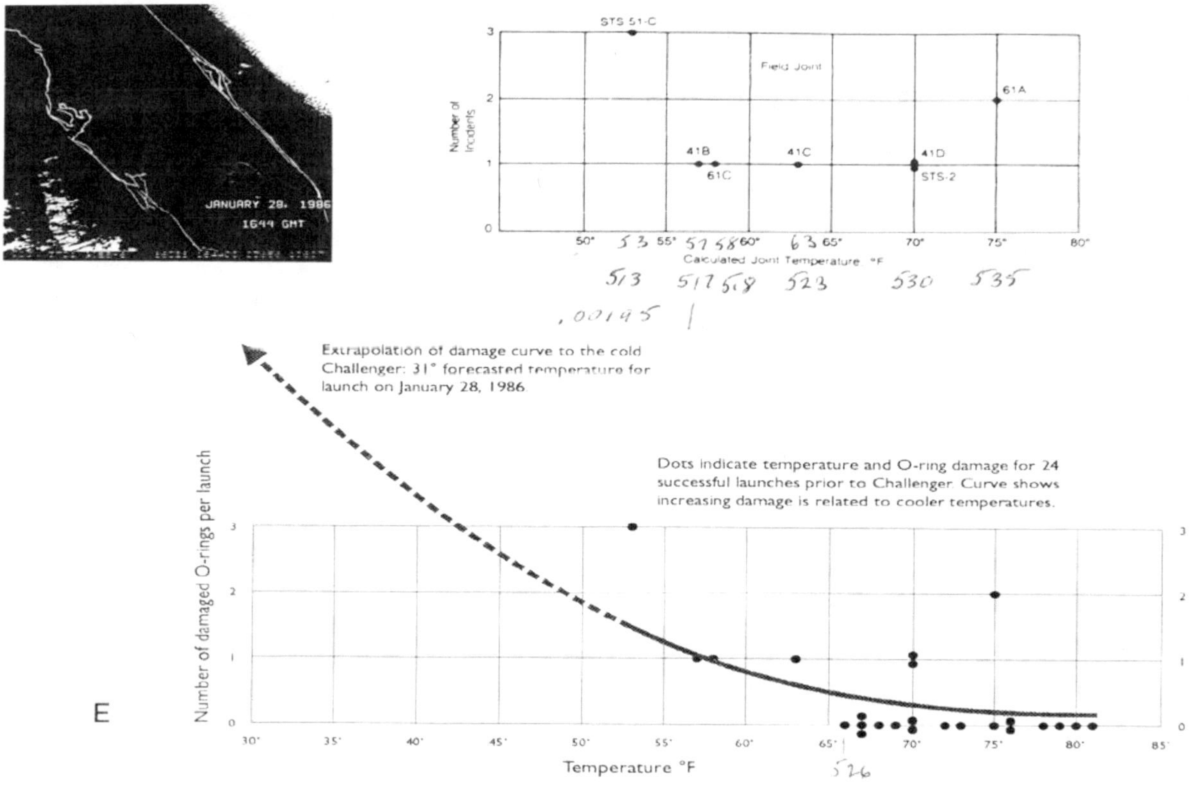

Figure 11-36 NASA Rocket Motor Case O-Ring Data Damage From Dunar & Waring

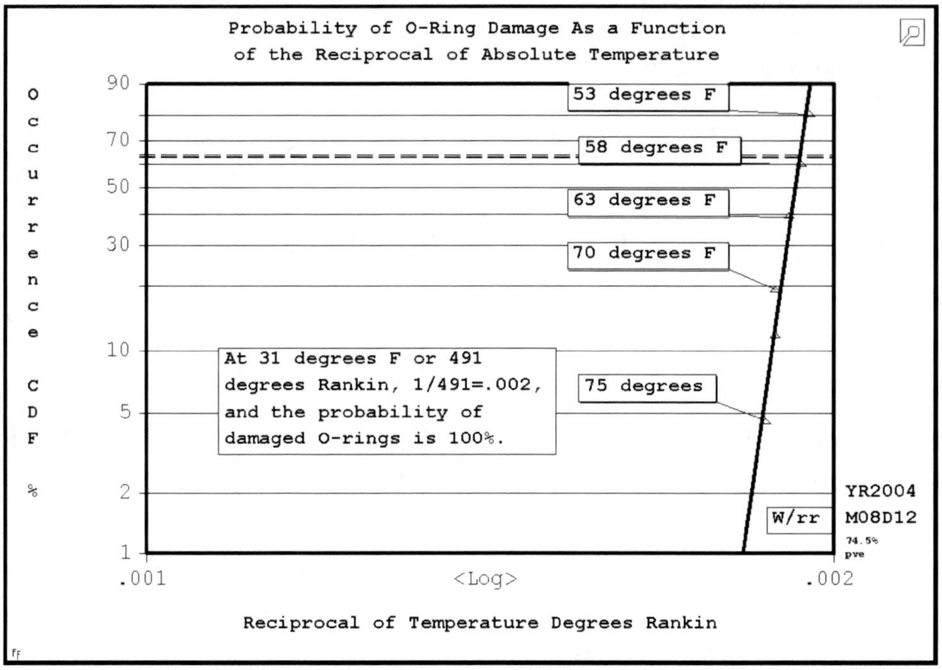

Figure 11-37 O-Ring Damage

References: 1. A. J. Dunar & Stephen P. Waring, "Power To Explore: History of Marshall Space flight Center 1960-1990," Government Printing Office, 033-000-01221-7, Chapter IX, "The Challenger Accident."

2. J. H. Wujek, "Challenger O-Ring Data Analysis," OnLine Ethics Center: http://onlineethics.org/edu/ncases/EE18.html

Appendix A: Glossary

Aggregate Cumulative Hazard (ACH) – **ACH is** the best method for detecting batch problems and compliments the Now Risk method. See Appendix F. See also Section 8.9 for more on batch problems.

Assurance Index - A single percent value that simultaneously specifies a lower one-sided demonstrated confidence bound for reliability and an equal confidence level as a function of age. Also called the Reliability Assurance Index. See Section 7.3.5.

Batch – A data set that contains some fraction of units that have a problem or failure mode that the other units do not have. Batch problems are often caused by quality problems, some units are defective compared to others. See Section 8.9.

β (Beta) - The parameter of the Weibull distribution that determines the pdf shape and implies the category of failure mode (infant mortality, random, or wear out). It is also called the slope parameter as it is estimated by the slope of the straight line on 1 x 1 scale Weibull probability paper. See Section 2.7.

Bias - The difference between the true value of a population parameter and the median of many estimates calculated from random samples drawn from the parent population. Also called systematic or fixed error. For symmetrical distributions the mean is used instead of the median but of course they are equal in that case.

Censored Units - Units that have not failed by the failure mode in question. They may have failed by another failure mode or they have not failed at all. Left or early censored means the unit's age is less than the earliest failure age; right censored means the age is greater than the oldest failure. Also called **suspensions**.

Characteristic Life - See Eta.

Confidence - The relative frequency that the (statistically derived) interval estimate contains the true unknown value being estimated.

Correlation Coefficient - A statistical measure of the strength of a linear relationship between two variables. Herein, the correlation coefficient, r, is used as a measure of goodness of fit of the Weibull or lognormal probability plots. For comparison with known Weibull distributions, the tenth percentile of the distribution of r, is called the critical correlation coefficient. The p-value is the percentile of r from an ideal Weibull with the same parameters and sample size and will have p-values greater than the critical value 90% of the time. If the p-value of an observed r is less than the critical value, (10%), it implies that the distribution is significantly different from the Weibull at the 90% level. Therefore, if the observed correlation coefficient is greater than critical value of r, the inference is that the fit is acceptable. The correlation coefficient squared, r^2, sometimes called the coefficient of determination, is also used as a measure of goodness of fit. See Section 3.2.1.

Distribution - A mathematical function giving the probability that a random quantity (e.g., a component's life) will be less than or equal to any given value. A bar chart or histogram may approximate a distribution.

Distribution Analysis - A comparison of the data fit to several distributions to determine which has the best fit. The author recommends the p-value of the correlation coefficient as the **best practice** for determining the best fit distribution with regression solutions. The likelihood ratio test is recommended for MLE-RBA solutions. There are alternative statistical goodness-of-fit tests which are not as good [Liu]. With life data, the Weibull 2 parameter, 3 parameter, and the lognormal are often contenders for the best fit.

Crow-AMSAA Model - A model for tracking the growth of reliability in a development program and recently applied to tracking maintainability for fleets of repairable systems and also tracking significant management events. See Chapter 9. The method is very robust; it handles mixtures of failure modes, and missing portions of data.

η (Eta) - The characteristic life of the Weibull distribution. 63.2% of the lifetimes will be less than the characteristic life, regardless of the value of β, the Weibull slope parameter. When $\beta = 1$, η = MTTF. Section 2.4.

Extreme Value - The statistics of rare events and minimum and maximum values. See Section 8.8.

Hazard Rate - The instantaneous failure rate. For the Weibull distribution, $\beta < 1$, infant mortality, will have decreasing hazard rates, as age increases. $\beta = 1$, the exponential distribution, has a constant hazard rate. $\beta > 1$, wear out distributions, have increasing hazard rates. If $\beta = 2$, the Rayleigh distribution, the hazard rate is a linearly increasing function.

Infant Mortality - A failure mode characterized by a hazard rate that decreases with age, i.e., new units are more likely to fail than old units.

Inspection Data - Failures are only found when the system is inspected at intervals of age. These are benign failures that do not produce catastrophic failure. The uncertainty of the analysis is increased because the exact time of the failure is not known; the failure is found at the next inspection. This is interval data. Coarse data, where the times-to-failure have been recorded imprecisely, is a related similar form of interval data. For example, failures may have been recorded in months instead of days or hours or minutes. When there is additional uncertainty related to detecting or finding the failure at the inspection and every part is inspected at every inspection, the data are called **probit data**. Eddy current inspections and destructive inspections are examples of probit data. The inspection option is used if the failures are obvious when inspected. The interval MLE solution is the most accurate for interval data. [Marquart] See Section 5.8.

Kaplan-Meier Model - A survival model developed in the medical industry and recently applied to life data analysis and warranty claims forecasting. See Chapter 8, Section 8.5.

Likelihood Function - The likelihood function measures the relative credibility of parameter values, given the data or observations. For continuous distributions, the probability density function appears in a product, once for each failure point. The log likelihood is usually used. The maximum value of the likelihood defines a point estimate of the parameter(s) called the maximum likelihood estimate, which has excellent statistical properties. See Chapter 5 and Appendix C. The likelihood contours are considered by many statisticians to be extremely important. The likelihood ratio test and the likelihood contour test are used to determine whether two or more data sets are significantly different. See Chapter 7.

Lognormal Distribution - If the log of x is normally distributed, x is said to be lognormally distributed. Common parameters of the lognormal are the mean and standard deviation logarithmic transforms. SSW uses "Med" and "SigF" instead, as they are in original measuring units (no transform needed). See Chapter 3.

MLE - Maximum Likelihood Estimate, the maximum value of the likelihood function. Chapter 5.

Monte Carlo Simulation - Mathematical model of a system with random elements, usually computer-adapted, whose outcome depends on the application of randomly generated numbers. See Chapter 4.

MTBF - Mean Time Between Failure is estimated by the total time on all systems divided by the total number of failures on all systems. It is used with <u>repairable</u> systems. **MTBF is often confused with MTTF. . See Figure A-1 below**. With complete samples, no suspensions, the MTTF equals MTBF, but with suspensions this is rarely true. MTBF is the average operating time between failures

MTTF - Mean Time To Failure is the expected or average time to failure. Before Weibull analysis it was the most popular measure of reliability. It is the parameter of the exponential distribution. The Weibull η and MTTF are related by β and the gamma function. For $\beta = 1$, η and MTTF are equal. See Section 2.4.

Outlier - A wild or different point from the others in a data set. See Chapter 3.

Parameter - A constant associated with a population (such as the characteristic life of a Weibull population or the mean of a normal population).

Precision - The degree of agreement or repeatability from repeated measurements. The precision is usually measured by the standard deviation or the range of the estimates. For repeated measurements of a constant parameter, say a dimension or a pressure, the scatter of the measurements is an indication of the precision error. For statistical estimates, a confidence interval defines the statistical precision of the estimate. See Chapter 7.

Production Process Control - The control, analysis and improvement of production processes using Weibull analysis. See Section 8.7.

pve% - p-value rank of fit in percent (0 – 100) of actual correlation among all possible from MC simulation.

Random Failure Mode - A failure mode that is independent of time, in the sense that an old unit is as likely to fail as a new unit. There is no aging. A Weibull with β = 1.0 is the exponential distribution and describes random failure modes. The hazard rate is constant.

Reliability - The probability that a system will perform its intended function satisfactorily for a specified interval of time when operating in the manner and environment intended.

Risk Analysis - The prediction of the number of failures expected to occur in some future time period. It is often called failure forecasting or predictive analysis.

SuperSMITH® Software - The software package including SuperSMITH Weibull, SuperSMITH Visual and YBath.

Suspension - A test or operational unit that has not failed by the mode under consideration at the time of the life data analysis. Often used as a synonym for "censored."

Sudden Death Testing - The units to be tested are randomly divided into subsets of perhaps 3, 4 or 5 units. Each subset is tested until the first unit fails. The resulting data, failures and suspensions, is used to estimate the Weibull distribution and measure compliance with a contract or design requirement.

t_0 - Zero age for the failure mode. It is known as the guaranteed minimum life parameter in the three-parameter Weibull, normal, and lognormal distributions. However the three parameter normal and lognormal distributions are almost never used. Units have zero probability of failure prior to t_0. It defines a guaranteed failure free period.

Uncertainty - An estimate of the largest expected error in the result. It includes both bias and precision errors while confidence intervals only include the statistical precision error.

Uncertainty Interval Estimate - An interval around the estimate of the result that should include the true unknown result.

Wear Out Failure Mode - The hazard rate increases with age, i.e., old units are more likely to fail than new units. For the Weibull distribution, a failure mode characterized by a slope, β, greater than one. Reliability decreases with age.

Weibayes/Weibest - A method for constructing a Weibull distribution based on assuming a value of β, the Weibull slope parameter. It is used when there are certain deficiencies in the data (for example, small samples of failures or no failures). With small samples, the elimination of the uncertainty in β improves the accuracy of the results compared to Weibull analysis. See Chapter 6.

Weibull Analysis - Procedure for finding the Weibull distribution that best describes a sample of life data, in order to estimate reliability, determine failure mode characteristics, and predict the occurrences of future failures.

Weibull plot - A plot of time-to-failure data on Weibull probability paper against probability of failure.

SSW - SuperSMITH Weibull Software

SSV - SuperSMITH Visual Software

SSY - YBath™ - SuperSMITH Software for analyzing mixtures of 2 or 3 failure modes. See Appendix J.

Figure A-1 is a comparison of MTBF and MTTF for a fleet of 1000 components subjected to a Weibull failure mode defined below. Throughout the life of the components, MTTF = 55. MTBF varies from over 1500 to 55 as the failed units are replaced many times and after 5 years becomes asymptotic to MTTF. Planned parts replacement will increase MTBF for wear out failure modes but it has no effect on MTTF. See Section 2.4.

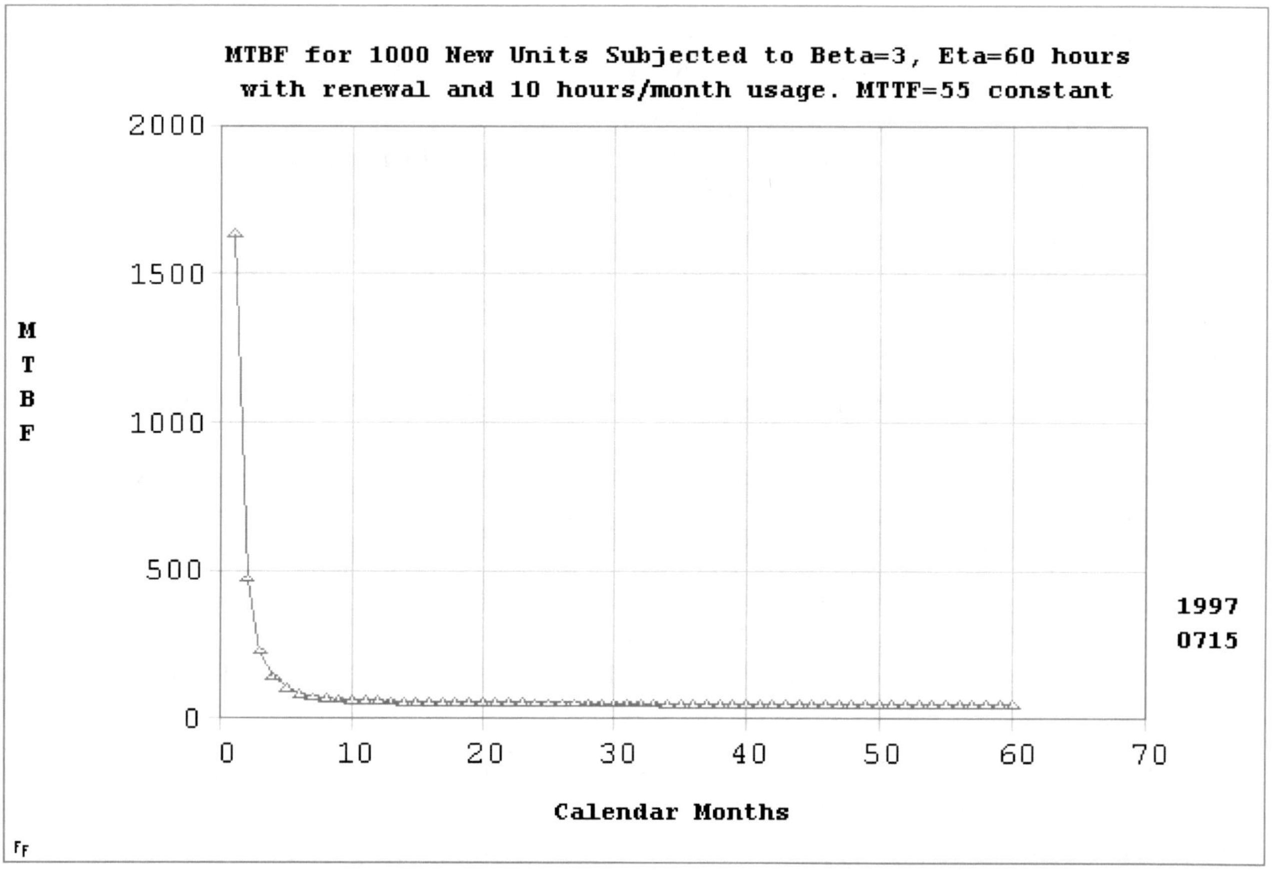

Figure A-1 MTBF Versus MTTF

Appendix B: Rank Regression and Correlation Method of Weibull Analysis

B.1 Method

Median rank regression uses a best-fit straight line, through the data plotted on Weibull paper, to estimate the Weibull parameters β and η. The "best-fit" line is found by the method of least squares.

First, the failure times and median ranks (see Chapter 2 and Appendix I for the calculation of median ranks) are transformed, as follows:

$$Y = \ln(\text{failure time}) \qquad X = \ln(\ln(1/(1 - \text{Median Rank of Y})))$$

(The median rank is expressed in decimal form.)

Least squares is then used to estimate A and B in the equation Y = A + BX. These estimates will be referred to as A and B, respectively. The median rank regression estimates of the Weibull parameters are:

$$\beta = 1/B \qquad \eta = e^{A}$$

B.2 Example and Step-By-Step Procedure

The median rank regression method will be illustrated with the censored data below:

Time	*Status*
1500	Failure
1750	Suspension
2250	Failure
4000	Failure
4300	Failure
5000	Suspension
7000	Failure

Step 1: Calculate the median ranks of the failure times using the methods of Chapter 2 or look up the values in Appendix I.

Failure Time	*Adj. Rank Order Number*	*Median Rank (decimal)*
1500	1.0000	0.0946
2250	2.1667	0.2523
4000	3.3333	0.4099
4300	4.5000	0.5676
7000	6.2500	0.8041

Step 2: For each failure, calculate the natural (base e) logarithm of the failure times (Y = ln(failure times)) and X = ln(ln(1/(1-Median Rank of Y))).

Failure Time	*Median Rank*	*Y*	*X*
1500	0.0946	7.3132	-2.3088
2250	0.2523	7.7187	-1.2353
4000	0.4099	8.2940	-0.6397
4300	0.5676	8.3664	-0.1763
7000	0.8041	8.8537	0.4887

Step 3: Calculate the least squares estimates A and B in the equation Y = A + BX, where Y bar is the average of the Y's and X bar is the average of the X's.

$$\mathbf{A = \bar{Y} - B\bar{X}}$$

and in the above example,

$$\sum x_i y_i = -28.8735 \qquad \sum y_i = 40.5460 \qquad \bar{Y} = 8.1092$$

$$\sum x_i = -3.8714 \qquad \sum x_i^2 = 7.5356 \qquad \bar{X} = -0.7743$$

$$B = \frac{\sum_{i=1}^{n} x_i y_i - \dfrac{\sum_{i=1}^{n} x_i \sum_{i=1}^{n} y_i}{n}}{\sum_{i=1}^{n} x_i^2 - \dfrac{\left(\sum_{i=1}^{n} x_i\right)^2}{n}}$$

$$B = \frac{-28.8735 - \dfrac{(-3.8714)(40.5460)}{5}}{7.5356 - \dfrac{(-3.8714)^2}{5}}$$

$$\mathbf{B} = \frac{2.5205}{4.5381} = 0.5554$$

$$\mathbf{A = 8.1092 - (0.5554)(-0.7743) = 8.5392}$$

The correlation coefficient can be calculated from the following and (more importantly) r^2, the proportion of the variation in the data that can be explained by the Weibull hypothesis, is then easily calculated.

$$r = \frac{\sum_{i=1}^{n} x_i y_i - \frac{\sum_{i=1}^{n} x_i \sum_{i=1}^{n} y_i}{n}}{\sqrt{\left(\sum_{i=1}^{n} x_i^2 - \frac{\left(\sum_{i=1}^{n} x_i\right)^2}{n}\right)\left(\sum_{i=1}^{n} y_i^2 - \frac{\left(\sum_{i=1}^{n} y_i\right)^2}{n}\right)}}$$

$$r = \frac{-28.8735 - \left(-3.8714 \frac{40.5460}{5}\right)}{\sqrt{\left(7.5356 - \frac{(-3.8714)^2}{5}\right)\left(330.2361 - \frac{(40.5460)^2}{5}\right)}} = 0.985722$$

In our example,

$$r^2 = 0.9716$$

Step 4: Calculate the median rank regression estimates of β and η:

$$\beta = \frac{1}{\hat{B}} = \frac{1}{0.5554} = 1.80$$

$$\eta = e^{A} = e^{8.5392} = 5111.25$$

The Weibull equation used to calculate the probability of failure before time t is then:

$$F(t) = 1 - e^{-\left(\frac{t}{5111.25}\right)^{1.8}}$$

Figure B-1 shows the data plotted on Weibull paper using SSW. (Note the small difference in η between what was calculated above and from SSW due to the round-off in the above calculations).

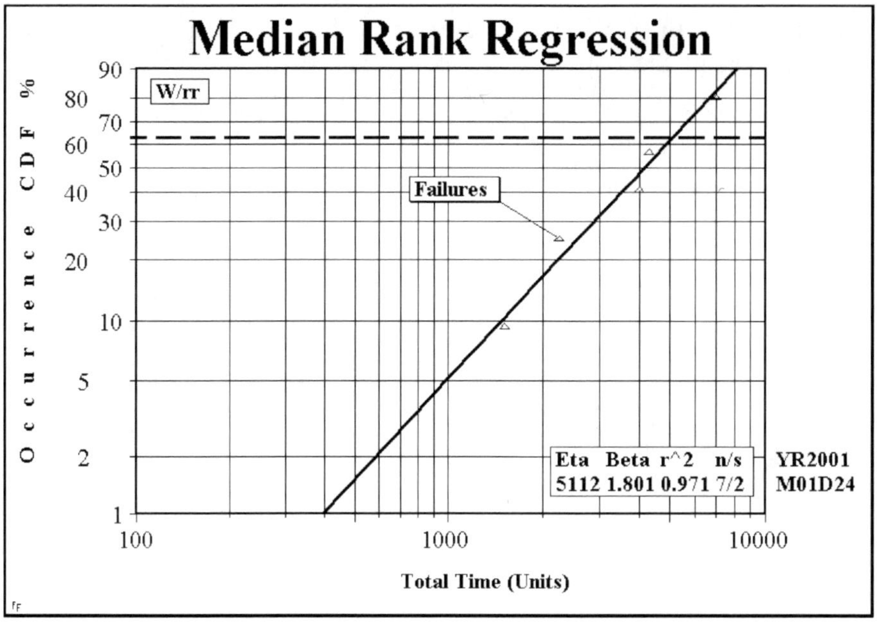

Figure B-1. Median Rank Regression

Appendix C: Maximum Likelihood Estimation

C.1. Foreword

Weibull analysis consists of fitting failure data to the Weibull distribution by estimating the parameters, β and η. The median rank regression method was presented in Chapter 2 and Appendix B; this Appendix presents an explanation and an example of maximum likelihood Weibull analysis. Further details of the maximum likelihood method can be found in [Nelson 1982], and [Lawless].

The maximum likelihood Weibull analysis method consists of finding the values β and η which maximize the "likelihood," of obtaining β and η, given the observed data. The likelihood is expressed in Weibull probability density form. It is a function of the data and the parameters β and η. If we are given the probability distribution and its parameters, and asked to calculate the probability of some event, that is a problem in probability. In a statistical problem, the data is employed in the distribution to make inferences about the parameters. In this statistical case, the product of the probability distributions for each respective data point is called the likelihood function.

C.2 Statistics, Probability and Likelihood

Statisticians prefer maximum likelihood estimates, MLE, to all other methods, because MLE have excellent statistical properties. As a result, statisticians recommend MLE as the primary method for fitting distributions to data. In contrast, the author and most engineers recommend the median rank regression (MRR) method as the primary method and MLE-RBA (Reduced Bias Adjustment) as an alternative. MLE is not recommended for life data analysis unless the data sets are enormous, 500 or more failures. Statisticians have different objectives for Weibull analysis which may explain the difference in opinion. There are advantages and disadvantages to both MLE and MRR methods which are described in Chapter 5, and later in this appendix. MLE's three major deficiencies are (1) that it is highly biased for small and medium size samples ... that is 1 to 100 failures, (2) it is very complex, requiring a computer solution, and (3) there is no graphical display of the data. Engineers like to look at a plot or graphic of the data in every case. Because there is no MLE data plot, there is no correlation coefficient. However, MLE does have advantages, particularly for non-life data. For example, it is often precise for estimating β and η. On the other hand, for small and medium size samples, β and η will be more biased than rank regression estimates. (See comparisons in Chapter 5.)

C.3 The Likelihood Function

The likelihood function is the statistical expression of the probability of β and η, given the observed data. It was first introduced as a part of the Bayes function. Sir Ronald Fisher later proved the great value of MLE, derived from the likelihood function. He showed the log likelihood function would be approximately normal in shape for moderate sample sizes. Many statisticians believe the likelihood function is the most important single statistic that exists and the author agrees.

When the sample is complete (all units are run to failure), the likelihood function is:

$$L = \prod_{i=1}^{n} f(x_i) = f(x_1)f(x_2)...f(x_n)$$

where n = sample size.

$$f(x) = \frac{dF(x)}{dx}$$

and for the Weibull distribution

$$F(x) = 1 - e^{-(x/\eta)^{\beta}}$$

In reliability terms, F(x) is the probability that a unit will fail before it acquires x units of operating time. F(x) is often called the "unreliability" at time x, and R(x) is the reliability at time x.

$$F(x) = 1 - R(x)$$

The Weibull probability density function is,

$$f(x) = \frac{\beta}{\eta}\left(\frac{x}{\eta}\right)^{\beta-1} e^{-(x/\eta)^{\beta}}$$

and the Weibull likelihood function is:

$$L = \prod_{i=1}^{n} \left(\frac{\beta}{\eta}\right)\left(\frac{x_i}{\eta}\right)^{\beta-1} e^{-(x_i/\eta)^{\beta}}$$

Note that the "likelihood" of the sample failure data,

$$x_1, x_2, \ldots x_n$$

is a function of the Weibull parameters β and η. The general form of the likelihood function for censored samples (where not every unit has been run to failure) is:

$$L = \prod_{i=1}^{r} f(x_i) \prod_{j=1}^{k} (1 - F(T_j))$$

where r = number of units run to failure and k = number of unfailed units

$$x_1, x_2, \ldots x_r = \textbf{known failure time}$$

$$T_1, T_2, \ldots T_k = \textbf{operating time on each unfailed unit}$$

When the time-to-failure distribution is Weibull,

$$L = \prod_{i=1}^{r} \left(\frac{\beta}{\eta}\right)\left(\frac{x_i}{\eta}\right)^{\beta-1} e^{-(x_i/\eta)^{\beta}} \prod_{j=1}^{k} e^{-(T_j/\eta)^{\beta}}$$

C.4 Maximizing the Likelihood Function

The maximum likelihood finds the values of β and η which maximize the log likelihood function. The method differentiates the logarithm of the likelihood function with respect to β and η, equates the resulting expressions to zero, and simultaneously solves for β and η. The maximum likelihood estimate of β is obtained from: (The "hat", ∧, signifies maximum likelihood).

$$\frac{\sum_{i=1}^{n} x_i^{\hat{\beta}} \ln x_i}{\sum_{i=1}^{n} x_i^{\hat{\beta}}} - \frac{1}{r}\sum_{i=1}^{r} \ln x_i - \frac{1}{\hat{\beta}} = 0$$

The maximum likelihood estimate of β is found using iterative procedures, where n = number of failures (r) + number of suspensions (k). Units censored at times T_i are assigned the values $x_{i+1} = T_i$. The second term in the equation sums the logarithms of the failure times only.

Analogous to the complete sample case, the maximum likelihood estimate of η is, in censored samples,

$$\hat{\eta} = \left(\frac{\sum_{i=1}^{n} x_i^{\hat{\beta}}}{r} \right)^{\frac{1}{\hat{\beta}}}$$

There are curious characteristics in these functions. An MLE estimate of β and η based on a single failure is possible if there are some right or late suspensions. This is surprising as it determines two parameters from a single failure (plus some suspensions). Further, if a plot of the MLE Weibull is made, and the single failure is plotted (using median ranks), the line does not go through the point! This is caused by our lack of a method for plotting data on an MLE Weibull.

C.5 Maximum Likelihood Example

The maximum likelihood method will be illustrated with the censored data listed below. (Note this is the same data employed in Appendix B to illustrate rank regression.)

Cycles	Status
1500	Failure
1750	Suspension
2250	Failure
4000	Failure
4300	Failure
5000	Suspension
7000	Failure

First use rank regression to make the Weibull plot as shown in Appendix B and Figure C-1.

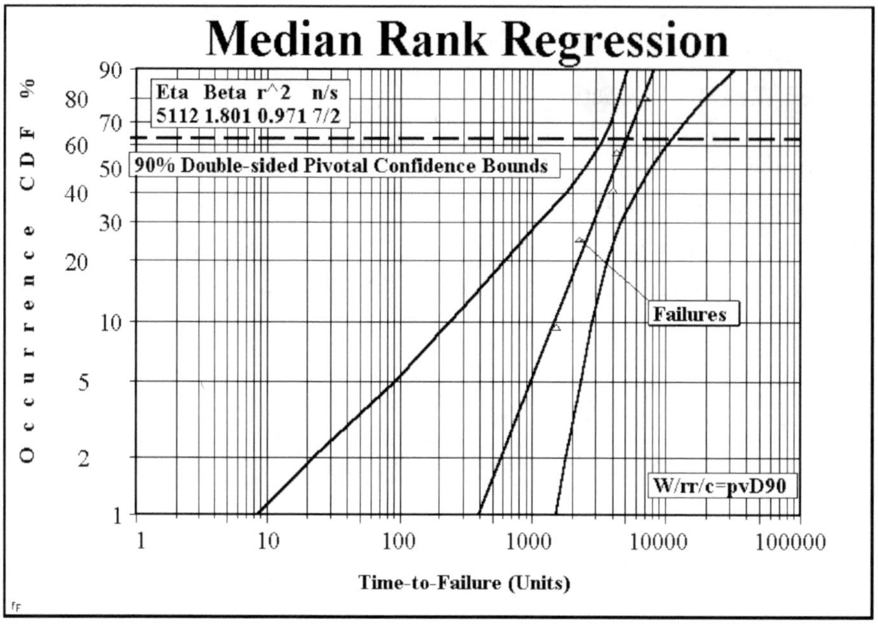

Figure C-1. Regression Solution from Appendix B

The confidence intervals shown are based on the pivotal method **best practice** for rank regression small samples.

The maximum likelihood estimate of β is the root of the equation below.

$$G(\beta) = \frac{\sum_{i=1}^{7} x_i^{\beta} \ln x_i}{\sum_{i=1}^{7} x_i^{\beta}} - \frac{1}{5}\sum_{i=1}^{5} \ln x_i - \frac{1}{\beta} = 0$$

An estimate of β of 1.8, was used as the initial value of β. This and subsequent estimates of β are listed below with the corresponding value of G(β).

β	G(β)
1.800	-0.1754
1.802	-0.1746
2.179	-0.0255
2.182	-0.0248
2.255	-0.0007
2.256	-0.00005
2.257	-0.0000

With the maximum likelihood estimate of $\hat{\beta}$ is 2.257, the maximum likelihood estimate of $\hat{\eta}$ = 4900.1 from:

$$\hat{\eta} = \left(\frac{\sum_{i=1}^{7} x_i^{2.257}}{5} \right)^{\frac{1}{2.257}} = 4900.1$$

Note that the MLE solution for η is identical to the Weibayes procedure based on a known β. Plotting the resulting MLE Weibull with data plotted with median ranks often indicates a poor "fit" but in fact, the plotted data has no relationship to MLE solution. See Figure C-2. Most analysts do not plot the data on MLE plots. SSW will make the MLE plot with or without the data.

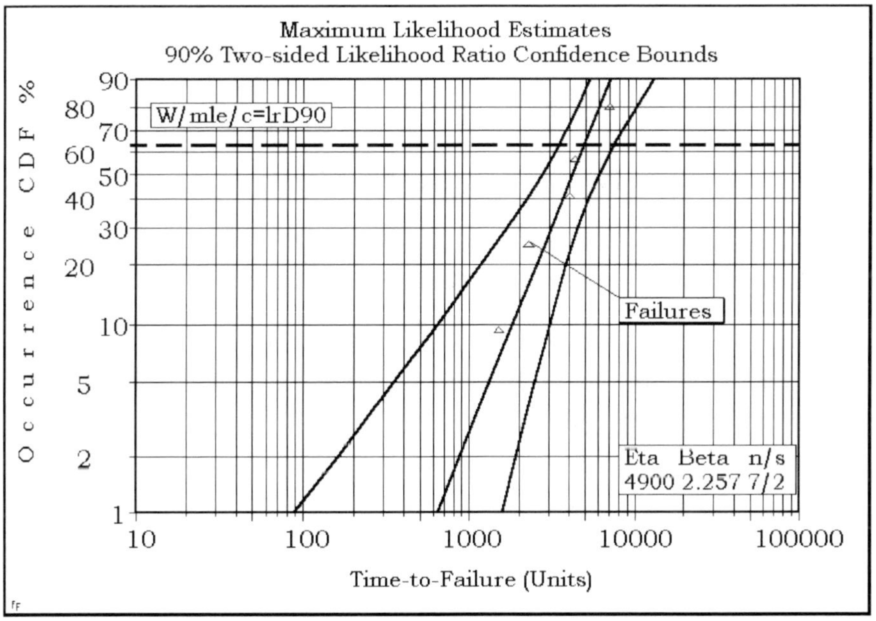

Figure C-2. Maximum Likelihood Estimates 90% Two-Sided Likelihood Ratio Confidence Bounds

There are MLE solutions for both the three parameter Weibull and lognormal, but they are not presented herein as they are overly complex. However, SSW provides this capability. The popular approach for three parameter solutions in other software is to determine t_0 using rank regression and then shift to MLE to estimate β and η. The likelihood function is also provided.

The Weibull likelihood function is a three dimensional surface. Horizontal slices through the surface form contours of equal likelihood. They are triangular shaped for small samples, and elliptically shaped for large samples. The function resembles a mountain with the MLE at the top. Using SuperSMITH Visual, the likelihood contours from SuperSMITH Weibull may be plotted. See Figure C-3 for a top-down view of the three dimensional mountain. The shape and width of the likelihood function are measures of the precision uncertainty of determining β and η. For example, the horizontal width of the 90% contour provides the 90% confidence intervals for η, and the vertical width shows 90% confidence intervals for β. These are likelihood ratio confidence intervals. Steep, narrow likelihood functions are preferred as they are more precise than broad likelihood functions. See Chapter 5 and the material below for comparisons of MLE and median rank regression (MRR).

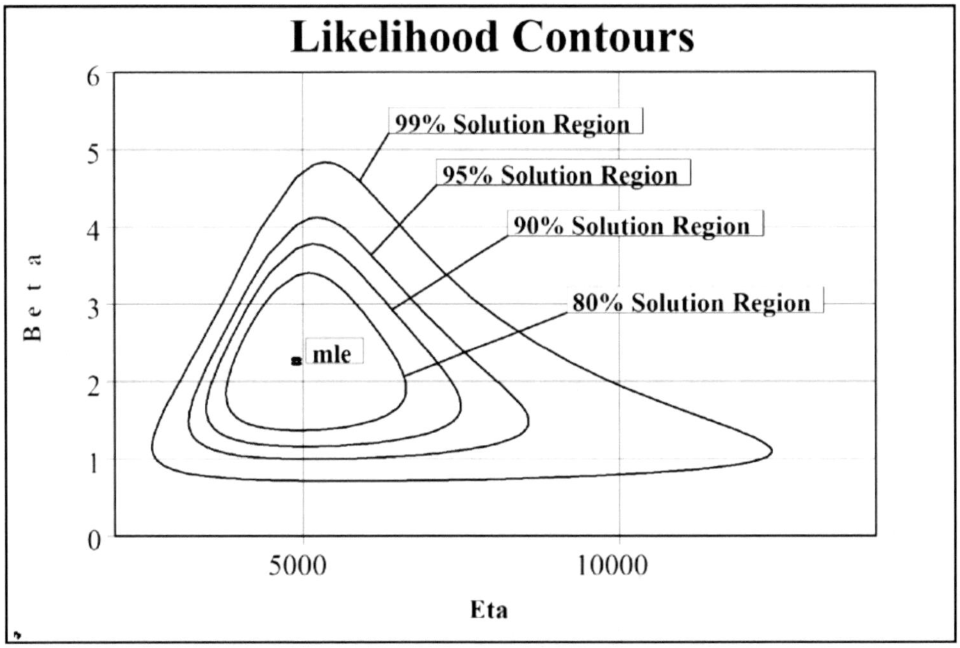

Figure C-3. Likelihood Contours

As described in Chapter 5, MLE betas are biased on the steep side with small samples. The author's new MLE-RBA eliminates this bias producing much more accurate results. This method was used to produce Figure C-4 which may be compared with Figure C-2. Note that the likelihood ratio contours may be adjusted using research developed by Wes Fulton to eliminate the small sample bias. The confidence bounds are wider, more conservative and the plot line is shallower. For small and medium size samples, (1-100 failures), MLE-RBA is recommended instead of MLE.

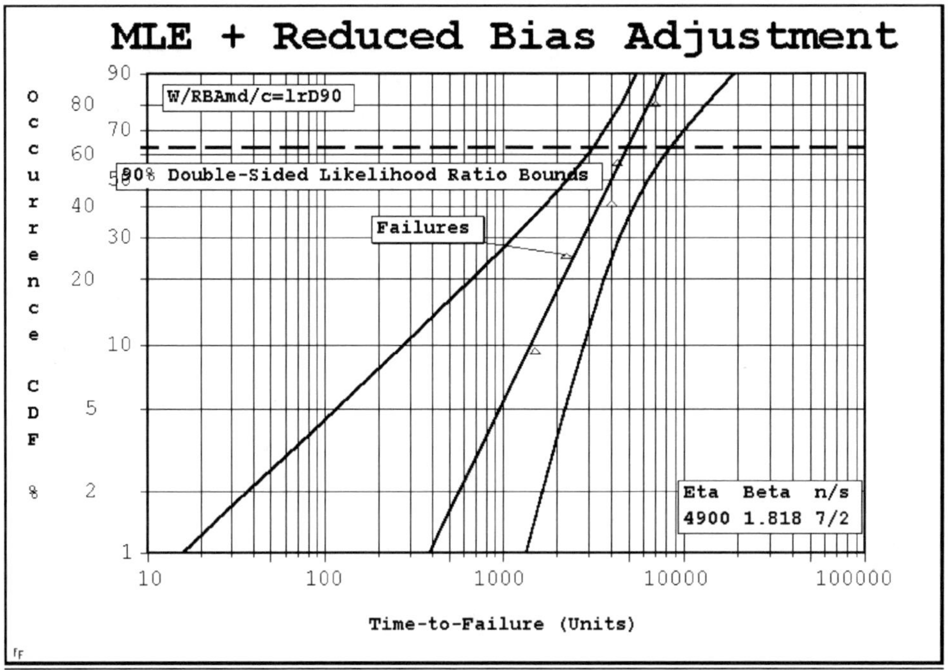

Figure C-4. MLE + Reduced Bias Adjustment

C.6 Interval MLE

There is also an MLE solution for interval or grouped data available in SSW. It will even analyze data when the inspection intervals vary from unit to unit. In fact it can be useful for data which is not grouped, where each unit has its own inspection schedule. Late research indicates interval MLE is slightly more accurate than either the inspection option or the Probit method. [Marquart- Appendix D]

Interval MLE has **four main data types, 1) occurrence, 2) suspension, 3) discovery, and 4) interval.** A failure age or event that is recorded with a precise counter or gauge is an example of an **occurrence** represented by a point on the time scale (such as t). The age of a unit that has not failed is an example of a **suspension** where the true time to failure is greater than (>t) the age attained so far. The analytical opposite of a suspension, called a **discovery**, occurs when a benign (or dormant) failure is discovered upon first inspection. For a discovery, the true time to failure is less than (<t) the age recorded at first inspection. Finally, all other inspection data for benign failures falls in the **interval** between the last inspection time (t1) and the next inspection time (t2) where failure was detected. In this case the true failure time is greater than the previous inspection age and less than the detection age (>t1<t2). Additionally each data type entry can have a frequency number associated. For example, ">t1<t2x5" indicates that there were five failures found between the ages of t1 and t2.

Interval MLE equations can be solved for problems involving all of these data types intermixed. This is the most general method for handling interval data. Many real data situations include a combination of all of these data types.

The general form of the likelihood function for failures and suspensions is given before in this Appendix. The general form of the likelihood function for discoveries and intervals is:

$$L_{Discoveries} = \prod_{m=1}^{m=p} \left(F\left(T_m\right)\right)$$

$$L_{Intervals} = \prod_{u=1}^{u=v} \left(F\left(T_u\right) - F\left(T_u - D_u\right)\right)$$

where:

- p = number of discovered samples
- v = number of interval samples
- T_M = Age for the m(th) discovered unit at its first inspection time
- T_u = Age for the u(th) interval sample at its inspection where failure was found
- D_u = Duration between T_u and age at last inspection where no failure was found
- F = CDF value

C.7 Maximum Likelihood Versus Median Rank Regression Estimates

Background. Safety considerations in some applications such as aerospace, automotive safety, and heart implants may require corrective action based on small samples of failure data, with or without large numbers of suspensions. Of prime importance is the accuracy and precision of the failure forecasts and the B.1 life estimates. Whenever there are alternative methods like MRR versus MLE, the author uses Monte Carlo simulations to identify the best practice. See Chapters 5 and 7, also [Wenham] and [Liu]. In this appendix we will summarize our conclusions on MRR versus MLE versus MLE-RBA.

Graphical Display. Rank regression, MRR, provides a **graphical display** of the data. This helps identify instances of poor fitting Weibull distribution plots perhaps suggesting another distribution, more than one failure mode affecting the units, mixtures of failure modes, batch problems, outliers. MLE does not provide a graphical display of the data. If the data is plotted with median ranks, the MLE line will usually be a bad fit. It is difficult to explain that the data as plotted does not relate to the MLE line. Engineers insist on a plot of the data while statisticians do not.

More Accurate. Rank regression and MLE-RBA provide more accurate and conservative estimates of "low" percentiles like the B.1 to B1 life, from small and medium sample sizes. These low B lives and corresponding high reliabilities may be extremely important for safety problems, warranties, guaranties and contract obligations. Maximum likelihood B lives tend be optimistically biased with small numbers of failures. See Chapter 5.

Risk Analysis. Rank regression and MLE-RBA failure forecasts are more accurate for small samples. MLE failure forecasts tend to be optimistic. MLE "now risks" cannot detect a batch problem. However, Appendix F describes the use of both MLE and median rank regression to detect batch problems using the aggregate cumulative hazard (ACH) function.

Beta & Eta. Maximum likelihood estimates tend to overestimate β with small samples. The slope of the Weibull plot is too steep. The slope of the lognormal and normal plots is similarly biased, too steep, as MLE standard deviation is underestimated. MRR and MLE-RBA provide more accurate estimates of the Weibull Beta and the normal/lognormal Sigma/SigF.

Confidence Intervals. Likelihood ratio interval estimates for MLE are rigorous and with MLE-RBA, they are adjusted for small sample size. Pivotal interval estimates are recommended for MRR. See Chapter 7

Convergence. Rank regression always provides a solution, while MLE will sometimes fail. However, the latest SSW has very high reliability, well over 99%.

Batch Problems. Rank regression "expected failures now" predictions will reflect the presence of batch problems. MLE is insensitive to these abnormalities in the data set.

One Failure. MLE may provide a solution with one failure and some right or late suspensions. The capability has large uncertainties, but there are situations where it cannot be avoided as it provides the only solution if beta is unknown. Weibayes is preferred if there is prior knowledge of beta.

Mathematical Rigor. There is a mathematical objection to the use of the regression least-squares method for rank regression. The residual scatter about the line is not uniform. The results are that the lower end of the line tends to be overweighed compared to the upper end. However, as all engineering interest is in the lower end of the curve, this is acceptable to engineers, actually preferred, while it is unacceptable to statisticians. MLE does have attractive mathematical qualities. Some authors never use regression on probability plots [Nelson, Meeker, Tobias]. Dr. Suzuki, of the University of Tokyo, suggests discarding the lower third of the data when fitting the line. The author strongly disagrees. See Chapter 5.

Presentations. For presentations to management, it is best to keep it simple and concise to improve communications. Rank regression plots are preferred for this purpose. MLE plots with data points located with median ranks are not recommended as they inspire comments about the poor fit of the Weibull line.

Summary. Assuming the data set is input to a computer, the author recommends the analyst employ both MRR and MLE-RBA for small samples. In most cases the two sets of results will be in reasonably good agreement providing some assurance of a good Weibull fit. MRR is still recommended as the engineering standard for all the reasons above but MLE-RBA based on the median bias described in Chapter 5 is a very good alternative. MLE-RBA based on mean bias instead of median bias requires a much larger correction (C_4^6 instead of $C_4^{3.5}$) but may be preferred by statisticians as it is consistent with Gosset's C_4 correction for the normal and lognormal. (However, Gosset was working with a symmetrical distribution.)

The author recommends MLE-RBA based on median bias as the distribution of beta is highly skewed. Table C-1 shows the mean bias results. The median bias results are in Chapter 5, Table 5-4A

Table C-1. Comparison of Methods for the Weibull Distribution
Mean Values from 20,000 Monte Carlo Trials, True Values Eta = 1000, Beta = 3.0, B1 = 215.8
...Bias = Mean – True value ...
Most accurate are in Bold, Italics

Sample Size	Median Ranks			MLE			Reduced Bias Adjustment		
	Eta	Beta	B1	Eta	Beta	B1	Eta	Beta	B1
4	***998.56***	3.817	257.33	981.03	4.942	336.95	981.03	***3.006***	***192.97***
6	***996.14***	3.354	366.66	986.97	4.002	292.67	986.97	***2.975***	***201.26***
10	***996.92***	3.145	***223.88***	993.64	3.506	260.58	993.64	***2.962***	206.53
30	***999.10***	***3.005***	***214.26***	997.90	3.149	230.47	997.90	2.991	213.70
100	1000.47	2.984	213.47	***999.67***	3.042	220.12	***999.67***	***3.000***	***215.50***

(Root Mean Square Error)

Sample Size	Median Ranks			MLE			Reduced Bias Adjustment		
	Eta	Beta	B1	Eta	Beta	B1	Eta	Beta	B1
4	175.98	3.011	185.45	***174.23***	4.130	221.95	***174.23***	***2.093***	***159.79***
6	143.70	1.649	142.09	***142.74***	2.131	163.53	***142.74***	***1.409***	***127.19***
10	112.10	1.063	107.88	***110.74***	1.162	113.17	***110.74***	***0.884***	***96.02***
30	64.84	0.552	62.41	***64.33***	0.499	57.63	***64.33***	***0.454***	***54.18***
100	35.70	0.302	34.82	***35.11***	0.245	29.90	***35.11***	***0.239***	***29.38***

C.8 Sir Ronald Fisher

Fisher on his desk calculator at Whittinghome Lodge 1952, from "R. A. Fisher, The Life of a Scientist," John Wiley & Sons 1978

The author thought Fisher was the greatest statistician of all time when he went to England to study under Professor George Barnard, Fisher's disciple. Sir Ronald Fisher was a genius and dominated both statistics and genetics during his lifetime. He developed most of what we know as engineering statistics including maximum likelihood estimation, the design of experiments, the theory of randomization, and the analysis of variance. Inspired by Gosset, he developed many of the distributions of small sample statistics. His lifelong war with Karl Pearson and later his son Egon Pearson continues today as a friendly rivalry among the disciples. He was friends with Egon Pearson until Egon (with Jerzy Neyman) invented confidence intervals and tests of hypothesis. He was so opposed to these concepts that the friendship ended. He was outspoken and arrogant, unfriendly by all accounts. Egon Pearson's book "Student" is an excellent documentary of the trilogy: Karl Pearson, Ronald Fisher and William Gosset.. The author now rates Gosset the greatest, with Fisher second greatest.

Sir Ronald Fisher

Professor George Barnard wrote in Fisher's obituary about his contributions to the theory and practice of statistics: "to assess the contributions made to a subject by one largely responsible for its creation would be futile."

The author has read much of Fisher's papers and books and highly recommends his first book, "The Design of Experiments." This little monologue can be read in an hour, there is no math and it provide the basic principles of DOE which Fisher invented. The author also recommends a derivative book, not by Fisher, "A Lady Tasting Tea." By way of criticism the author believes Fisher made two mistakes in his lifetime; he assumed mortality due to improper dosages would be normally distributed and he supported the tobacco companies in the fifties in the great debate that high correlation of lung cancer with smoking did not imply that smoking causes lung cancer. "Correlation is not causation." However, his many contributions overwhelm, indeed, totally outweigh, these mistakes.

Appendix D: Goodness of Fit

D.1 Introduction

Goodness of fit is a complex topic. There are many measures of goodness-of-fit available. The author believes the three best statistical methods are discussed herein and are available in SSW. There are two questions to answer with all medium and large data sets (21 failures or more): (1) how good does the data fit the distribution and (2) which distribution is the best fit? With small data sets (20 or less failures) the two parameter Weibull is recommended as there is not enough information to produce a credible distribution analysis. In Chapter 3 the correlation coefficient, r, and r^2, the coefficient of determination, were introduced as measures of goodness of fit for median rank regression solutions. The **Critical Correlation Coefficient squared** (CCC^2), the tenth percentile of r^2, was introduced as a lower 90% confidence bound for r^2. If the observed r^2 is less than CCC^2, a bad fit is indicated. Here, "**pve%**" denotes a smoothed simulation estimate of the p-value of r and r^2 and is discussed along with the fresh simulation estimate of the same thing, "**prr.**"

In Chapter 5 the **likelihood function and likelihood ratio** were presented as goodness of fit measures applicable to MLE and MLE-RBA methods. The **Anderson-Darling** is another method recommended by some experts. With experience, the **eyeball is a very good** measure of fit. Confirmed outliers should be removed before using any measures of goodness of fit. (See Handbook Section 3.4)

The author has been concerned with how good are these goodness-of-fit measures? To ensure the correct selection of best practices and to measure the **"power"** of these methods, the author asked Wes Fulton to program a Monte Carlo simulation within the most recent versions of SSW. "Power" is the probability that the goodness of fit method will correctly identify the parent distribution of the data as the best distribution, i.e., if the data came from a two parameter Weibull will the goodness of fit indicate the two parameter Weibull is the best distribution? This research is ongoing, a work in progress. The Monte Carlo simulation is now available in SSW. This means that you can now measure the power of the goodness of fit method you have chosen for your data set. Paul Barringer, Todd Marquart and Carl Tarum are major contributors in this research and Wes Fulton has incorporated their results in the software. Each of the three methods, p-value of r^2, likelihood ratio, and Anderson-Darling are open for possible improvement.

It may not be obvious that the results of distribution analysis depend on how many distributions are considered. For example, if four distributions are considered, the probability of selecting any of the distributions as best will be less than a three-distribution comparison. This was one of the reasons why the normal distribution was moved out of the life data comparisons years ago. The other reason is because all normally distributed data sets are fit well by the Weibull. Also, if a method has low power for one distribution it will have higher power for other distributions. For example, the r^2- CCC^2 has low power for Weibull 3P so it will have higher power for Weibull 2P.

P-Value of r^2. As presented in Chapter 3, pve% is an estimate of the true p-value from many Monte Carlo (MC) simulations, and prr% is a single simulation result provided with pivotal MC confidence bounds. The pve% estimate is based on the Critical Correlation Coefficient squared (CCC^2). So, the first problem for our research was to improve the accuracy of CCC^2. Paul Barringer calculated the values in Table D-1 based on ten million trials each compared to 1,000 trials used in earlier editions. These values were confirmed by Carl Tarum and are now incorporated in SSW.

Table D-1 CCC^2 Values as a function of the number of failures

Weibull	2P	Weibull	3P	Lognormal	
Failures	CCC^2	Failures	CCC^2	Failures	CCC^2
3	0.7920	3	0.8914	3	0.7939
4	0.7991	4	0.8905	4	0.7992
5	0.8076	5	0.8874	5	0.8144
6	0.8203	6	0.9003	6	0.8286
7	0.8325	7	0.9088	7	0.8416
8	0.8425	8	0.9154	8	0.8531
9	0.8515	9	0.8914	9	0.8630
10	0.8594	10	0.9254	10	0.8718

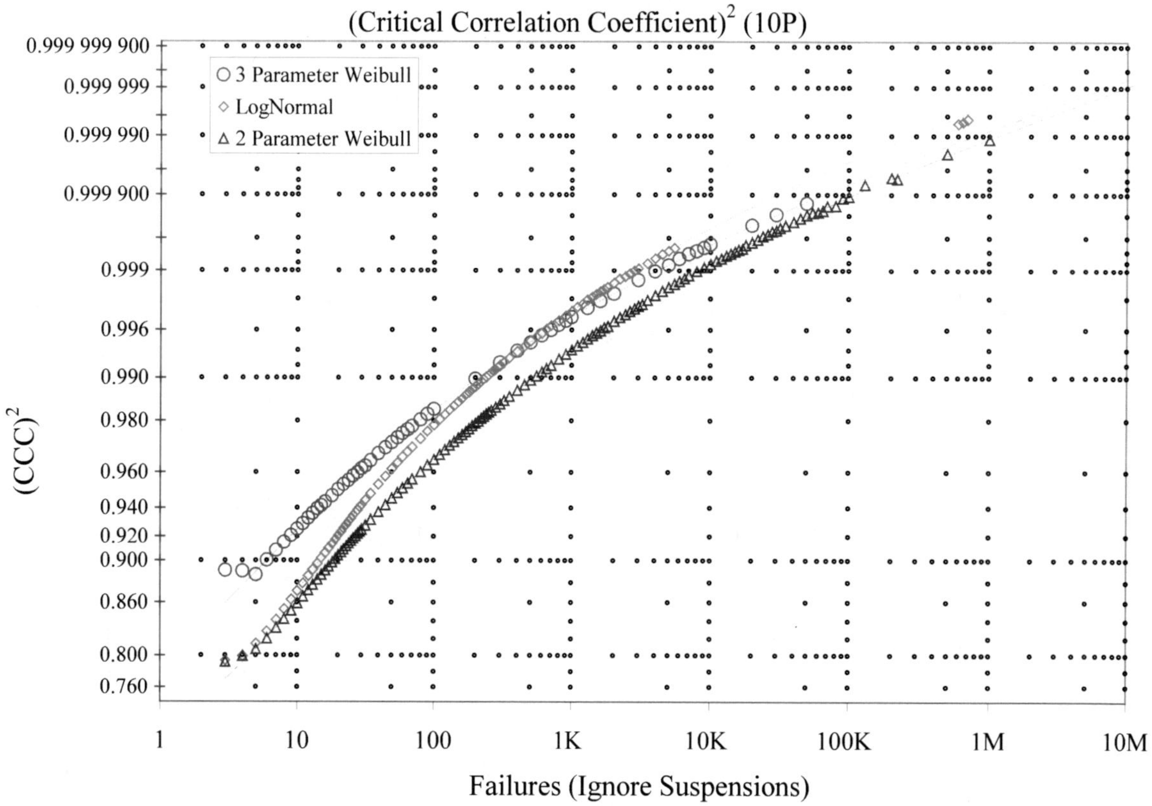

Figure D-1. CCC2 Versus Failures

The original Figure 3-4 in earlier editions was based on 1000 Monte Carlo trials and extended out to 100 failures. Figure D-1 is based on millions of trials and Carl Tarum has extended this to a million failures.

Power. With the new CCC^2 values, the next question is how well do the four methods; $r^2 - CCC^2$, pve%, MLE/LR, and Anderson-Darling, ("pA" on these plots, but "pAD" in software), identify the Weibull two parameter distribution as the best choice when the data is Weibull two parameter? What is the probability of a correct choice with this data? What is the "power" of these methods of goodness of fit? The results are shown in Figure D-2, D-3, and D-4. Notice the small sample power for Weibull 2P and lognormal 2P samples is about 50% or even less. For Weibull 3P samples it is much less. This substantiates our recommendation to require at least 21 failures for *credible distribution analysis results.*

"pve%" and "prr%". The older method, $r^2 - CCC^2$, was recommended before Wes Fulton developed pve%, and before Carl Tarum greatly improved pve% accuracy. Our new research shows that r^2- CCC^2 fails for Weibull 3P detection as suspected. The prr% (like pve% but using only a quick simulation) can be useful for special data sets with non-occurrences (non-failures), but in general pve% is more accurate.

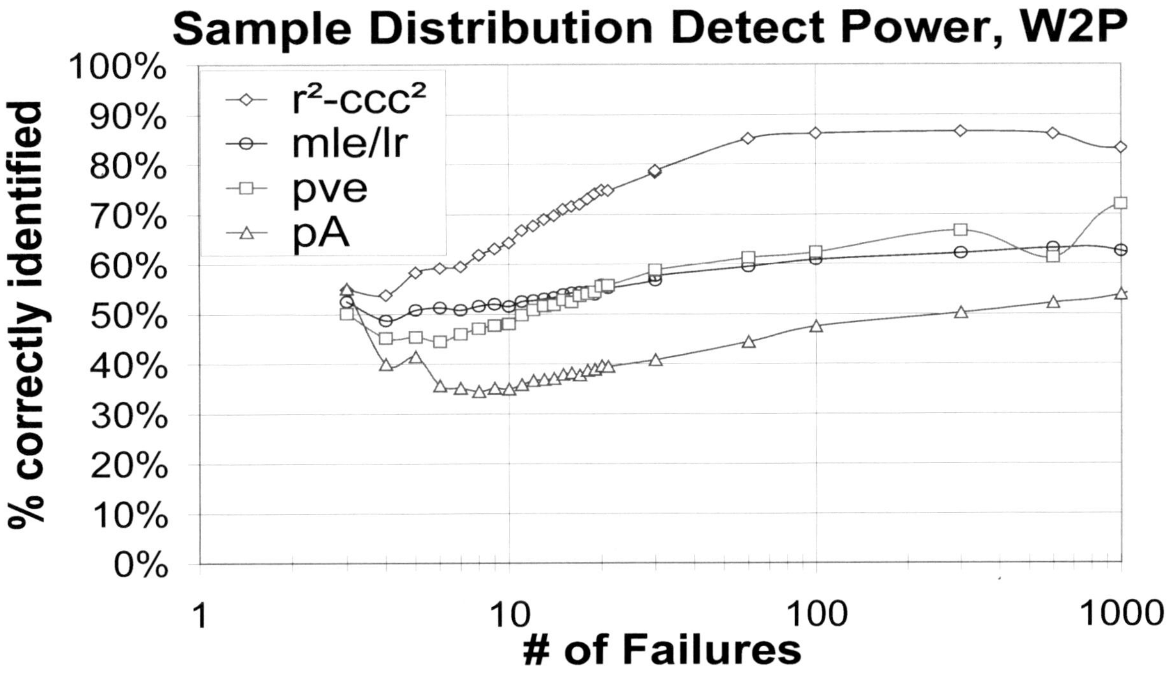

Figure D-2. Power for Weibull 2 Parameter

Likelihood Ratio based on Maximum Likelihood Estimate (mle/lr). If MLE-RBA is selected as the method, goodness of fit will utilize the likelihood function corrected for small samples with the Fulton Factor. For the two parameter solutions, Weibull 2P and lognormal 2P, the distribution maximum likelihood solution with the highest likelihood is the best choice between those two. This simple method cannot be employed with the Weibull 3P as the degrees of freedom are different from the two parameter distributions. The likelihood ratio (lr) is employed then to compare the best of the two parameter distributions with the Weibull 3P. If the p-value is greater than 50%, the W3P is recommended. The p-value may be varied to values greater than 50% if desired. The power of this method is shown in these figures. As [Chi Chao Liu] indicated in his thesis this method and the pve% method are the two best measures of goodness of fit compared to the other statistical measures.

Anderson – Darling ("pA" on these plots, but "pAD" in software). Statisticians often recommend this method. It is available in SSW under the computer icon. There are various corrections for small samples suggested in the literature. By comparison with the above two methods it does not show great power.

Research in Progress. The current technology in SSW is the state of the art for the best goodness-of-fit methods. For example, you may exercise the SSW Monte Carlo simulation for your data set and your distribution to estimate the power of the goodness of fit by these methods.

Thank You. The author would like to thank Paul Barringer, Wes Fulton, Carl Tarum and Todd Marquart for their extraordinary efforts on this project. Carl Tarum's paper on Goodness of Fit may be accessed through SAE (Paper 1999-01-057).

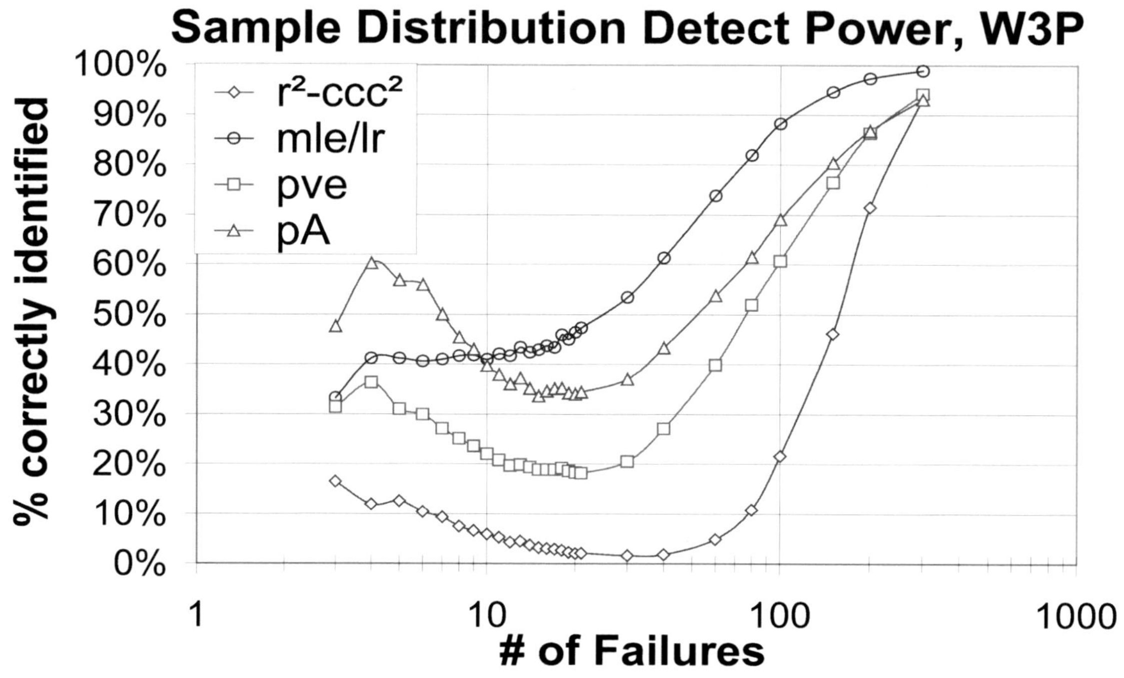

Figure D-3. Power for the Weibull 3 Parameter

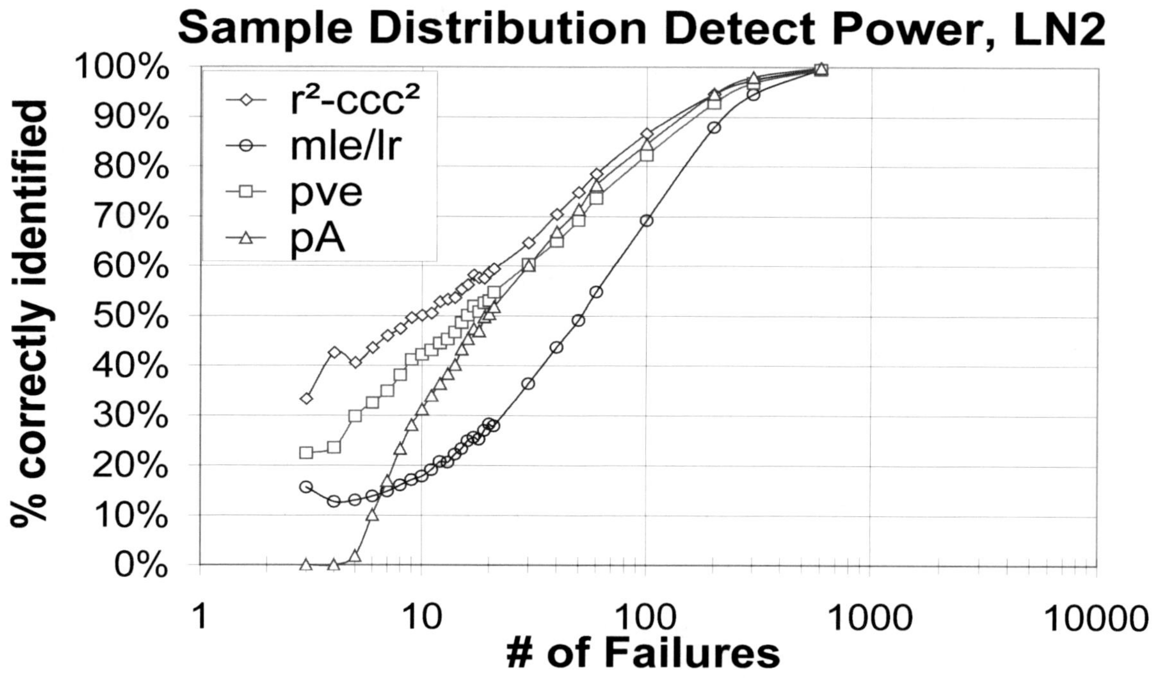

Figure D-4. Power for the Lognormal

Appendix E: Weibayes Analysis

E.1 Foreword

Weibayes is a method for constructing a Weibull distribution based on knowing or assuming a value of β, the Weibull slope parameter. It is used when there are certain deficiencies in the data. For instance, when there are no failures or few failures. Weibayes analysis is significantly more accurate than Weibull when beta is known from prior experience [Abernethy & Fulton 1992]. **Weibayes is best practice** for small samples if a reasonable estimate of β is available. Chapter 6 describes many applications of this method as well as the historical background.

In a Weibayes analysis, the slope/shape parameter β is assumed from historical failure data, prior experience, or from engineering knowledge of the physics of the failure. Given β, an equation may be derived using the method of maximum likelihood to determine the characteristic life, η.

$$\eta = \left[\sum_{i=1}^{n} \frac{t_i^{\beta}}{r} \right]^{1/\beta} \qquad \text{(E-1)}$$

where t_1 = time/cycles on unit_i
r = number of failed units
η = maximum likelihood estimate of characteristic life
n = total number of failures + suspensions

With β assumed and η calculated from Equation E-1, a Weibull distribution is derived. Alternatively Weibayes may use regression analysis to determine eta, given beta. A Weibayes line is plotted on Weibull probability paper. The Weibayes plot is used exactly like any other Weibull plot. Estimates of B lives, risk predictions, and reliability are available from Weibayes analysis.

If no failures have occurred, r is assumed to be one, i.e., the first failure is imminent. The most popular application is to test several redesigned units with no failures to show that the incidence of a particular Weibull failure mode is significantly reduced. Assuming one failure in this case, when there have been none, is obviously conservative, and the resulting Weibayes is also. This argument was employed for many years to explain Weibayes (and may still be preferred to the statistical derivation that follows). Later, Grant Reinman, of Pratt & Whitney Aircraft, working with the author, showed that Weibayes with no failures provides a lower one-sided confidence bound. With zero failures, $\hat{\eta}$ is a conservative 63% lower confidence bound on the true value of η if we assume the first failure is imminent. If failures have occurred and Weibayes is used, $\hat{\eta}$ is the maximum likelihood estimator of the true value of η [Nelson 1985]. Further, quantities calculated from maximum likelihood estimates are also MLE. Therefore, the Weibull line, B lives and reliabilities are MLE, if η is MLE. As mentioned earlier, Weibayes is a one parameter Weibull.

E.2 Weibayes Equation With no Failures

If no failures have occurred, the Weibayes equation with r = 1 gives a conservative 63% lower confidence bound on the true value of η. The lower bound is derived using two facts from statistics:

a. If $t_1, t_2, t_3, \ldots t_n$ represent failure times drawn from a Weibull population with slope parameter β and characteristic life η, then $t_1^{\beta}, t_2^{\beta}, \ldots, t_n^{\beta}$, represents a random sample from an exponential population with mean life $\theta = \eta^{\beta}$. The exponential distribution function is $F(t) = 1 - e^{-t/\theta}$.

b. If no failures have occurred in a fleet with n units having operating times $t_1, t_2, \ldots, t_n$, and the units are susceptible to an exponential failure mode, then a conservative 100(1 - α)% one-sided lower confidence limit on the mean life θ is:

$$\theta \geq \frac{\sum_{i=1}^{n} t_i}{-\ln \alpha} \tag{E-2}$$

Conservative here means that the true confidence level is unknown, but is at least 100(1 - α)%. Thus, if no failures have occurred in a fleet with n units having operating times $t_1, ..., t_n$ and the units are susceptible to a Weibull failure mode with known β (and unknown η), then a conservative 100(1 - α)% one-sided lower confidence limit $\theta = \eta^\beta$ is:

$$\theta = \eta^\beta \geq \frac{\sum_{i=1}^{n} t_i^\beta}{-\ln \alpha} \tag{E-3}$$

$$\eta \geq \left(\frac{\sum_{i=1}^{n} t_i^\beta}{-\ln \alpha} \right)^{1/\beta}$$

The Weibayes lower bound on η as shown in Equation E-1 is equal to the lower confidence bound in Equation E-3 with the denominator, -ln α = 1.0. Solving for α, we find: $-\ln\alpha = 1.0$ and therefore, $\alpha = e^{-1.0} = 0.368$. Thus, the Weibayes lower bound on η is a 100(1 - 0.368)% = 63.2% conservative lower confidence bound on η. The confidence level can be increased by increasing α in the denominator of the expression on the right-hand side of the inequality. Other confidence levels and corresponding denominators are given below. Note that with β assumed, determining a lower bound for η also determines a lower bound for the Weibull line.

Confidence	50%	80%	90%	95%	99%
-ln α	0.693	1.61	2.3	3.0	4.6

"Weibest" was an early Pratt & Whitney version based on r = 0.693 assumed failures instead of the usual Weibayes assumption of r = 1.0. 0.693 is less conservative. The "Weibest" line is a 50% confidence bound, a median estimate of the true Weibull characteristic life. Weibest is useful for apples-to-apples, median-to-median, comparisons with Weibull lines. With r = 1.0 assumed, Weibayes provides 63.2%, but Weibayes may be used at any confidence level.

E.3 Weibayes With Failures

If r failures have occurred, and Weibayes is used, $\hat{\eta}$ is the maximum likelihood estimator of the true value of η. This is shown by finding the value of η that maximizes the Weibull likelihood equations from Appendix C, while assuming that β, the Weibull slope parameter, is known. These calculations, similar to those discussed in Appendix C, result in the following equation for the maximum likelihood estimator of η (assuming that β is known):

$$\hat{\eta} = \left(\frac{\sum_{i=1}^{n} t_i^\beta}{r} \right)^{1/\beta} \tag{E-4}$$

[Nelson 1985] provides methods for determining confidence bounds with Weibayes when failures are present. See Chapter 6. These confidence bounds are available in SSW. 2008: New studies show regression analysis is more accurate than MLE for small samples, 20 failures or less. SSW has this capability.

Appendix F: Batch Failures Using the Aggregate Cumulative Hazard (ACH) Function

Contributed by: G. K. Cole (Specialist - Reliability Modeling Rolls Royce plc) & D. A. Nevell (Operational Research Specialist - EDS)

Key Words: Cumulative Hazard Function, Batch problems, Load-Strength interaction. See also Section 4.4.1 and Appendix J.

F.1 Batch Failures on Weibull Plots

Batch problems occur when "sub-sets" or "sub-groups" have markedly different failure characteristics from the rest of the population. Batch sub-groups can arise for many reasons (e.g., manufacturing errors, misuse or abuse, or unforeseen increases in service usage). Experience indicates that batch problems are very common in practice. Batch problems may sometimes be revealed by the shape of the Weibull plot.

One type of batch problem occurs due to an imperfection in the quality control of the production process. A small proportion of parts are sub-standard and exhibit a lower strength curve than required by the design. They may have machining flaws, inclusions, small cracks, or deficient material. If the strength of the sub-standard parts is significantly reduced, they will fail much earlier than the rest. If a Weibull plot is performed under this scenario, then a beta value of less than 1 usually results, implying an infant-mortality problem. There is a high risk of early failure for the sub-standard parts in the batch. After all the sub-standard parts have failed, the standard parts fail with longer lives. Figure F-1.

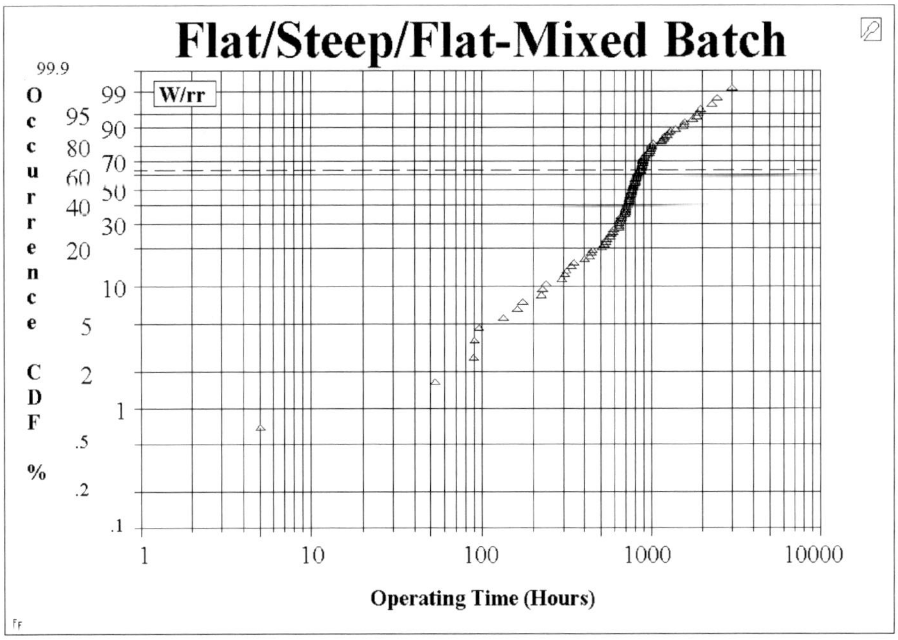

Figure F-1. Flat/Steep/Flat Mixed Batch Model

Another class of batch problems occurs when there is an increase in load or a reduction in the strength of some of the parts due to a process, operating usage, or handling error. In this case very early in-service failures may not occur. There are two onset points for failure, one for the batch or sub-family and one for the rest. See Figure F-2. The failure distribution for the batch can be represented by one Weibull line, and the rest by another one shifted to the right. A distinctive shape is observed, three distinct sections with a "steep-flat-steep" profile. The initial steep section comes from batch failures, the flat section from a mixture of batch and non-batch failures, the final steep section from the upper "non-batch" parts. In practice, when sufficient failure data is present, these "cusps" are quite common features observed in Weibull plots. Time and data constraints usually preclude getting both ends of the curve displayed in one problem. Sometimes we only observe the initial transition from steep-to-flat, because we are forced to undertake the analysis prior to the later upward swing from flat to steep.[1] Often, we do not even get as far as the first transition, getting a potentially misleading straight line plot for an undiscovered batch problem. Many variations on this theme are possible. Figure F-2 is just one variation among many.

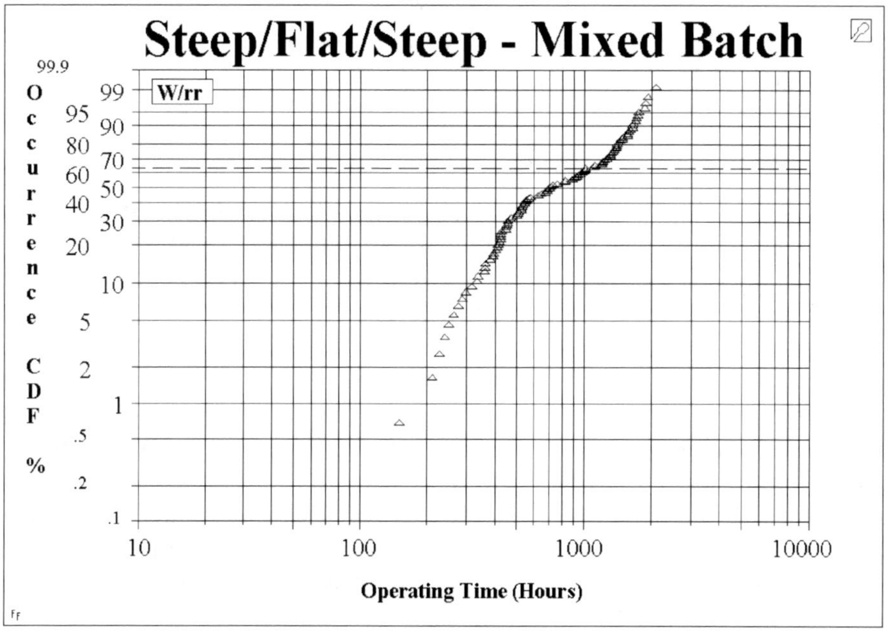

Figure F-2. Steep/Flat/Steep – Mixed Batch

F.2 Batch Problems With the "Present-Risk" Method

Chapter 4, Section 4.4.1, outlines the technique for validating the Weibull line by calculating the "present-risk". For service data, the present-risk calculation is performed by taking the fitted β and η parameters for the Median Rank Regression (MRR) line and aggregating the risk that each failed and unfailed (or suspended) component has accumulated under the fitted line:

$$\textbf{Expected failures} = \textbf{present risk} = \sum_{i=1}^{n} F(t_i) \qquad \text{(F-1)}$$

where:

$$F(t) = 1 - \exp(-(t/\eta)^{\beta})$$

and n is the number of failures and suspensions in the population.

For batch problems, the present-risk calculation will overestimate the number of failures that have occurred. If no batch problem exists, then the present-risk will be in reasonable agreement with the number of failures to date. Unfortunately, the same approach cannot be used for MLE. It is a characteristic of MLE

[1]Sometimes, this initial transition can be wrongly assumed to indicate the presence of a "t_0-shift" in the failure distribution.

that adjusts the fitted β and η to produce a present-risk that is close to the number of observed failures, regardless of whether a batch effect is present or not. If a batch effect is present, this property of MLE produces a noticeable disparity with the MRR line. The MLE line will usually cant down, have a shallower slope, with respect to that produced by MRR. Disagreement between the methods becomes more acute as the magnitude of the batch effect increases. For "interval-grouped" failure data, the inspection option in SuperSMITH Weibull may be used if the intervals are the same for all units. If not the analyst is forced to use the modified MLE function for interval data (covered by the "int5" option in SSW). The Likelihood function is given by:

$$\mathbf{L} = \sum_{\mathbf{i}} \ln\left[\mathbf{F}(\mathbf{T}_{\mathbf{ij_i}}) - \mathbf{F}(\mathbf{T}_{\mathbf{ij_i}-1})\right] \qquad \text{(F-2)}$$

The two terms inside the logarithm represent the likelihood function for failure of the ith component inside the jth and j-1th inspection interval. Suspension terms are dealt with in the same way as multi-censored data. For a particular problem, if MLE is employed, we need another approach, ideally graphical, which illustrates the goodness of fit of our model regardless of the fitting approach. Aggregate Cumulative Hazard (ACH) is offered as just such an approach.

F.3 The ACH Method

The hazard function h(t) for any continuous CDF, F(t), is:

$$\mathbf{f(t) = d(F(t)) / dt} \qquad \text{(F-3)}$$

$$\mathbf{R(t) = 1 - F(t)} \qquad \text{(F-4)}$$

$$\mathbf{h(t) = f(t) / R(t)} \qquad \text{(F-5)}$$

[Nelson 1982] states that "the hazard function is the instantaneous failure rate at age y", or more precisely "in the short time Δ from y to y + Δ, a proportion Δ h(y) of the population that reached age y fail". The Cumulative Hazard Function (CHF or H(t)) is derived directly from the hazard function by the following:

$$\mathbf{H(t) = \int_0^t h(t)dt} \qquad \text{(F-6)}$$

For Weibull:

$$\mathbf{H(t) = (t / \eta)^{\beta}} \qquad \text{(F-7)}$$

At Rolls Royce we have taken the concept of the Cumulative Hazard Function (or CHF) and extended it to detect batch problems using both MLE and MRR methods via the Aggregate Cumulative Hazard (or ACH) function. After fitting a Weibull to the data, ACH at time t is calculated by aggregating the H(t) value for all survivors at time t, and adding that to the sum of CHF values at the time of each failure occurring prior to t. This is best demonstrated by an example.

Column 1 of Table F-1 lists (in ascending order) a sub-set of 10 failure times sampled from a precise Weibull distribution with parameters β = 3 and η = 1000 hours. The next three columns illustrate the calculation of the hazard and corresponding CHF. The procedure is to estimate the hazard function via the reciprocal of the inverse rank of each failure, and then aggregate this value at each successive failure to obtain the CHF value at that life. The CHF reaches a value of 10 at 1000 hours (i.e., the characteristic life of the distribution). By contrast, the ACH function is derived at the first failure point as the aggregation of the CHF for each survivor at 411.5 hours (i.e., 9*(411.5/1000)^3=0.627), together with the one failure at 411.5 hours (i.e., 1*(411.5/1000)^3=0.07). At the second failure time the sum now becomes:

ACH = CHF (1st failure) + CHF (2nd failure) + Sum (ACH of 8 survivors)
= 0.07 + 0.178 + (8*0.178)
= 0.248 + 1.428 = 1.676 ---- (compared to 2 actual failures!)

This example Weibull illustrates the ACH method for complete failure data. ACH is also applicable when suspensions are present or when the format of the data is "Interval-grouped". For "Interval-grouped" data the ACH function increments in discrete steps at each inspection time, the actual failures only incrementing at those inspections where the failures are uncovered. ACH closely tracks the actual number of failures from zero hours right up to the present part time if the model is a good fit. If batch problems exists then ACH will deviate significantly from the cumulative observed failures. ACH can visually uncover batch, and is consistent and repeatable regardless of the form of the failure data or the fitting method employed. ACH can be used for complete failure data (as might be encountered on a test rig where all the components are run until they fail) as well as the more usual incomplete data.

Table F-1. Sample Calculations to Compare the Hazard, ACH, and Present Risk Methods

	Cumulative Hazard Method				ACH Method			
Failure Times	**Inverse Rank (k)**	**Hazard (10/k)**	**Cumulative Hazard**	**Survivors**	**Hazard (Failures)**	**Hazard (Suspensions)**	**ACH**	**F(t)**
411.50	10	1.000	1.000	9	0.070	0.627	0.697	0.067
563.03	9	1.111	2.111	8	0.248	1.428	1.676	0.163
669.87	8	1.250	3.361	7	0.549	2.104	2.653	0.260
760.42	7	1.429	4.790	6	0.988	2.638	3.627	0.356
844.06	6	1.667	6.456	5	1.590	3.007	4.596	0.452
926.09	5	2.000	8.456	4	2.384	3.177	5.561	0.548
1011.04	4	2.500	10.956	3	3.418	3.100	6.518	0.644
1104.81	3	3.333	14.290	2	4.766	2.697	7.463	0.740
1218.95	2	5.000	19.290	1	6.577	1.811	8.388	0.837
1392.22	1	10.000	29.290	0	9.276	0.000	9.276	0.933
								Total=5

F.4 A Case Study: Aero-Engines – (LP Turbine Strap Failures)

In the low-pressure turbine of one of our aero-engines, two adjacent turbine shrouds are welded together via a support strap. Cracking was experienced in the weld joining the support-strap to the turbine shrouds. After the initial crack developed it ran around the outside of the support strap. Eventually, either the support-strap parted company with the blade shrouds, or it was left partially connected to one blade. At this point its function in helping to retain the two blades together was lost.

The loss of the support-strap itself does not present a problem. However, with the strap's function lost an overload situation presents itself at the top of each of the two blades in the segment, and eventually one of them fails from a creep crack which emanates from the trailing edge near the shroud. When the blade eventually fails, an In-flight-shut-down (IFSD) occurs. Thus, we have multi-censored failure data.

The MLE and MRR plots are given in Figure F-3. The characteristic "canting-down" of the MLE line in relation to MRR will be noticed. This is a strong indication of a batch problem (i.e., a few engines have a turbine blade segment with a defective weld, the rest do not). The present-risk obtained from the MRR line was 16.5 compared to 8 actual failure events. Figures F-4 and F-5 show the corresponding ACH plots produced for both MRR and MLE. Both show significant deviation supporting the batch hypothesis. In this particular case it was established that larger platform gaps at top tolerance on some LPT segments during manufacture induced high stresses on the electron beam weld during assembly of those particular blade segments into their disk dovetails. Subsequently, some cracks progressed benignly along this weld whilst on other occasions they "joined-up" with the strap TIG weld which then unzipped and released the strap, eventually resulting in blade failure. Once the true cause was established, later risk mitigation included lower platform gaps on manufacture, blade inspection periods halved to catch cracks in time, and the imposition of a lower overall blade-set release life.

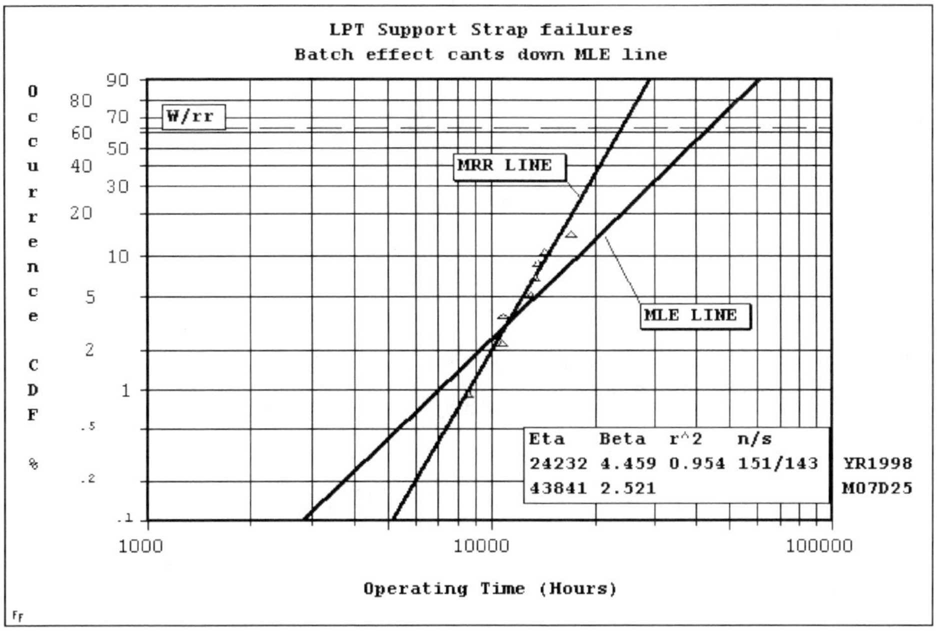

Figure F-3. LPT Strap

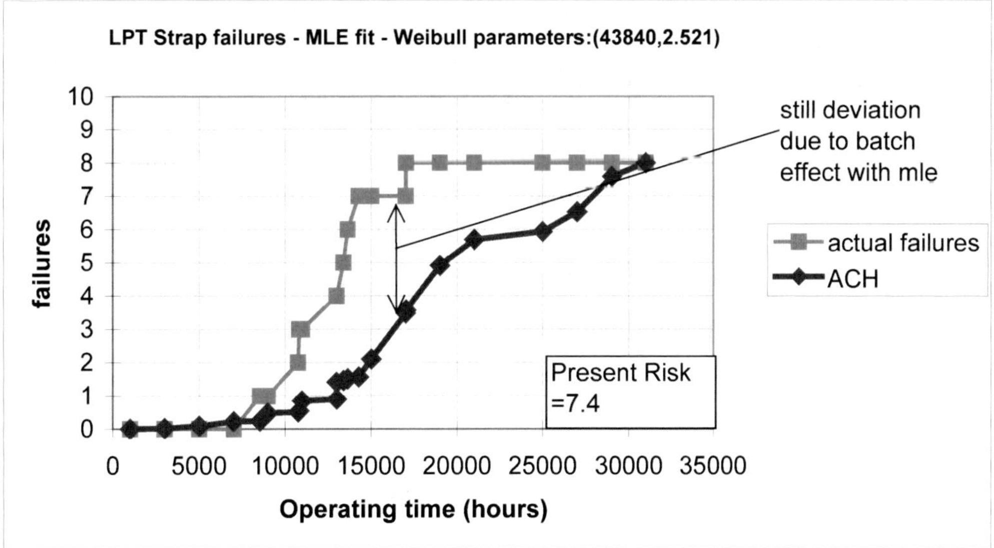

Figure F-4. LPT Strap Failures – MLE Fit – Weibull Parameters

F.5 Concluding Remarks

The ACH method has provided a useful visual fitting routine in order to spot batch problems, and is a valuable addition to the existing Weibull toolbox. It appears to consistently reveal the presence of batch problem regardless of the form of the available failure data (e.g., Multi-Censored or Interval-grouped).

A couple more examples (including an application for Interval-data with burnt Nozzle Guide Vanes) are covered in a more detailed paper given at the February '98 SAE Weibull Users Conference in Detroit.

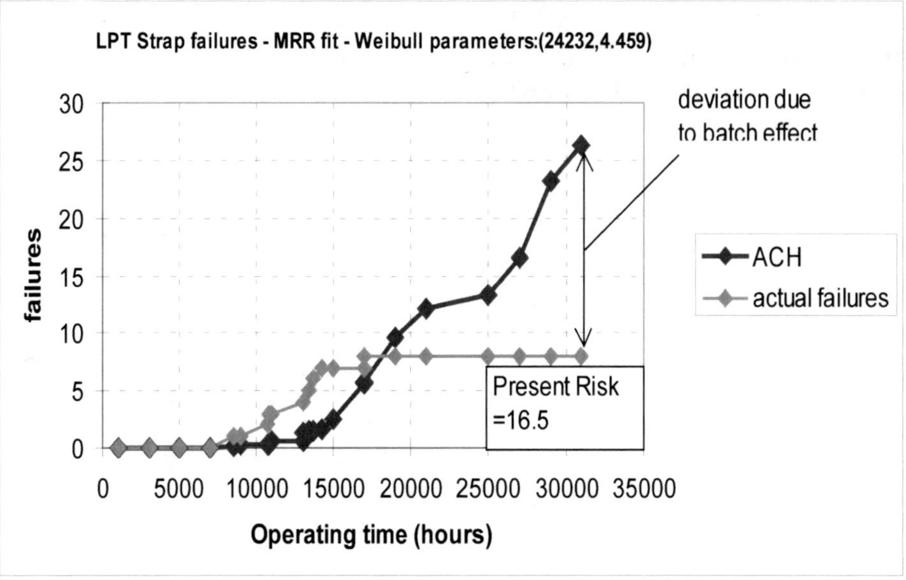

Figure F-5. LPT Strap Failures – MRR Fit – Weibull Parameters

Author's Comments: The ACH MRR method has proven to be a significant contribution to Weibull technology. It enhances the present risk method described in Chapter 4 by providing good graphics. It also makes some use of the MLE method even for these small sample batch problems. It works for complete samples which the present risk method cannot handle. The median rank regression aggregate cumulative hazard plot should be considered when a batch is suspected. For example, in separate correspondence, the authors at Rolls Royce studied the bleed valve problem at Base D shown in Figures 4-5 and 4-6. Note that the present risk of 22 compared to the ten observed failures does hint at a batch problem. Here is the ACH MRR plot of those same data. It confirms the suspected batch problem.

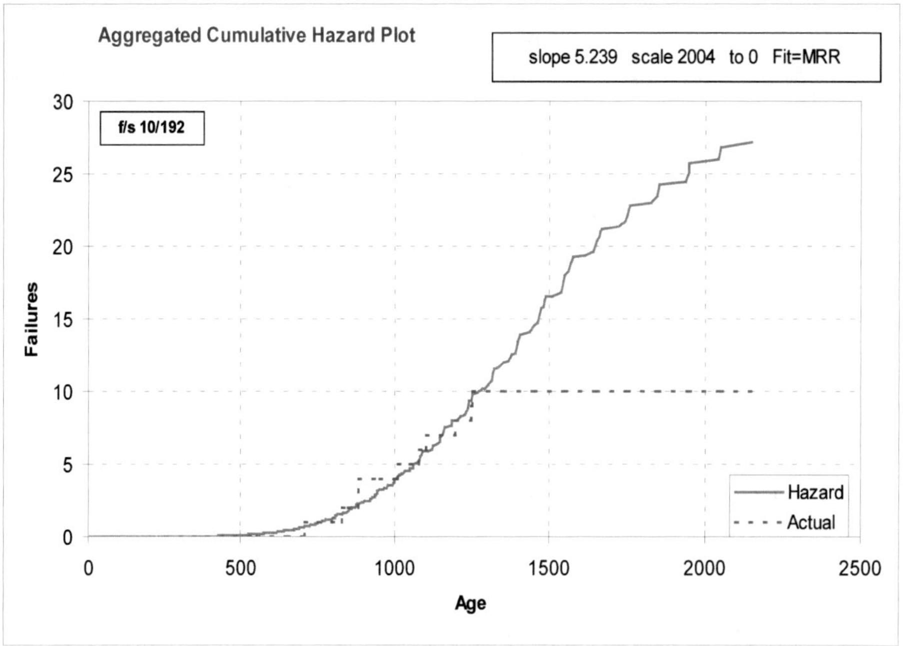

Figure F-6. Bleed Valve Shows Batch Effect

Appendix G: Weibull and Lognormal Mean and Variance

G.1 Rth Moments

In general, the rth moment about the origin is defined as:

$$\mu_r = \int_0^\infty t^r f(t)dt$$

where f(t) is the probability density function for the Weibull distribution, and $\beta > 0$, $t_0 = 0$, and $\eta > 0$. so, for the Weibull:

$$\mu_r = \frac{\beta}{\eta^\beta} \int_0^\infty t^{r+(\beta-1)} e^{-\left(\frac{t}{\eta}\right)^\beta} dt$$

Let $y=(t/\eta)^\beta$ and differentiate both sides:

$$\frac{\beta t^{\beta-1}}{\eta^\beta} dt = dy$$

and note that $t = \eta y^{1/\beta}$.

$$\therefore \frac{dy}{dt} = \frac{\beta t^{\beta-1}}{\eta^\beta}, \text{and } dt = \frac{\eta^\beta}{\beta t^{\beta-1}} dy$$

Substituting, we get

$$\mu_r = \frac{\beta}{\eta^\beta} \int_0^\infty \left(\eta y^{\frac{1}{\beta}}\right)^{r+(\beta-1)} e^{-y} \frac{\eta^\beta}{\beta\eta\left(y^{\frac{1}{\beta}}\right)^{\beta-1}} dy$$

removing parentheses, we have

$$\mu_r = \frac{\beta}{\eta^\beta}\int_0^\infty \eta^r \eta^{\beta\text{-}1}\left(y^{\frac{1}{\beta}}\right)^r \left(y^{\frac{1}{\beta}}\right)^{\beta\text{-}1} e^{-y}\ \frac{\eta^\beta}{\beta\eta^{\beta\text{-}1}\left(y^{\frac{1}{\beta}}\right)^{\beta\text{-}1}} d$$

∴ "canceling", we have

$$\mu_r = \int_0^\infty \eta^r \left(y^{\frac{1}{\beta}}\right)^r e^{-y}\,dy$$

noting that $(X^{\frac{a}{b}})^r = (X^{\frac{ar}{b}})$, this becomes

$$\mu_r = \int_0^\infty \eta^r \left(y^{\frac{r}{\beta}}\right) e^{-y}\,dy$$

Therefore, substituting this we obtain:

$$\mu_r = \eta^r \int_0^\infty y^{\frac{r}{\beta}} e^{-y}\,dy$$

G.2 Weibull Mean

The Weibull mean can be derived using the fact that the mean is the first moment about the origin; so, setting r = 1 in the last equation:

$$\mu_1 = \eta \int_0^\infty y^{\frac{1}{\beta}} e^{-y}\,dy$$

but the integral is the well known gamma function, (See Figure 2-3):

$$\mu_1 = \eta\,\Gamma\left(1+\frac{1}{\beta}\right) = \text{MTTF}$$

Also note that if $t_0 \neq 0$,

$$\text{MTTF} = t_0 + \eta\,\Gamma\left(1+\frac{1}{\beta}\right)$$

G.3 Weibull Variance

For the Weibull variance, first note that the variance is the second moment about the origin less the first moment squared. Then, the second moment about the origin is:

$$\mu_2' = \eta^2 \int_0^\infty y^{2/\beta} e^{-y} dy$$

where, taking advantage of the gamma function, we have:

$$\mu_2' = \eta^2 \,\Gamma\left(1 + \frac{2}{\beta}\right)$$

since,

$$\sigma^2 = \mu_2' - (\mu_1)^2$$

For the Weibull, we have:

$$\sigma^2 = \eta^2 \left(\Gamma\left(1 + \frac{2}{\beta}\right) - \left[\Gamma\left(1 + \frac{1}{\beta}\right)\right]^2 \right)$$

G.4 Weibull Mode

If $\beta \leq 1$, there is no mode; but for $\beta > 1$:

$$\text{Mode} = t_0 + \eta\left(1 - \frac{1}{\beta}\right)^{1/\beta}$$

G.5 Weibull Median

The Weibull distribution median is:

$$\text{Median} = t_0 + \eta\left[(\ln 2)^{1/\beta}\right]$$

G.6 Lognormal Mean and Standard Deviation

The Lognormal is more confusing than the Weibull because the parameters, the mean and standard deviation of the normal distribution can be in logarithms. Using SSW symbols, there is interest in the mean of the log of x, "muL", and the standard deviation of the log of x, "sigL", and their antilogs, "Log Mean Antilog" which is simply the median "Med", and the antilog of sigL which is called std. dev. Factor "SigF". In addition, the mean of x is "mu":

$$\mu = e^{\left(\mathrm{muL} + \left(\frac{\mathrm{sigL}^2}{2}\right)\right)}$$

Within SSW, the parameters, Med, and SigF, are estimated either by median rank regression or maximum likelihood methods.

G.7 Lognormal Variance

Using SuperSMITH symbols, the variance of (x) is:

$$(\mathbf{Sig})^2 = \left(\mathbf{e}^{\mathbf{sigL}^2} - 1\right) * \left(\mathbf{e}^{2\mathbf{muL} + \mathbf{sigL}^2}\right)$$

The antilog of sigL is SigF. When maximum likelihood is used to estimate sigL, a less biased estimate of sigma, "SigF1," is provided by taking the antilog of:

(sigL) x (square root of (N/N - 1) = sigF1

SuperSMITH® Lognormal Distribution Parameters

	Antilog Values of Log x Parameters	**Log x (Log Base e)**	**x**
Mean	MuAL "Med" (Log Mean Antilog)	muL	mu
Standard Deviation	SigAL "SigF" (Std. Dev. Factor)	sigL	sig

NOTE: "Med" is now used to designate the MuAL parameter shown in SuperSMITH® software for the lognormal solution, because MuAL and B50 (the median) are identical. "SigF" remains the designation of variability parameter shown. With Med and SigF as the lognormal parameters (both are in the original measurement scale domain), the base of logarithm used does not matter.

Appendix H – Weibull Graph Paper

WEIBULL PAPER
(3 Cycles)

WEIBULL DISTRIBUTION
BETA =
ETA =
SAMPLE SIZE =
FAILURES =

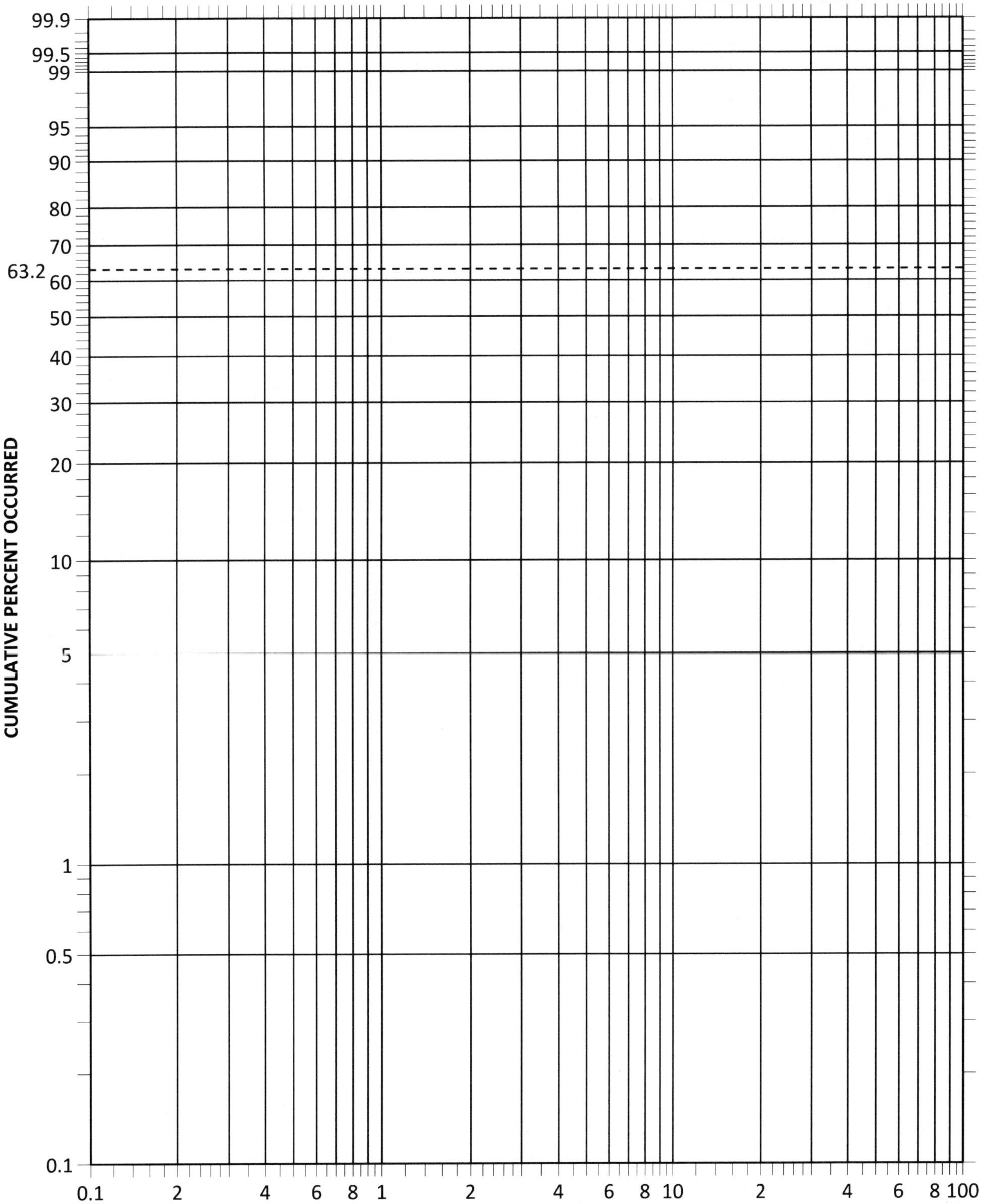

Customize the X-axis to your data. Start at the far left with a scale less than your actual data. If your smallest data is 85 then start your left horizontal scale at 10. Remember the vertical axis is % of cumulative failures.

WEIBULL PAPER
(3 Cycles)

WEIBULL DISTRIBUTION
BETA =
ETA =
SAMPLE SIZE =
FAILURES =

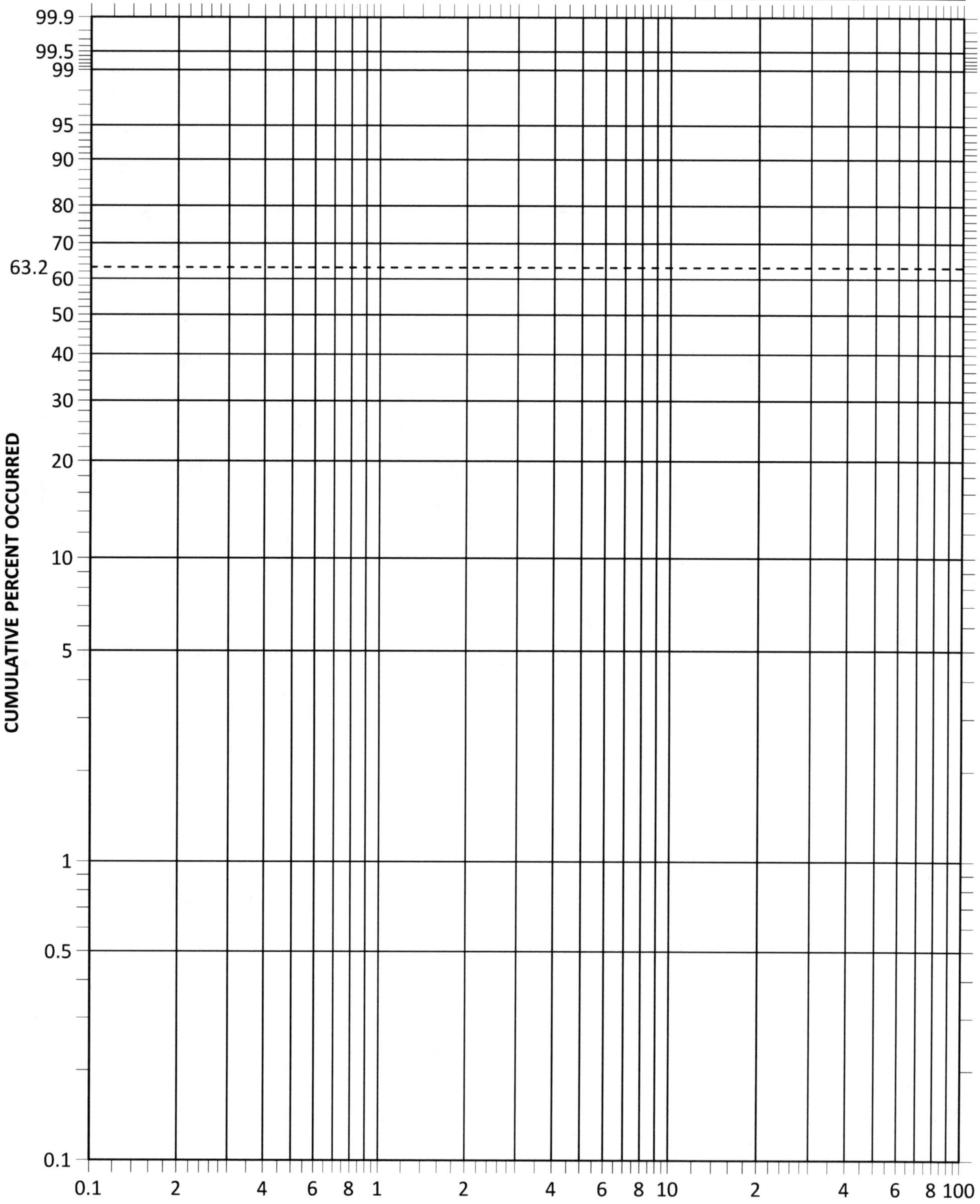

Appendix I: Beta-Binomial Ranks

I.1 Beta-Binomial 50% Ranks (Median)

Rank Order	Median Ranks (50%) Sample Size									
	1	2	3	4	5	6	7	8	9	10
1	50	29.29	20.63	15.91	12.94	10.91	9.43	8.30	7.41	6.70
2		70.71	50.00	38.57	31.38	26.44	22.85	20.11	17.96	16.23
3			79.37	61.43	50.00	42.14	36.41	32.05	28.62	25.86
4				84.09	68.62	57.86	50.00	44.02	39.31	35.51
5					87.06	73.56	63.59	55.98	50.00	45.17
6						89.09	77.15	67.95	60.69	54.83
7							90.57	79.89	71.38	64.49
8								91.70	82.04	74.14
9									92.59	83.77
10										93.30

Rank Order	Median Ranks (50%) Sample Size									
	11	12	13	14	15	16	17	18	19	20
1	6.11	5.61	5.19	4.83	4.52	4.24	4.00	3.78	3.58	3.41
2	14.80	13.60	12.58	11.70	10.94	10.27	9.68	9.15	8.68	8.25
3	23.58	21.67	20.04	18.65	17.43	16.37	15.42	14.58	13.83	13.15
4	32.38	29.76	27.53	25.61	23.94	22.47	21.18	20.02	18.99	18.05
5	41.19	37.85	35.02	32.58	30.45	28.59	26.94	25.47	24.15	22.97
6	50.00	45.95	42.51	39.54	36.97	34.71	32.70	30.92	29.32	27.88
7	58.81	54.05	50.00	46.51	43.48	40.82	38.47	36.37	34.49	32.80
8	67.62	62.15	57.49	53.49	50.00	46.94	44.23	41.82	39.66	37.71
9	76.42	70.24	64.98	60.46	56.52	53.06	50.00	47.27	44.83	42.63
10	85.20	78.33	72.47	67.42	63.03	59.18	55.77	52.73	50.00	47.54
11	93.89	86.40	79.96	74.39	69.55	65.29	61.53	58.18	55.17	52.46
12		94.39	87.42	81.35	76.06	71.41	67.30	63.63	60.34	57.37
13			94.81	88.30	82.57	77.53	73.06	69.08	65.51	62.29
14				95.17	89.06	83.63	78.82	74.53	70.68	67.20
15					95.48	89.73	84.58	79.98	75.85	72.12
16						95.76	90.32	85.42	81.01	77.03
17							96.00	90.85	86.17	81.95
18								96.22	91.32	86.85
19									96.42	91.75
20										96.59

For those of you that would like to generate your own median ranks, Carl Tarum suggests the following **Excel method:** You can generate the ranks in Excel with the following function
= BETA.INV(C, J, N-J+1)
where:

C is the confidence level
J is the rank order
N is the sample size

Rank Order	Median Ranks (50%) Sample Size 21	22	23	24	25	26	27	28	29	30
1	3.25	3.10	2.97	2.85	2.73	2.63	2.53	2.45	2.36	2.28
2	7.86	7.51	7.19	6.90	6.62	6.37	6.14	5.92	5.72	5.53
3	12.53	11.97	11.46	10.99	10.55	10.15	9.78	9.44	9.11	8.81
4	17.21	16.44	15.73	15.09	14.49	13.94	13.43	12.96	12.52	12.10
5	21.89	20.91	20.01	19.19	18.43	17.74	17.09	16.48	15.92	15.40
6	26.57	25.38	24.30	23.30	22.38	21.53	20.74	20.01	19.33	18.69
7	31.26	29.86	28.58	27.41	26.32	25.32	24.40	23.54	22.74	21.99
8	35.94	34.33	32.86	31.51	30.27	29.12	28.06	27.07	26.14	25.28
9	40.63	38.81	37.15	35.62	34.22	32.92	31.71	30.59	29.55	28.58
10	45.31	43.29	41.43	39.73	38.16	36.71	35.37	34.12	32.96	31.87
11	50.00	47.76	45.72	43.84	42.11	40.51	39.03	37.65	36.37	35.17
12	54.69	52.24	50.00	47.95	46.05	44.31	42.68	41.18	39.77	38.46
13	59.37	56.71	54.28	52.05	50.00	48.10	46.34	44.71	43.18	41.76
14	64.06	61.19	58.57	56.16	53.95	51.90	50.00	48.24	46.59	45.06
15	68.74	65.67	62.85	60.27	57.89	55.69	53.66	51.76	50.00	48.35
16	73.43	70.14	67.14	64.38	61.84	59.49	57.32	55.29	53.41	51.65
17	78.11	74.62	71.42	68.49	65.78	63.29	60.97	58.82	56.82	54.94
18	82.79	79.09	75.70	72.59	69.73	67.08	64.63	62.35	60.23	58.24
19	87.47	83.56	79.99	76.70	73.68	70.88	68.29	65.88	63.63	61.54
20	92.14	88.03	84.27	80.81	77.62	74.68	71.94	69.41	67.04	64.83
21	96.75	92.49	88.54	84.91	81.57	78.47	75.60	72.93	70.45	68.13
22		96.90	92.81	89.01	85.51	82.26	79.26	76.46	73.86	71.42
23			97.03	93.10	89.45	86.06	82.91	79.99	77.26	74.72
24				97.15	93.38	89.85	86.57	83.52	80.67	78.01
25					97.27	93.63	90.22	87.04	84.08	81.31
26						97.37	93.86	90.56	87.48	84.60
27							97.47	94.08	90.89	87.90
28								97.55	94.28	91.19
29									97.64	94.47
30										97.72

I.2 Beta-Binomial 5% Ranks

Rank Order	5% Ranks Sample Size 1	2	3	4	5	6	7	8	9	10
1	5.00	2.53	1.70	1.27	1.02	0.85	0.73	0.64	0.57	0.51
2		22.36	13.54	9.76	7.64	6.28	5.34	4.64	4.10	3.68
3			36.84	24.86	18.93	15.32	12.88	11.11	9.77	8.73
4				47.29	34.26	27.13	22.53	19.29	16.88	15.00
5					54.93	41.82	34.13	28.92	25.14	22.24
6						60.70	47.93	40.03	34.49	30.35
7							65.18	52.93	45.04	39.34
8								68.77	57.09	49.31
9									71.69	60.58
10										74.11

Rank Order	5% Ranks Sample Size 11	12	13	14	15	16	17	18	19	20
1	0.47	0.43	0.39	0.37	0.34	0.32	0.30	0.28	0.27	0.26
2	3.33	3.05	2.81	2.60	2.42	2.27	2.13	2.01	1.90	1.81
3	7.88	7.19	6.60	6.11	5.68	5.31	4.99	4.70	4.45	4.22
4	13.51	12.29	11.27	10.40	9.67	9.03	8.46	7.97	7.53	7.14
5	19.96	18.10	16.57	15.27	14.17	13.21	12.38	11.64	10.99	10.41
6	27.12	24.53	22.40	20.61	19.09	17.78	16.64	15.63	14.75	13.96
7	34.98	31.52	28.70	26.36	24.37	22.67	21.19	19.90	18.75	17.73
8	43.56	39.09	35.48	32.50	30.00	27.86	26.01	24.40	22.97	21.71
9	52.99	47.27	42.74	39.04	35.96	33.34	31.08	29.12	27.39	25.87
10	63.56	56.19	50.54	46.00	42.26	39.10	36.40	34.06	32.01	30.20
11	76.16	66.13	58.99	53.43	48.92	45.17	41.97	39.22	36.81	34.69
12		77.91	68.37	61.46	56.02	51.56	47.81	44.60	41.81	39.36
13			79.42	70.33	63.66	58.34	53.95	50.22	47.00	44.20
14				80.74	72.06	65.62	60.44	56.11	52.42	49.22
15					81.90	73.60	67.38	62.33	58.09	54.44
16						82.93	74.99	68.97	64.06	59.90
17							83.84	76.23	70.42	65.63
18								84.67	77.36	71.74
19									85.41	78.39
20										86.09

Rank Order	5% Ranks Sample Size 21	22	23	24	25	26	27	28	29	30
1	0.24	0.23	0.22	0.21	0.20	0.20	0.19	0.18	0.18	0.17
2	1.72	1.64	1.57	1.50	1.44	1.38	1.33	1.28	1.24	1.20
3	4.01	3.82	3.65	3.50	3.35	3.22	3.10	2.98	2.88	2.78
4	6.78	6.46	6.17	5.90	5.66	5.43	5.22	5.03	4.85	4.69
5	9.88	9.41	8.98	8.59	8.23	7.90	7.59	7.31	7.05	6.81
6	13.24	12.60	12.02	11.49	11.01	10.56	10.15	9.77	9.42	9.09
7	16.82	15.99	15.25	14.57	13.95	13.38	12.85	12.37	11.92	11.50
8	20.57	19.56	18.63	17.80	17.03	16.33	15.68	15.09	14.53	14.02
9	24.50	23.27	22.16	21.16	20.24	19.40	18.62	17.91	17.25	16.63
10	28.58	27.13	25.82	24.64	23.56	22.57	21.66	20.82	20.05	19.33
11	32.81	31.13	29.61	28.24	26.99	25.84	24.79	23.83	22.93	22.11
12	37.19	35.25	33.51	31.94	30.51	29.21	28.01	26.91	25.89	24.95
13	41.72	39.52	37.54	35.76	34.14	32.66	31.31	30.07	28.93	27.87
14	46.41	43.91	41.68	39.68	37.86	36.21	34.70	33.31	32.03	30.85
15	51.26	48.45	45.95	43.71	41.68	39.84	38.16	36.62	35.20	33.89
16	56.30	53.15	50.36	47.86	45.61	43.57	41.71	40.00	38.44	36.99
17	61.56	58.02	54.90	52.13	49.64	47.38	45.34	43.46	41.75	40.16
18	67.08	63.09	59.61	56.53	53.78	51.30	49.05	47.00	45.12	43.39
19	72.94	68.41	64.51	61.09	58.05	55.32	52.86	50.62	48.57	46.69
20	79.33	74.05	69.64	65.82	62.46	59.46	56.77	54.33	52.10	50.06
21	86.71	80.19	75.08	70.77	67.04	63.74	60.79	58.13	55.71	53.49
22		87.27	80.98	76.02	71.83	68.18	64.94	62.03	59.40	57.01
23			87.79	81.71	76.90	72.81	69.24	66.06	63.20	60.61
24				88.27	82.39	77.71	73.73	70.23	67.11	64.30
25					88.71	83.02	78.47	74.58	71.16	68.10
26						89.12	83.60	79.18	75.39	72.04
27							89.50	84.15	79.84	76.14
28								89.85	84.66	80.47
29									90.19	85.14
30										90.50

I.3 Beta-Binomial 95% Ranks

Rank Order	95% Ranks Sample Size 1	2	3	4	5	6	7	8	9	10
1	95.00	77.64	63.16	52.71	45.07	39.30	34.82	31.23	28.31	25.89
2		97.47	86.46	75.14	65.74	58.18	52.07	47.07	42.91	39.42
3			98.30	90.24	81.07	72.87	65.87	59.97	54.96	50.69
4				98.73	92.36	84.68	77.47	71.08	65.51	60.66
5					98.98	93.72	87.12	80.71	74.86	69.65
6						99.15	94.66	88.89	83.12	77.76
7							99.27	95.36	90.23	85.00
8								99.36	95.90	91.27
9									99.43	96.32
10										99.49

Rank Order	95% Ranks Sample Size 11	12	13	14	15	16	17	18	19	20
1	23.84	22.09	20.58	19.26	18.10	17.07	16.16	15.33	14.59	13.91
2	36.44	33.87	31.63	29.67	27.94	26.40	25.01	23.77	22.64	21.61
3	47.01	43.81	41.01	38.54	36.34	34.38	32.62	31.03	29.58	28.26
4	56.44	52.73	49.46	46.57	43.98	41.66	39.56	37.67	35.94	34.37
5	65.02	60.91	57.26	54.00	51.08	48.44	46.05	43.89	41.91	40.10
6	72.88	68.48	64.52	60.96	57.74	54.83	52.19	49.78	47.58	45.56
7	80.04	75.47	71.30	67.50	64.04	60.90	58.03	55.40	53.00	50.78
8	86.49	81.90	77.60	73.64	70.00	66.66	63.60	60.78	58.19	55.80
9	92.12	87.71	83.43	79.39	75.63	72.14	68.92	65.94	63.19	60.64
10	96.67	92.81	88.73	84.73	80.91	77.33	73.99	70.88	67.99	65.31
11	99.53	96.95	93.40	89.60	85.83	82.22	78.81	75.60	72.61	69.80
12		99.57	97.19	93.89	90.33	86.79	83.36	80.10	77.03	74.13
13			99.61	97.40	94.32	90.97	87.62	84.37	81.25	78.29
14				99.63	97.58	94.69	91.54	88.36	85.25	82.27
15					99.66	97.73	95.01	92.03	89.01	86.04
16						99.68	97.87	95.30	92.47	89.59
17							99.70	97.99	95.55	92.86
18								99.72	98.10	95.78
19									99.73	98.19
20										99.74

Rank Order	95% Ranks Sample Size 21	22	23	24	25	26	27	28	29	30
1	13.29	12.73	12.21	11.73	11.29	10.88	10.50	10.15	9.81	9.50
2	20.67	19.81	19.02	18.29	17.61	16.98	16.40	15.85	15.34	14.86
3	27.06	25.95	24.92	23.98	23.10	22.29	21.53	20.82	20.16	19.53
4	32.92	31.59	30.36	29.23	28.17	27.19	26.27	25.42	24.61	23.86
5	38.44	36.91	35.49	34.18	32.96	31.82	30.76	29.77	28.84	27.96
6	43.70	41.98	40.39	38.91	37.54	36.26	35.06	33.94	32.89	31.90
7	48.74	46.85	45.10	43.47	41.95	40.54	39.21	37.97	36.80	35.70
8	53.59	51.55	49.64	47.87	46.22	44.68	43.23	41.87	40.60	39.39
9	58.28	56.09	54.05	52.14	50.36	48.70	47.14	45.67	44.29	42.99
10	62.81	60.48	58.32	56.29	54.39	52.62	50.95	49.38	47.90	46.51
11	67.19	64.75	62.46	60.32	58.32	56.43	54.66	53.00	51.43	49.94
12	71.42	68.87	66.49	64.24	62.14	60.16	58.29	56.54	54.88	53.31
13	75.50	72.87	70.39	68.06	65.86	63.79	61.84	60.00	58.25	56.61
14	79.43	76.73	74.18	71.76	69.49	67.34	65.30	63.38	61.56	59.84
15	83.18	80.44	77.84	75.36	73.01	70.79	68.69	66.69	64.80	63.01
16	86.76	84.01	81.37	78.84	76.44	74.16	71.99	69.93	67.97	66.11
17	90.12	87.40	84.75	82.20	79.76	77.43	75.21	73.09	71.07	69.15
18	93.22	90.59	87.98	85.43	82.97	80.60	78.34	76.17	74.11	72.13
19	95.99	93.54	91.02	88.51	86.05	83.67	81.38	79.18	77.07	75.05
20	98.28	96.18	93.83	91.41	88.99	86.62	84.32	82.09	79.95	77.89
21	99.76	98.36	96.35	94.10	91.77	89.44	87.15	84.91	82.75	80.67
22		99.77	98.43	96.50	94.34	92.10	89.85	87.63	85.47	83.37
23			99.78	98.50	96.65	94.57	92.41	90.23	88.08	85.98
24				99.79	98.56	96.78	94.78	92.69	90.58	88.50
25					99.80	98.62	96.90	94.97	92.95	90.91
26						99.80	98.67	97.02	95.15	93.19
27							99.81	98.72	97.12	95.31
28								99.82	98.76	97.22
29									99.82	98.80
30										99.83

Appendix J: Mixtures of Populations and Failure Modes & YBath™ Software

By: Carl Tarum, Editor's Update 2006, 2021

When analyzing a failure distribution, be aware that there may be subpopulations or a mixture of failure modes. Subpopulations are often called "batches." There is a special section in Chapter 8 devoted to batch clues and analysis. Appendix F is devoted to the Aggregated Cumulative Hazard method for analyzing batches. Batch problems are common and often are caused by some change in the production process. Carl Tarum developed the software package, YBath, to analyze mixtures of failure modes. YBath is now part of the SuperSMITH® software.

If no significant difference is seen when two batches or subpopulations are plotted, then they are combined. But if there is a significant difference between the subpopulations, the mixture must be analyzed appropriately. Environment may also contribute to a mixture of subpopulations. Products used in Florida may have different failure modes and different failure rates when compared to products used in Sweden. This may be due to the weather, or it may be due to the usage of the individual operators.

It is always preferable to separate the failure modes based on engineering analysis of the parts (and environment) and analyze them separately rather than rely on statistical methods. Suppose you have a data set of 50 parts, and 20 of them have one failure mode and the other 30 have a different failure mode. You would analyze the first set as 20 failures (of F_1) and 30 suspensions (for F_2). The second set would be 30 failures (of F_2) and 20 suspensions (for F_1). These two plots and sets of parameters are then used to predict the cumulative failure distributions.

When parts are not available for physical analysis, the data may be split into groups based on the apparent "corner" on the plot. Although this is common practice it can cause errors, because a small percentage of wear out failures will occur at an "early" life, and a percentage of "infant mortality" failures will occur at "later" life. If the two distributions overlap, grouping parts based on failure time might result in improper classification. YBath uses either least squares regression or maximum likelihood methods to find the best fit based on the entire sample. This may be more accurate than grouping points based on the corner. Separating failure modes based on physical broken part analysis is superior.

A minimum of at least 21 failures is needed for credible results from the competing risk model, and 50 or more failures for the other mixtures. You still must use judgment. Suppose that you have 100 failures to analyze, and the YBath model indicates that 1% of these are in an infant mortality population. One percent of one hundred is only one failure, so out of the 100 failures, there is only one associated with infant mortality. It seems unreasonable to conclude that this one failure can somehow provide enough information to generate a percentage defective, as well as a slope and characteristic life of the defective population! It could just be an outlier. Mixture models have interactions between parameters, and two different solutions may both appear to have a good fit, even though they have drastically different values. Whatever model you use, there must be enough failures to justify each section.

Following are brief descriptions of the more common methods for handling mixtures. In the descriptions, u indicates the portion or batch of the total population that has a particular failure distribution (F_1 in the simple mixture). F_1, F_2, and F_3 indicate failure distributions. The population cumulative failure distribution is F. The descriptions are given without describing the particular distribution shape (e.g., Weibull, lognormal, normal, or exponential.) An appropriate distribution shape needs to be substituted for each F_n. For example, using the Weibull distribution in the competing risk mixture results in the WeiBath equation. (Both of these are described later.)

J.1 Competing Risk – Competing Failure Modes

$F = 1 - (1-F_1)(1-F_2)$ or $R = R_1 \times R_2$ An example of competing risk is an automobile tire. It can fail due to puncture, or it can fail due to wear out. Just because a tire is about to wear out does not negate its probability of puncture! SSW uses a likelihood ratio technique to estimate the parameters of the two distributions. If this SSW technique is used, the Abernethy Risk can produce a failure forecast for the mixture. YBath has both MLE and regression solutions for the mixture problems.

Competing risk occurs when a population has two or more failure modes, and the entire population is at risk from either failure mode. Even though a Weibull plot of this data will show a cusp or corner or appear curved, it is a homogeneous population subject to a mixture of competing failure modes. Note that this is simply a series reliability problem. If the first slope is shallow and the second slope steeper, this is called the Classic BiWeibull.

J.2 Noncompeting Failure Modes

$F = uF_1+(1-u)F_2$. The data comes from two independent subpopulations with no common failure modes of interest. Each subpopulation has its own unique failure mode.

There are very few data sets that fit this model. It is quite rare. Most sets have at least one common failure mode. The simple mixture may be used as an approximation for more complex distributions, such as the competing risk mixture, described below. An example would be resistor life in a circuit, where one population is wire-wound resistors, and the other population is resistive film resistors. While the failure mode (open circuit) appears to be the same, the mechanisms and failure distributions may be different. The proper analysis is to separate the two resistor types and analyze them separately. If there is no significant difference, then the resistor type does not matter, and the sets can be combined.

J.3 Competing Risk with a Batch Problem

$F = u[1-(1-F_1)(1-F_2)]+(1-u)F_2$. Many mixtures contain a batch problem. There is at least one failure mode (F_1) that is unique to one batch - subpopulation, and there is a failure mode (F_2) that is common to both subpopulations.

In this case, one subpopulation is subject to failure modes 1 and 2, as indicated by the portion of the equation in brackets [ ]. This subpopulation by itself has competing risk. If the tire example is considered, this would be a case where some tires may be out of round. They would have a failure due to wobble (F_1). However, the tire could still get a puncture (F_2) on the way to the dealer to have the tire replaced. YBath will attempt a solution for this mixture.

J.4 Compound Competing Risk with a Batch Problem

$F = u[1-(1-F_1)(1-F_2)(1-F_3)]+(1-u)[1-(1-F_2)(1-F_3)]$. This is an expansion of the Competing Risk Mixture to include a third failure mode. This distribution can model the classic "Bathtub Curve" of reliability. (See Figures 2-6 and 2-7.) The model can address infant mortality, chance failures, and wear out. This distribution is seldom observed in practice. Usually, one of the failure modes dominates and the data are adequately analyzed using a competing risk with or without a batch.

In the tire example, the defective subpopulation, u, has risk due to wheel wobble (F_1), puncture (F_2), or wear out (F_3). Wear out is a possibility in the batch subpopulation because the driver may not know anything is wrong with the tire, or the defect may be barely noticeable. The remaining subpopulation (1 - u) is subject to risks due to puncture or wear out.

J.5 WeiBath Model

$F = 1-(u)\mathbf{exp}[-(t/h)^b-(t/\eta)^\beta-(t/H)^B]-(1-u)\mathbf{exp}[-(t/\eta)^\beta-(t/H)^B]$. When a Weibull distribution, $F = 1-\mathbf{exp}[-(t/\eta)^\beta]$, is used for each failure mode in the compound competing risk model, it is called a WeiBath (**Weibull** + **Bathtub**) model. The author (Tarum) proposed this model and has independently produced YBath software to solve for the parameters.

u	is the portion of the total population in the subpopulation with an additional failure mode
t	is the time to failure,
b, h	are the slope and characteristic life of the subpopulation failure mode (Infant mortality),
β, η	are the slope and characteristic life of (Chance failures), and
B, H	are the slope and characteristic life of (Wear out).

Reliability texts often refer to a "bathtub curve" for a product's life. This curve has 3 distinct areas: 1) A period of initially high, decreasing failure rate (infant mortality); 2) A period of low, constant failure rate (chance failures), 3) A period of increasing failure rate as the product wears out. When the instantaneous failure rate, the hazard rate, is plotted over time, it has a bathtub shape. Some texts also refer to these three stages as QRD (Quality, Reliability, and Durability). Until recently, it has been impossible to fit a bathtub curve model to data unless the data can be separated into failure modes by broken parts inspection and engineering analysis. YBath software now provides distribution analysis to a bathtub curve model when failure modes cannot be analyzed with separate Weibulls. SuperSMITH Visual™ does a nice job of adding the pdf together to get the bathtub curve.

The WeiBath distribution has 7 variables, so it is also called a 7-parameter model. Using the nomenclature shown, Weibull equations for the various mixtures are:

Competing Risk: (4 Parameter) $F = 1-\mathbf{exp}[-(t/\eta)^\beta-(t/H)^B]$

Competing Risk (with a batch problem) Mixture: (5 Parameter)

$F = 1-(u)\mathbf{exp}[-(t/h)^b-(t/\eta)^\beta]-(1-u)\mathbf{exp}[-(t/\eta)^\beta]$

Simple Mixture: (5 Parameter) $F = 1-(u)\mathbf{exp}[-(t/h)^b]-(1-u)\mathbf{exp}[-(t/\eta)^\beta]$.

J.6 Considerations

The Weibull distribution can be used for the cumulative failure distributions, F_n, shown above. There may also be other models with additional subpopulations, t_0 shifts, or additional failure modes. The ones shown here are the most common. A literature search shows that these models are known by other names. YBath uses the Competing Risk Mixture instead of the Simple Mixture.

J.7 Curve shape

The shape of the data points on the Weibull plot is often informative. Figure J-1 shows the various distributions that may be encountered. Your data may not show all of the characteristics demonstrated, depending on your data. The curves are labeled as follows:

7	(7 Parameter),	Compound Competing Risk Mixture with a batch problem
5	(5 Parameter),	Competing Risk Mixture with a batch problem.
4	(4 Parameter),	Competing Risk
3-	(3 Parameter),	Minimum Life, t_0 is negative. Part of life is used up before $t = 0$.
3+	(3 Parameter),	Minimum Life, t_0 is positive. All parts live for at least t_0.
LN	(Lognormal),	Lognormal Distribution, plotted on Weibull Paper
2	(2 Parameter),	Weibull Distribution

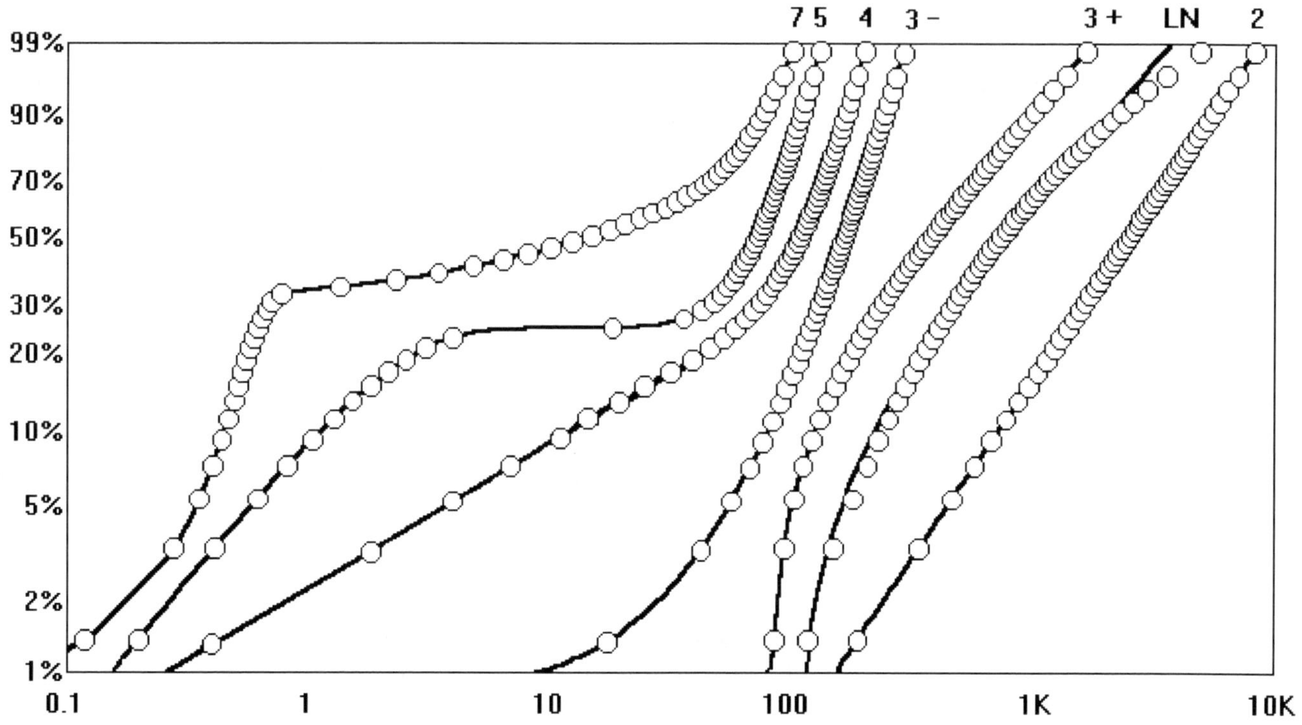

Figure J-1. General Shapes on Weibull Plots

To determine what distribution you may have, consider the following general shapes:

- Straight line: A standard Weibull distribution. (2 parameter)
- Continuously curved line: Minimum life (3-, 3+) or a lognormal (LN) distribution. The competing risk model (5) may appear to be curved and look like (3- or 3+), depending on the value of p and the relative slope and intercepts of the two populations.
- Straight line, curving into a second line with steeper slope: Competing risk (4) or competing risk mixture (5). Consider the first four failures in Figure 3-19. This is the "Classic BiWeibull."
- Straight line, curving to a decreasing slope: Competing risk mixture (5) or perpetual survivors, both are batch problems. Some competing risk mixtures may appear as a line, with a "bump" of increased failures, followed by an asymptotic return to the original failure line. Depending on the relative values of the slopes and characteristic lives, the bump may not even appear on the graph.

  A single failure mode that does not apply to all units in the fleet may also have this characteristic; the units that are immune to the failure mode are called "perpetual survivors." This has one inflection point. For example, consider steel bars designed for fatigue life. The design stress would be below the endurance limit. Now suppose that a subpopulation was defective and failed in fatigue. The only failures that will ever be observed are the defective ones. The standard population is made up of perpetual survivors that will never fail due to fatigue.
- Line, curving to an increasing slope, a decreasing slope, then an increasing slope: This curve can be represented with a compound competing risk with a batch problem (7). A compound competing risk mixture may have up to three slope changes, combining features of competing risk and competing risk with batch problems.

Author's Comments: Carl Tarum developed the YBath software. It has taken years and he is commended for the effort. The results include both regression and MLE solutions for these complex models. Large sample sizes are required to estimate the many parameters. Hopefully most of the readers will rarely need these solutions but on the other hand, when they are needed, the YBath software is the best solution. YBath software is now bundled with the SuperSMITH software. It is easily accessible using the mixture icon menu in SSW. Large samples, 50 or more failures as a minimum, are needed to estimate all these parameters. 100 failures would produce more credible results.

Appendix K: Answers to Problems

Chapter 2: Problem 2-1.

Problem 2-1: Rank Order Number	Failure Time	Median Rank*
1	85	6.70%
2	135	16.23%
3	150	25.86%
4	150	35.51%
5	190	45.17%
6	200	54.83%
7	200	64.49%
8	240	74.14%
9	240	83.77%
10	250	93.30%

*From Median Rank Tables - Appendix I

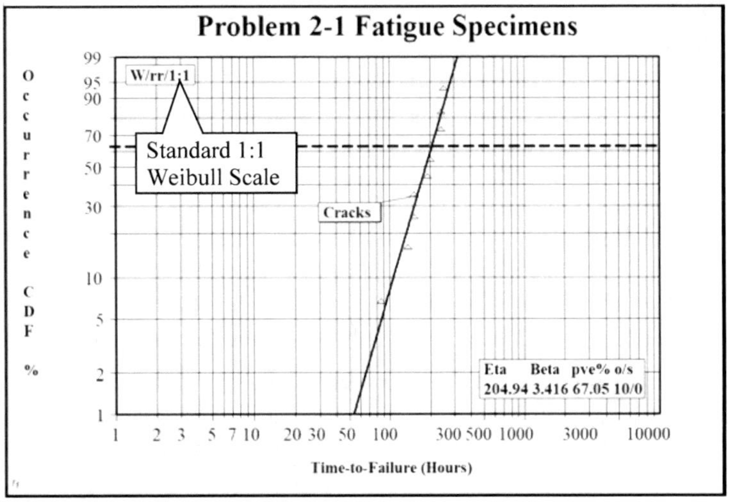

Figure K-1. Problem 2-1 Fatigue Specimens

a. See Figure Problem 2-1
b. β = 3.4 implying a wear out failure mode class
c. η = 204 hours
d. B1 = 54 hours
e. 63.2%
f. 2% (CDF = 98%, and R = 1-CDF)

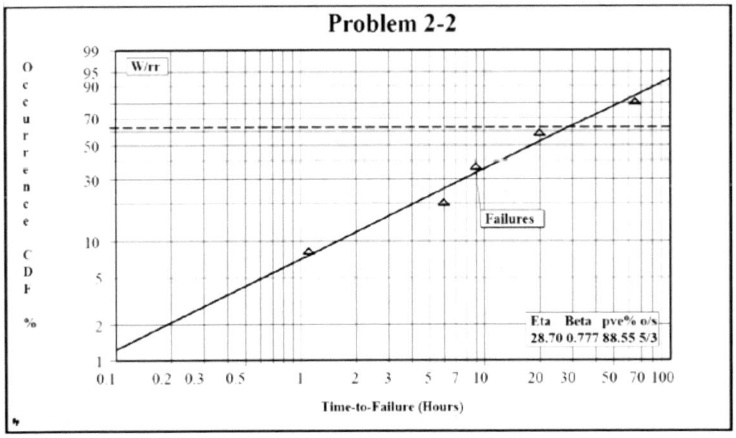

Figure K-2. Problem 2-2 Weibull with Suspensions

Problem 2-2: Plot a Weibull with suspensions.

a. See Figure Problem 2-2
b. See Figure Problem 2-2
c. β = 0.77 implies infant mortality.
d. Did you notice the serial numbers are close together? This is often the first clue that there is a batch problem.

Order	Time	Inverse Order	Adjusted Rank (i)	Median Rank Benard or App I
1	1.1F	8	[(8x0)+(8+1)]/(8+1)=1	8.33% or 8.30%
2	6.0F	7	[(7x1)+(8+1)]/(7+1)=2	20.23% or 20.11%
3	7.0S	6	----	-
4	8.0S	5	----	-
5	9.0F	4	[(4x2)+(8+1)]/(4+1)=3.40	36.90%
6	14.6S	3	----	-
7	20.0F	2	[(2x3.40)+(8+1)]/(2+1)=5.27	59.13%
8	65.0F	1	[(1x5.27)+(8+1)]/(1+1)=7.14	81.37%

Problem 2-3: Understanding Good Weibulls

a. The hardest part of Weibull analysis is getting good data. Good software has eliminated the drudgery of hand calculations and hand plotting.
b. Weibull plots of the air conditioning compressor data will show the goodness of fit. Figure 3-5 and SuperSMITH™ provide the critical correlation coefficient which may be compared to the observed value, to determine if the fit is good enough.
c. The slope parameter is also called the shape parameter because it indicates the shape of the probability density function shown in Figure 2-2.
d. η = MTTF when beta = 1.0. This is the exponential distribution.
e. The standard method for doing Weibull analysis is median rank regression, X on Y. It should always be used for small samples, 20 or less failures, and generally provides the most accurate results. More details will be found in Chapter 5.
f. What class of failure [infant mortality, random, or wear out] are the following?
    1. Failures from inadequate stress screening or burn-in are infant mortality, $\beta < 1$.
    2. Woodpecker damage to power poles in Florida are random, $\beta = 1$. Most of Mother Nature's failure modes are random in time, at least over short intervals. Foreign object damage in engines and aircraft, lightning strikes on transformers, even human error.
    3. Alzheimer's disease affects elderly people primarily. Therefore it is a wear out failure mode. Whooping cough afflicts children, therefore, infant mortality would be expected.
    4. Bearing failures from spalling and skidding balls are wear out modes.
    5. Turbine vanes fail early from plugged fuel nozzles, a quality problem. This implies infant mortality. Much later, stress corrosion cracking appears, which implies a wear out mode.

Chapter 3:

Problem 3-1: A Curved Weibull

a. A value of about 80 would be appropriate. See Figure K-3. SuperSMITH™ uses an accurate method, iterating on t_0 until the correlation coefficient reaches a maximum value. The optimum value is about 78.
b. Subtracted.

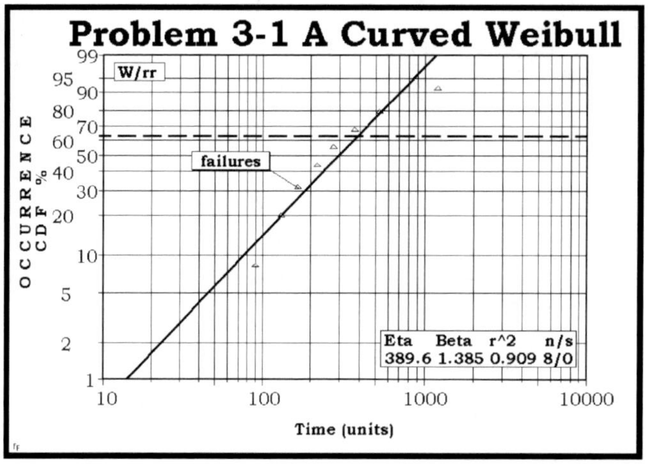

Figure K-3. Problem 3-1 A Curved Weibull

Problem 3-1: Interpreting the Plot

Figure 3-21 Garbage???? Terrible Weibull! A mixture of failure modes is indicated. Check the broken parts and failure reports. See if you can categorize the failures into groups and do a Weibull for each group. The resulting Weibulls should have a better fit. If not, there may be curvature, implying the need for t_0 or lognormal.

Figure 3-22 Broken Units This looks like a mixture of two competing failure modes. It is a curious mixture as the first steep mode should "cover" or prevent the second. In correspondence with Dimitri Kececioglu, he suggests there may be a subpopulation, a batch, that is subject to the first mode. This is a reasonable explanation. Pay no attention to beta as mixtures will drive it toward one. Another possibility is that this is a single failure mode with some perpetual survivors.

Figure 3-23 Roller Bearing Fatigue A good Weibull! Expect to see lots of these. β is typically 1.5 for roller bearings and 2.0 for ball bearings. t_0 corrections are often indicated although not in this case.

Figure 3-24 What Kind of Data? Vertical columns of points indicate this is either interval inspection data or coarse data. The interval inspection option that fits the upper most point only when there is more than one point at the same time may help. Use Inspection Option #1 in SSW. Use r or r squared as a guide.

Figure 3-25 Accelerated Test Data Note all three sets show curvature hinting that the lognormal or the t_0 distributions may be a better choice.

Figure 3-26 California Earthquakes As I don't know when California was created, this is interval data. Intervals between events will tend to have a beta of one regardless of the true unknown beta. It could even be lognormal data. It is also coarse data (the intervals should be in days instead of years). Inspection Option #1 may give a better fit.

Figure 3-27 Laboratory Connector Failures Steep β's tend to hide problems. This looks like a great Weibull. It isn't. The three parameter Weibull or the lognormal should be investigated. There may even be outliers present. (One trick to get a better look at the anomalies is to input a t_0 just below the first failure, in this case, a value of 20. This will spread out the data.) A distribution analysis may be needed. Laboratory tests often have very steep betas as conditions are closely controlled and measurements are made as accurately as possible.

Figure 3-28 Cable Wire Fatigue Samples This data was taken in our Weibull Workshops. A liter of water is suspended from the 120 pound test fishing leader wire held by locking pliers. The curvature and some coarseness in the data is obvious. A distribution analysis with or without the inspection option is suggested. As the pliers are rotated back and forth, the wire fatigues and ruptures, strand by strand. This type wire was selected with the expectation that total rupture would be obtained by progressive deterioration. As each strand fails the stress on the remaining strands increases progressively. This should produce lognormal data for half twists to total rupture. On the other hand, physically it is impossible to rupture all strands on the first half twist. This would imply a t_0 three parameter Weibull. With small samples, 15-25, in each class, both results have been produced. With large samples, the data is lognormal.

Figure 3-29 Name?? This is a "Classic BiWeibull", the most popular mixture of two modes. The recommended action is to use engineering skills, to categorize the data into separate modes. This is best done by examining the broken parts, perhaps fracture surfaces under a microscope. Second best is reappraisal of the failure data or reports. If these steps are impossible (and they often are impossible,) the SuperSMITH™ software will provide a solution as illustrated in Chapter 3.

Figure 3-30 Yield Strength - Steel Typical curvature is shown for materials properties. A distributional analysis is recommended. In all of these examples, prior information is invaluable as an aid to improving the analysis.

Figure 3-31 What is This? The curvature implies the need for either t_0 or a lognormal solution. In this case the data is truly lognormal as shown in **Figure 3-32**. There the data on Figure 3-31 is plotted on lognormal probability paper.

Chapter 4:

Problem 4-1: A Curved Weibull

Risk Analysis: Use the Weibull Equation 2-1 or read F(t) off the plot

Engine Time	F(t)	# Engines	Product
150	0.0033	20	0.067
200	0.008	20	0.16
250	0.0155	20	0.31
300	0.0266	20	0.533
350	0.042	20	0.84

Now Risk = Sum (1) = 1.909

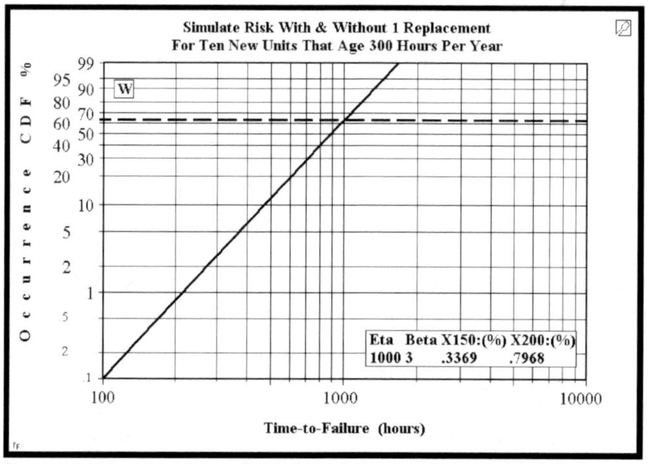

Figure K-4. Simulate Risk With & Without 1 Replacement For Ten New Units That Age 300 Hours Per Year

u = 6 months x 25 hours/month = 150, FF = {F(t + u) - F(t)}/{1 - F(t)}

t + u	F(t + u)	FF = {F(t + u) - F(t)}/1 - F(t)	20 x FF
300	0.0266	0.0234	0.468
350	0.0420	0.0343	0.685
400	0.0620	0.0472	0.945
450	0.0871	0.0622	1.243
500	0.1175	0.0788	1.576

Expected Failures Next 6 Months = Sum = 4.92

Problem 4-2: Turbine Airfoils

a. Comparing Figure K-5 with K-6, Location A appears to be different. The two data sets were tested with both the likelihood contour method and the likelihood ratio test described in Chapter 7. Both tests agree that the sets are not significantly different at 90% confidence.

b. The risk for the entire fleet is shown in Figure K-7, generated from SSW output using SuperSMITHVisual.

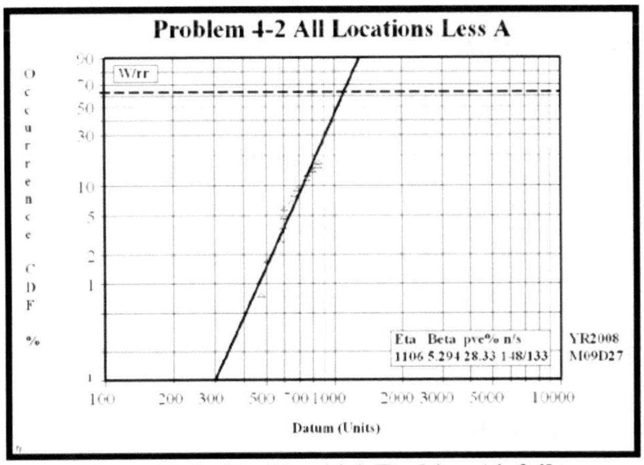

Figure K-5. Problem 4-2 Turbine Airfoils

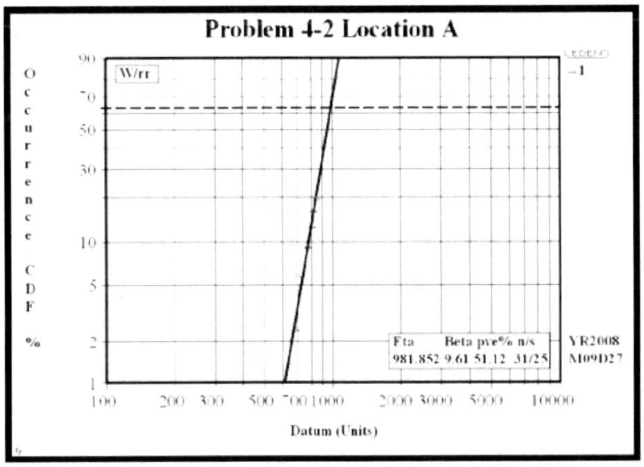

Figure K-6. Problem 4-2 Location A

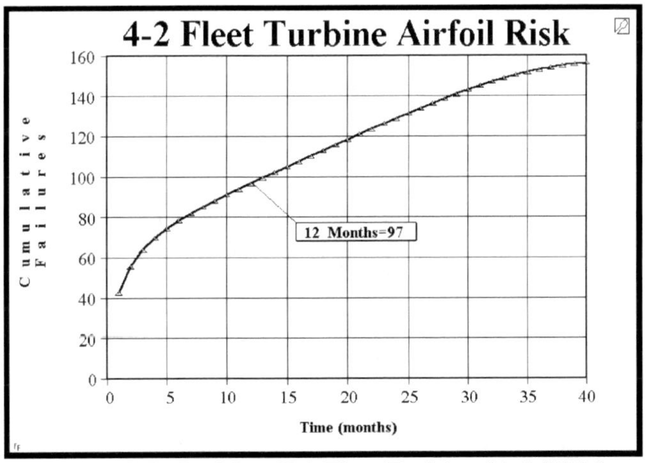

Figure K-7. Problem F-2 Fleet Turbine Airfoil Risk

Problem 4-3: Control Failure Mode

a. t = 1000,

$$F(1000) = 1 - e^{-(1000/19735)^{1.26}} = 0.0231$$

Number of failures = (0.0231)(1,308) = 30

b. t = 2000,

$$F(2000) = 1 - e^{-(2000/19735)^{1.26}} = 0.0544$$

Number of failures = (0.0544)(1,308) = 71

c. t = 4000,

$$F(4000) = 1 - e^{-(4000/19735)^{1.26}} = 0.1253$$

F(4000) = 0.1253

Number of failures = (0.1253)(1,308) = 164

d. Inspection at 1000 hours makes units "good as new"

P(failure at 4000 hours) = 4(F(1000)) = 0.0924

Number of failures = (0.0924)(1308) = 121

Problem 4-4: Monte Carlo 3 Failure Modes 4 Engines

a. Engine at 100 hours using Equation 4-4:

$$T_a = 96587\left[\ln\left(\frac{1}{1-0.329}\right)\right]^{1/0.76}$$

= 28831 hours = 1149 months ahead

$$T_b = 4996\left[\ln\left(\frac{1}{1-0.604}\right)\right]^{1/2.6386}$$

= 4853 hours = 190 months ahead

$$T_c = 1126\left[\ln\left(\frac{1}{1-0.615}\right)\right]^{1/7.4}$$

= 1119 hours = 40 months ahead

The 1000 hour inspection is 900 hours or 36 months ahead. Therefore, reset mode a and c to "0".

$$T_a = 96587\left[\ln\left(\frac{1}{1-0.3}\right)\right]^{1/0.76}$$

= 24877 hours = 995 months ahead

$$T_c = 1126\left[\ln\left(\frac{1}{1-0.07}\right)\right]^{1/7.4}$$

= 24877 hours = 31 months ahead

The 2000 hour inspection is 40 months from the last inspection which was at 900 hours, or at 76 months. Therefore there are no failures on the first engine. Similarly, for the engines at 200 hours, 700 hours and 900 hours there are no failures within the next 48 with this string of random numbers. (B) Five additional engines fail without the 1000 hour inspection (which is a block replacement).

Chapter 5:

(Earlier editions had a Monte Carlo problem, which was removed around 2013)

Chapter 6: Weibayes

Problem 6-1: Five redesigned units

a. Construct a Weibayes line for the new units. $\beta = 3$, $\eta = 2387$

b. What is the minimum improvement demonstrated in the characteristic life? [2387/500] = 4.77 = $[MTBF_{NEW}/MTBF_{OLD}] = [B1_{NEW}/B1_{OLD}]$

Engine Times (t)	$(t)^{\beta}$
800	$5.12x10^8$
1100	$1.33x10^9$
1400	$2.74x10^9$
1600	$4.1x10^9$
1700	$4.9x10^9$
	Sum = $1.36x10^{10}$

$(Sum)^{(1/\beta)} = 2387$

Problem 6-2:

$\eta = 3846$, $\beta = 3$, Therefore, B.1 = 385, using the Weibayes Equation 5-1 and Equation 4-4.
Conclusion: the success data is insufficient to increase the predicted design life.

Problem 6-3:

Do a Weibayes with the six bolt times and Equation 5-1. For $\beta = 2$, $\eta = 187$. Now do a risk calculation, setting the times on the two high time bolts to zero assuming they were the failures. 2.18 failures are predicted in the next year.

Problem 6-4: Plot a Weibayes

These two data sets produce the same Weibayes line. $\eta = 2389$, $\beta = 3$. However, the interpretation is different. With one failure the Weibayes is an MLE estimate of the true Weibull. With no failures, it is a lower, one-sided, 63% confidence bound for the true Weibull.

Problem 6-5: Test Substantiation

A new component has been designed which has 2 times the characteristic life of the old one. The old component fails by a Weibull, $\beta = 4.0$, $\eta = 300$ hours.

90% Confidence Multipliers from Table 6-1 x Life = 600 Hours

Units	Multiplier Times	Expected Life	Test Hours/Unit	Total Hours
4	0.871	600	522.6	2090
8	0.732	600	439.2	3514

Problem 6-6: One or Less Test

How much time should the new unit in Problem 6-1, be tested with one or zero failures, to demonstrate with 90% confidence that it has 2 times the life with 4 times units to test? From Table 6-3, the factor is 1.033 x 600 = 619.8 hours.

Problem 6-7: Sudden Death vs. Zero Failure Tests

What are the advantages and disadvantages of these two types of Weibull substantiation tests? With good prior knowledge would you employ Weibayes with Sudden Death tests?

Answer: The zero failure tests have minimum test time and cost, but Sudden Death provides the Weibull plot which may have significant value. Prior knowledge of beta should almost always be employed with Sudden Death or otherwise. The exception would be the rare luxury of having large samples of failures available for inference.

***Problem 6-8: Sudden Death**

Your company tests 16 randomly selected roller bearings to failure from each batch of 100,000 bearings. You suggest a cost reduction based on Sudden Death bearing rig tests of sets of four simultaneously. Here is the last data set, failures in cycles from 16 roller bearings. For Sudden Death, assume the sets are contained by the columns and the first bearing failure ends the test for each set. For example, Set 1 would consist of one failure and three suspensions at a value of 47,385 cycles.

SET 1	SET 2	SET 3	SET 4
106575	32484	147191	177217
47385	35968	52531	38649
71710	112053	42114	70812
54546	176187	132621	113508

1. How does the Weibull and Weibayes for Sudden Death, compare to the Weibull for all 16 bearings to failure? Plot both Weibulls and a Weibayes assuming $\beta = 1.5$. In the box in the lower right hand corner are the solutions from top to bottom for Weibull-16 failures, Sudden Death Weibull 4F-12S, Sudden Death Weibayes, and the True Weibull. In this case, the Sudden Death Weibayes is by far the most accurate.
2. These data were generated by SuperSMITH Weibull sampling option from a true Weibull with $\beta = 1.5$ and $\eta = 100,000$ cycles. Plot the true Weibull. Which B10 life is the most accurate, Weibull-16F, Weibull-4F-12S, or Weibayes? The B10 lives are shown on the bottom of the plot.
3. Compare the Sudden Death total rig test time to the total life test time? Testing four bearings together, the total rig time would be = 47385 + 32484 + 42114 + 38649 = 160632. The test time for all 16 to failure = 1,411,552. 160,632/1,411,552 = 11.4%
4. Compare the Zero-Failure test time for 16 units to test, ß = 1.5, k = 0.275. Assume the design goal is 100,000. 16 x 0.275 x 100,000 = 440000 total with 27500 cycles on each bearing without failure. Note the 16 bearings would have passed in this case...unusual. This should only happen 10% of the time.

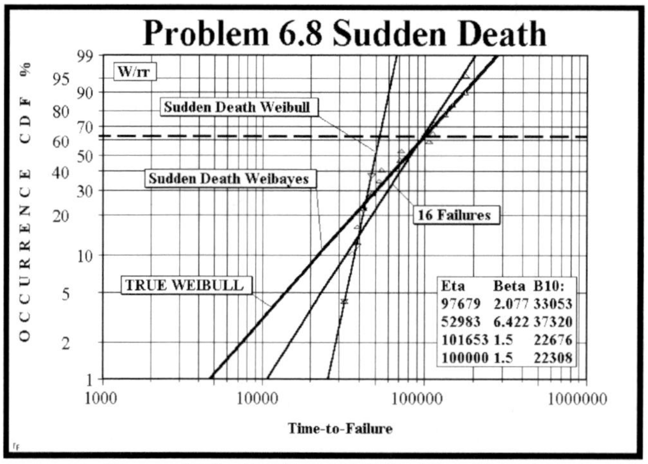

Figure K-8. Problem 6-8 Sudden Death

Chapter 7: Confidence Intervals

True or False:
7-1 False. The probability that the interval contains the true values is either zero or one. The frequency with which 90% confidence intervals contain the true value is 0.90.
7-2 False. Engineering uncertainties are not included.
7-3 False. 97.3% is a lower bound of an interval estimate.
7-4 True.
7-5 False. Even most statisticians do not understand the concept.
7-6 True.
7-7 True.
7-8 True.
7-9 True.
7-10 False

1. To make inferences from data we need data. There is none here. The data might have come from a Binomial zero failure test, a Binomial zero or one failure test, Weibull, Weibayes and others. Each of these data sets would produce different estimates of the system reliability.
2. The characteristic of confidence is variant under transformation. If used in a calculation, the confidence level changes. The equivalent system confidence cannot be predicted without the data and complex statistics or simulation.
3. The estimate of reliability is an interval estimate, the true value lies between 0.99 and 1.0, the estimate is not 0.99. We need the data or point estimates of reliability from data to estimate system reliability.

Chapter 8: Related Math Models

Problem 8-1: Binomial SQC

See Figures 8-3 and 8-4 for solutions to 8-1, 8-2, and 8-4. With a sampling plan that requires one or less defects in 100 pieces, entering the A-W Chart at Np = 1, and reading the C-2 curve, the probability of C-2 defects is 26%.

Problem 8-2: Binomial Reliability

Find the intersection of Cumulative Probability of 90% and $C \geq 1$ on Figure 8-4. $\mu = 2.3 = N\ (1\text{-}R)$. N = 230. Therefore, a reliability demonstration with 230 tests without failure, demonstrates 99% reliability with 90% confidence (lower one sided)....or use Table 8-1.

Problem 8-3: Exponential Reliability

The system MTBF is 164.

$$1/MTBF_{SYSTEM} = \sum_{i=1}^{i=n} (1/MTBF_i)$$

Problem 8-4: Poisson

On 8.3 the solution is the intercept of the 10% line and the $\mu = 6$ vertical, which falls in the "nine" space. Therefore, the upper one-sided 90% confidence bound = 9 for mu = 6.

Problem 8-5: Which Distribution?

a. Receiving inspection reports 2 defects in a sample of 100... = Binomial.
b. North Palm Beach averages 13.7 long outages per year...Poisson and the durations are probably exponential.
c. Transformer failure data by calendar time caused by lightning strikes ...Exponential but seasonal.
d. Bearing times-to-failure...3 parameter Weibull.
e. Professor Mischke says a specimen will rupture under fixed conditions as described by a Weibull, but in service, the time to failure is the product of the Weibull time and factors for temperature, vibration, acceleration, corrosion, etc...lognormal.
f. IQ scores..Normal, mean = 100, Standard Deviation = 13.
g. What was the reliability of the Jumo with MTBF = 25 hours for the typical 20 minute mission?...Exponential.
h. A reliability demonstration requires 2300 tests without failure...Binomial.
i. A component is subjected to a combined environmental test to force failures...a CERT test ....Crow-AMSSAA-CROW Model.
j. Mechanics sometimes make mistakes....so do engineers. Weibull. $\beta = 1$ or Exponential.
k. The duration of forced outages.......Exponential. The duration of telephone calls too.
l. Performance estimates. Normal.....like almost all measurements.
m. Foreign object damage as a function of operating time...Exponential=Weibull $\beta = 1$.
n. Uniform random numbers are the basis for Monte Carlo simulation...Rectangular distribution.
o. The reciprocals of uniform random numbers are useful...Exponential distribution.
p. Weibull times to failure raised to the beta power...Also Exponential.
q. Wire cable is made of multiple twisted strands of wire. Under load what is the expected distribution of first strand time to rupture? Weibull...total strands rupture? t_0 or lognormal.
r. Accident data for a plant or facility....Exponential.
s. The probabilistic statement that if two events are equally likely, the worst will happen is derived from who's law?....Murphy's Law.

Problem 8-6: Exponential...MTTF = 11415

Appendix L: The C_4 Factor

The C_4 factor is used to reduce the bias of the MLE standard deviation estimates for the normal and lognormal distributions. It is also used to reduce the bias in the Weibull MLE beta estimates. See Chapter 5.

N	C_4	$(C_4)^{3.520}$	$(C_4)^{-0.217}$	$(C_4)^{6.0}$
2	0.797885	0.451659	1.050231	0.258012
3	0.886227	0.653655	1.026564	0.484473
4	0.921318	0.749402	1.017947	0.611585
5	0.939986	0.804233	1.013525	0.689806
6	0.951533	0.839552	1.010842	0.742237
7	0.959369	0.864144	1.009044	0.779675
8	0.965030	0.882229	1.007756	0.807693
9	0.969311	0.896081	1.006789	0.829427
10	0.972659	0.907026	1.006035	0.846768
11	0.975350	0.915889	1.005432	0.860921
12	0.977559	0.923213	1.004939	0.872688
13	0.979406	0.929366	1.004527	0.882624
14	0.980971	0.934607	1.004179	0.891124
15	0.982316	0.939124	1.003880	0.898479
16	0.983484	0.943059	1.003622	0.904904
17	0.984506	0.946516	1.003395	0.910566
18	0.985410	0.949578	1.003195	0.915592
19	0.986214	0.952308	1.003018	0.920084
20	0.986934	0.954758	1.002859	0.924122
21	0.987583	0.956969	1.002716	0.927772
22	0.988170	0.958974	1.002586	0.931088
23	0.988705	0.960801	1.002469	0.934113
24	0.989193	0.962471	1.002361	0.936883
25	0.989640	0.964006	1.002263	0.939430
26	0.990052	0.965420	1.002172	0.941780
27	0.990433	0.966727	1.002089	0.943954
28	0.990786	0.967938	1.002011	0.945971
29	0.991113	0.969065	1.001940	0.947849
30	0.991418	0.970115	1.001873	0.949601
31	0.991703	0.971097	1.001810	0.951238
32	0.991969	0.972015	1.001752	0.952773
33	0.992219	0.972878	1.001697	0.954214
34	0.992454	0.973688	1.001645	0.955570
35	0.992675	0.974452	1.001597	0.956847
36	0.992884	0.975173	1.001551	0.958054
37	0.993080	0.975854	1.001508	0.959195
38	0.993267	0.976498	1.001468	0.960275
39	0.993443	0.977110	1.001429	0.961300
40	0.993611	0.977690	1.001392	0.962273
41	0.993770	0.978241	1.001357	0.963198
42	0.993922	0.978766	1.001324	0.964079
43	0.994066	0.979266	1.001293	0.964919
44	0.994203	0.979744	1.001263	0.965721
45	0.994335	0.980199	1.001234	0.966486
46	0.994460	0.980635	1.001207	0.967219
47	0.994580	0.981052	1.001180	0.967920
48	0.994695	0.981451	1.001155	0.968591
49	0.994806	0.981834	1.001131	0.969235
50	0.994911	0.982201	1.001108	0.969854

N	C_4	$(C_4)^{3.520}$	$(C_4)^{-0.217}$	$(C_4)^{6.0}$
51	0.995013	0.982554	1.001086	0.970447
52	0.995110	0.982893	1.001065	0.971018
53	0.995204	0.983219	1.001044	0.971568
54	0.995294	0.983533	1.001024	0.972096
55	0.995381	0.983836	1.001005	0.972606
56	0.995465	0.984127	1.000987	0.973097
57	0.995546	0.984409	1.000969	0.973571
58	0.995624	0.984680	1.000952	0.974029
59	0.995699	0.984942	1.000936	0.974471
60	0.995772	0.985195	1.000920	0.974898
61	0.995842	0.985440	1.000905	0.975311
62	0.995910	0.985677	1.000890	0.975711
63	0.995976	0.985907	1.000876	0.976098
64	0.996040	0.986129	1.000862	0.976473
65	0.996102	0.986344	1.000848	0.976836
66	0.996161	0.986553	1.000835	0.977188
67	0.996219	0.986755	1.000823	0.977530
68	0.996276	0.986951	1.000810	0.977862
69	0.996330	0.987142	1.000798	0.978183
70	0.996383	0.987327	1.000787	0.978496
71	0.996435	0.987507	1.000775	0.978800
72	0.996485	0.987682	1.000765	0.979096
73	0.996534	0.987852	1.000754	0.979383
74	0.996581	0.988017	1.000744	0.979662
75	0.996627	0.988178	1.000734	0.979934
76	0.996672	0.988335	1.000724	0.980199
77	0.996716	0.988488	1.000714	0.980457
78	0.996759	0.988636	1.000705	0.980709
79	0.996800	0.988781	1.000696	0.980953
80	0.996841	0.988922	1.000687	0.981192
81	0.996880	0.989060	1.000679	0.981425
82	0.996918	0.989194	1.000670	0.981652
83	0.996956	0.989325	1.000662	0.981874
84	0.996993	0.989453	1.000654	0.982090
85	0.997028	0.989578	1.000646	0.982302
86	0.997063	0.989700	1.000639	0.982508
87	0.997097	0.989819	1.000631	0.982710
88	0.997131	0.989936	1.000624	0.982907
89	0.997163	0.990050	1.000617	0.983099
90	0.997195	0.990161	1.000610	0.983288
91	0.997226	0.990270	1.000603	0.983472
92	0.997257	0.990376	1.000596	0.983652
93	0.997286	0.990480	1.000590	0.983828
94	0.997315	0.990582	1.000584	0.984001
95	0.997344	0.990682	1.000577	0.984169
96	0.997372	0.990779	1.000571	0.984335
97	0.997399	0.990875	1.000565	0.984497
98	0.997426	0.990969	1.000560	0.984655
99	0.997452	0.991060	1.000554	0.984811
100	0.997478	0.991150	1.000548	0.984963

$$C_4=\sqrt{\frac{2}{n-1}}\frac{\left(\frac{n-2}{2}\right)!}{\left(\frac{n-3}{2}\right)!}=\sqrt{\frac{2}{n-1}}\frac{\left(\Gamma\left(\frac{n}{2}\right)\right)}{\left(\Gamma\left(\frac{n-1}{2}\right)\right)}$$

Paul Barringer calculated these C_4 factors using the Gamma function in Excel where the equation is:

C₄ = ((2/(n-1))^0.5)*(EXP(GAMMALN(n/2)))/(EXP(GAMMALN((n-1)/2))) Revised equation

of course "n" would represent the location for the value of n = 2, 3…343 as you recognize "n" does not correspond to a literal cell location for Excel.

A further Excel simplification suggested by Carl Tarum for values of n>325 use the mathematical equivalence to reduce the size of calculated values ***exp***(A)/***exp***(B) = ***exp***(A-B) then

C₄ = ((2/n-1))^0.5)*(EXP((GAMMALN(n/2))-(GAMMALN((n-1)/2)))) New equation

which is valid up through n ≈ 5,000

The C_4 results in the table agree with ***ASTM*** "***Manual on Presentation of Data and Control Chart Analysis,***" *Sixth Edition*, which also contains excellent approximations for the C_4 factor (the approximation is unnecessary with the Excel equation which was not available when the table for ASTM was prepared). For example,

$$\mathbf{C_4 = \sqrt{(4n - 5)/(4n - 3)}}$$

which is accurate to 3 decimal places for n of 7 or more and to 4 decimal places for n of 13 or more.

Graphical Repair Analysis

Wayne Nelson, consultant
WNconsult@aol.com, 739 Huntingdon Drive,
Schenectady, NY 12309

SUMMARY. This expository article presents a simple and informative non-parametric plot of repair data on a sample of systems. The plot is illustrated with transmission repair data from cars on a road test.

KEY WORDS: repair data; reliability data; graphical analysis.

1. INTRODUCTION

Purpose. This article presents a simple and informative plot for analyzing data on numbers or costs of repeated repairs of a sample of systems. The plot provides a non-parametric graphical estimate of the population mean cumulative number or cost of repairs per system versus age. This estimate can be used to:

1. Evaluate whether the population repair (or cost) rate increases or decreases with age (this is useful for system retirement and burn-in decisions),
2. Compare two samples from different designs, production periods, maintenance policies, environments, operating conditions, etc.,
3. Predict future numbers and costs of repairs,
4. Reveal unexpected information and insight, an important advantage of plots.

Overview. Section 2 describes typical repair data. Section 3 defines the basic population model and its mean cumulative function (MCF) for the number or cost of repairs. Section 4 shows how to calculate and plot a sample estimate of the MCF from data from systems with a mix of ages. Section 5 explains how to use and interpret such plots.

2. REPAIR DATA

Purpose. This section describes typical repair data from a sample of systems.

Transmission data. Table 1 displays a small set of typical repair data on a sample of 34 cars in a preproduction road test. Information sought from the data includes (1) the mean cumulative number of repairs per car by 24,000 test miles (132,000 customer miles, design life) and (2) whether the population repair rate increases or decreases as the population ages. For each car the data consist of the car's mileage at each transmission repair and the latest observed mileage. For example, the data on car 024 are a repair at 7068 miles and its latest mileage 26,744+ miles; here + indicates this how long the car has been observed. Nelson (1988,1990,1995,2003) gives repair data on blood analyzers, residential heat pumps, window air-conditioners, power supplies, turbines, and other applications. The methods below apply to recurrence data from many fields.

Table 1. Transmission repair data.

CAR	MILEAGE		
024	7068	26744+	
026	28	13809+	
027	48	1440	29834+
029	530	25660+	
031	21762+		
032	14235+		
034	1388	21133+	
035	21401+		
098	21876+		
107	5094	18228+	
108	21691+		
109	20890+		
110	22486+		
111	19321+		
112	21585+		
113	18676+		
114	23520+		
115	17955+		
116	19507+		
117	24177+		
118	22854+		
119	17844+		
120	22637+		
121	375	19607+	
122	19403+		
123	20997+		
124	19175+		
125	20425+		
126	22149+		
129	21144+		
130	21237+		
131	14281+		
132	8250	21974+	
133	19250	21888+	

Censoring. A system's latest observed age is called its "censoring age", because the system's repair history beyond that age is censored (unknown) at the time of the data analysis. Usually, system censoring ages differ. The different censoring ages complicate the data analysis and require the methods here. A system may have no failures; then its data are just its censoring age. Other systems may have one, two, three, or more repairs before its censoring age.

Age. Here "age" (or "time") means any useful measure of system usage, e.g., mileage, days, cycles, months, etc.

3. THE POPULATION AND ITS MEAN CUMULATIVE FUNCTION

Model. Needed information on the repair behavior of a population of systems is given by the population mean cumulative function (MCF) versus age *t*. This function is a feature of the following model for the population without censoring. At a particular age *t* each population system has accumulated a total cost (or number) of repairs. These cumulative system totals usually differ. Figure 1 depicts the population of such uncensored system *cumulative cost history functions* as smooth curves for ease of viewing. In reality the histories are staircase functions where the rise of each step is a system's cost or number of repairs at that age. However, staircase functions are hard to view in such a plot. At age *t*, there is a population distribution of the cumulative cost (or number) of repairs. It appears in Figure 1 as a continuous density. This distribution at age *t* has a population mean *M*(*t*). *M*(*t*) is plotted versus *t* as a heavy line in Figure 1. *M*(*t*) is called the population "mean cumulative function" (MCF) for the cost (or number) of repairs. It provides most information sought from repair data.

Repair rate. When *M*(*t*) is for the *number* of repairs, the derivative

$$m(t) = dM(t)/dt$$

is assumed to exist and is called the population "instantaneous repair rate". It is also called the "recurrence rate" or "intensity function" when some other repeating occurrence is observed. It is expressed in repairs per unit time per system, e.g., transmission repairs per 1000 miles per car. Some mistakenly call *m*(*t*) the "failure rate", which causes confusion with the quite different failure rate (hazard function) of a life distribution for non-repaired units (usually components). The failure rate for a life distribution has an entirely different definition, meaning, and use, as explained by Ascher and Feingold (1984). Note that the hazard function for a time-to-repair distribution is also called a "repair rate" but is an entirely different concept.

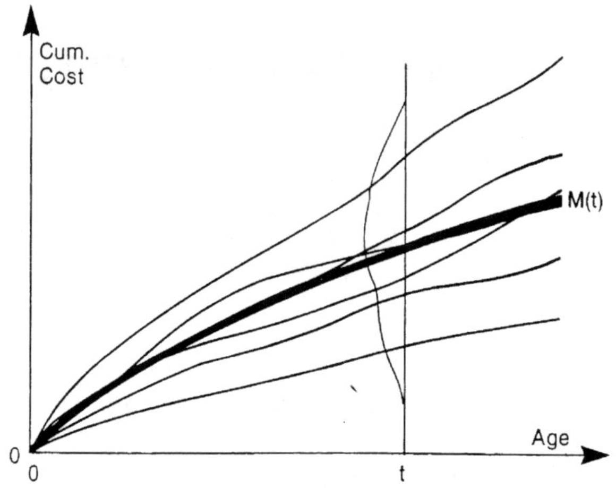

Figure 1. Population cumulative cost histories (uncensored), distribution

4. ESTIMATE AND PLOT OF THE MCF

Steps. The following steps yield a non-parametric estimate *M**(*t*) of the population MCF *M*(*t*) for the number of repairs from a sample of *N* systems; *N* = 34 cars.

1. List all repair and censoring ages in order from smallest to largest as in column (1) of Table 2. Denote each censoring age with a +. If a repair age of a system equals its censoring age, put the repair age first. If two or more systems have a common age, list them in a suitable order, possibly random.

2. For each sample age, write the number *I* of systems then in use ("at risk") in column (2) as follows. If the .earliest age is a censoring age, write *I* = *N* −1; otherwise, write *I* = *N*. Proceed down column (2) writing the same *I*-value for each successive repair age. At each censoring age, reduce the *I*-value by one. For the last age, *I* = 0.

3. For each repair, calculate its observed mean number of repairs at that age as 1/*I*. For example, for the repair at 28 miles, 1/34 = 0.03, which appears in column (3). For a censoring age, the observed mean number is zero, corresponding to a blank in column (3). However, the censoring ages determine the *I*-values of the repairs and thus are properly taken into account.

4. In column (4), calculate the sample mean cumulative function *M**(*t*) for each repair as follows. For the earliest repair age, this is the corresponding mean number of repairs, namely, 0.03 in Table 2. For each successive repair age this is the corresponding mean number of repairs (column (3)) plus the preceding mean cumulative number (column (4)). For example, at 19,250 miles, this is 0.04+0.26 = 0.30. Censoring ages have no mean cumulative number.

5. For each repair, plot on graph paper its mean cumulative number (column (4)) against its age (column (1)) as in Figure 2. This plot displays the non-

parametric estimate $M^*(t)$, which is a staircase function and is called the *sample MCF*. Censoring times are not plotted but are taken into account in the MCF estimate.

Table 2. MCF Calculations

(1) Mileage	(2) No. r obs'd	(3) mean no. $1/r$	(4) MCF
28	34	0.03	0.03
48	34	0.03	0.06
375	34	0.03	0.09
530	34	0.03	0.12
1388	34	0.03	0.15
1440	34	0.03	0.18
5094	34	0.03	0.21
7068	34	0.03	0.24
8250	34	0.03	0.27
13809+	33		
14235+	32		
14281+	31		
17844+	30		
17955+	29		
18228+	28		
18676+	27		
19175+	26		
19250	26	0.04	0.31
19321+	25		
19403+	24		
19507+	23		
19607+	22		
20425+	21		
20890+	20		
20997+	19		
21133+	18		
21144+	17		
21237+	16		
21401+	15		
21585+	14		
21691+	13		
21762+	12		
21876+	11		
21888+	10		
21974+	9		
22149+	8		
22486+	7		
22637+	6		
22854+	5		
23520+	4		
24177+	3		
25660+	2		
26744+	1		
29834+	0		

Plot. Figure 2 was plotted by Nelson and Doganaksoy's (1994) program MCFLIM, which does the calculations above. The program also calculates nonparametric approximate 95% confidence limits for $M(t)$; they are shown above and below each data point with a -. Nelson's (1995,2003) complex calculations of these limits requires a computer program like MCFLIM.

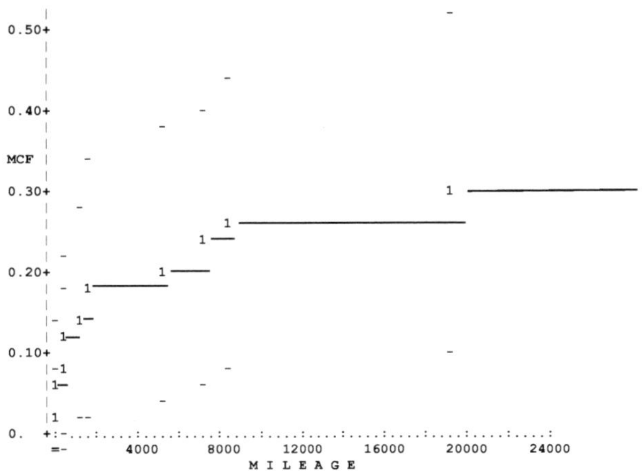

Figure 2. Transmission data MCF and 95% confidence limits - .

Software. In addition to the SuperSMITH software the following programs calculate and plot the MCF estimate and confidence limits from data with exact ages and right censoring. They handle the number and cost (or value) of recurrences and allow for positive or negative values.

- MCFLIM of Nelson and Doganaksoy (1989). This was used to obtain Figure 2.
- The RELIABILITY Procedure in the SAS/QC Software of the SAS Institute (1999), 947-951.
- The JMP software of SAS Institute (2005), 565-572.
- SPLIDA features developed by Meeker and Escobar (2004) for S-PLUS.
- A program for General Motors by Robinson (1995)

5. HOW TO INTERPRET AND USE A PLOT

MCF estimate. The plot displays a nonparametric estimate $M^*(t)$ of $M(t)$. That is, the estimate involves no assumptions about the form of $M(t)$ or the process generating the system histories. This nonparametric estimate is a staircase function that is flat between repair ages, but the flat portions need not be plotted. The MCF of a large population is usually regarded as a smooth curve, and one usually imagines a smooth curve through the plotted points. Interpretations of such plots appear below. See Nelson (1988,1995,2003) for more detail.

Mean cumulative number. An estimate of the population mean cumulative number of repairs by a specified age is read directly from such a curve through the plotted points. For example, from Figure 2 the estimate of this by 24,000 miles is 0.31 repairs per car, an answer to a basic question.

Repair rate. The derivative of such a curve

(imagined or fitted) estimates the repair rate $m(t)$. If the derivative increases with age, the population repair rate increases as systems age. If the derivative decreases, the population repair rate decreases with age. The behavior of the rate is used to determine burn-in, overhaul, and retirement policies. In Figure 2 the repair rate (derivative) decreases as the transmission population ages, the answer to a basic question.

Burn-in. Some systems are subjected to a factory burn-in. Systems typically are run and repaired until the instantaneous (population) repair rate decreases to a desired value m'. An estimate of the suitable length t' of burn-in is obtained from the sample MCF as shown in Figure 3. A straight line segment with slope m' is moved until it is tangent to the MCF. The corresponding age t' at the tangent point is suitable, as shown in Figure 3.

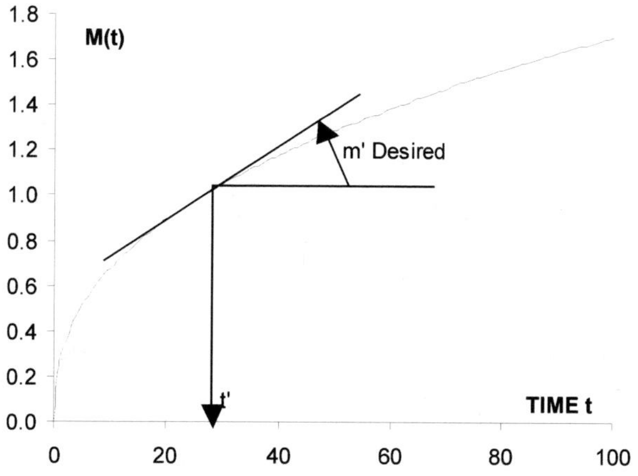

Figure 3. To determine age for burn-in t'.

Other information. Nelson (2003) gives other applications and information on

1. Predicting future numbers or costs of repairs for a fleet,
2. Analyzing repair cost or downtime data or other numerical values associated with repairs,
3. Analyzing availability data, including downtimes for repairs,
4. Analyzing data with more complex censoring where system histories have gaps with missing repair data,
5. Analyzing data with a mix of types of repairs,
6. The minimal assumptions on which the nonparametric estimate $M^*(t)$ and confidence limits depend.

Nelson and Doganaksoy (1994) and Nelson (2003) show how to statistically compare two data sets with respect to their MCFs. Such data sets may come from different designs, production periods, environments, maintenance policies, etc.

Literature. Most models and data analysis methods for repair data are parametric and involve more assumptions, which often are unrealistic. For example, Englehardt (1995) and Ascher and Feingold (1984) present such models, analyses, and assumptions for a single system, not for a sample of systems. The simplest such parametric model is the Poisson process. Also, the methods here extend to costs, downtimes, and other values associated with repairs, whereas previous methods apply only to simple counts of repairs.

Concluding remarks. The simple plot of the sample MCF is informative and widely useful. It requires minimal assumptions and is simple to make and present to others.

Acknowledgments. This updated version of Nelson (1998) appears here with the kind permission of Wiley, publisher of *Quality and Reliability Engineering International.* The author gratefully thanks Mr. Richard J. Rudy of Daimler-Chrysler, who generously granted permission to use the transmission data here.

Author. Dr. Wayne Nelson consults and trains privately on reliability data analysis. He is a Fellow of the Institute of Electrical and Electronic Engineers, the American Society for Quality, and the American Statistical Association. He has published 100+ articles, two Wiley classic books (*Applied Life Data Analysis* and *Accelerated Testing*), the SIAM book Nelson (2003), and various booklets.

REFERENCES

Abernethy, R.B. (2006), *The New Weibull Handbook,* 5th ed., available from Amazon.com. ISBN 0965306232.

Ascher, H. and Feingold, H. (1984), *Repairable Systems Reliability,* Marcel Dekker, New York.

Englehardt, M. (1995), "Models and Analyses for the Reliability of a Single Repairable System", in N. Balakrishnan (ed.), *Recent Advances in Life-Testing and Reliability,* CRC Press, Boca Raton, FL, 79-106.

Nelson, Wayne (1995), "Confidence Limits for Recurrence Data--Applied to Cost or Number of Repairs", *Technometrics* **37**, 147-157.

Nelson, Wayne B. (2003), *Recurrent Events Data Analysis for Product Repairs, Disease Recurrences, and Other Applications*, ASA-SIAM Series on Statistics and Applied Probability, www.siam.org/books/sa10 .

Nelson, Wayne and Doganaksoy, Necip (1994), *Documentation for MCFLIM and MCFDIFF -- Programs for Recurrence Data Analysis, available from Nelson, WNconsult@aol.com*

APPENDIX N: WALODDI WEIBULL

Dr. E.H. Waloddi Weibull
1887-1979
By: Dr. Robert B. Abernethy

The Weibull distribution is by far the world's most popular statistical model for life data. It is also used in many other applications, such as weather forecasting and fitting data of all kinds. It may be employed for engineering analysis with smaller sample sizes than any other statistical distribution. Having researched and applied this method for almost half a century, I was recently honored to be asked to write a short biography of this remarkable man from Sweden.

Waloddi Weibull was born on June 18, 1887 in Vittskoevle, a small village in southern Sweden. His family originally came from Schleswig-Holstein, at that time closely connected with Denmark. There were a number of famous scientists and historians in the family. His own career as an engineer and scientist is certainly an unusual one.

He was a midshipman in the Royal Swedish Coast Guard in 1904 was promoted to sublieutenant in 1907, Captain in 1916, and Major in 1940. He took courses at the Royal Institute of Technology where he later became a full professor and graduated in 1924. His doctorate is from the University of Uppsala in 1932. He worked in Swedish and German industries as an inventor (ball and roller bearings, electric hammer,) and as a consulting engineer. My friends at SAAB in Trollhatten Sweden gave me some of Weibull's papers. SAAB is one of many companies that employed Weibull as a consultant.

His first paper was on the propagation of explosive wave in 1914. He took part in expeditions to the Mediterranean, the Caribbean, and the Pacific Ocean on the research ship "Albatross" where he developed the technique of using explosive charges to determine the type of ocean bottom sediments and their thickness, just as we do today in offshore oil exploration.

In 1941 BOFORS, a Swedish arms factory, gave him a personal research professorship in Technical Physics at the Royal Institute of Technology, Stockholm.

Waloddi Weibull 1887-1979
Photo by Sam C. Saunders

He published many papers on strength of materials, fatigue, rupture in solids, bearings, and of course, the Weibull distribution. The author has identified 65 papers to date plus his excellent book on fatigue analysis (10), 1961. 27 of these papers were reports to the US Air Force at Wright Field on Weibull analysis. (Most of these reports to WPAFB are no longer available even from NTIS. The author would appreciate copies.) Dr. Weibull was a frequent visitor to WPAFB and many of our universities.

His most famous paper (7), at least in the USA, was given before the ASME in 1951, seven case studies using the Weibull distribution. Many including the author were skeptical that this method of allowing the data to select the most appropriate distribution from the broad family of Weibull distributions would work. However the early success of the method with very small samples at Pratt & Whitney Aircraft could not be ignored. Further, Dorian Shainin, a consultant for Pratt & Whitney, strongly encouraged the use of Weibull analysis. The author soon became a believer.

Robert Heller (1) spoke at the 1984 Symposium to the Memory of Waloddi Weibull in Stockholm, Sweden and said, in 1963, at the invitation of the Professor Freudenthal, he became a Visiting Professor at Columbia University's Institute for the Study of Fatigue and Reliability. I was with the Institute at that time and got to know Dr. Weibull personally. I learned a great deal from him and from Emil Gumbel and from Freudenthal, the three founders of Probabilistic Mechanics of Structures and Materials. It was interesting to watch the friendly rivalry between Gumbel, the theoretician and the two engineers, Weibull and Freudenthal.

"The Extreme Value family of distributions, to which both the Gumbel and the Weibull type belong, is most applicable to materials, structures and biological systems because it has an increasing failure rate and can describe wear out processes. Well, these two men, both in their late seventies at the time, showed that these distributions did not apply to them. They did not wear out but were full of life and energy. Gumbel went skiing every weekend and when I took Dr. and Mrs. Weibull to the Roosevelt Home in Hyde Park on a cold winter day, he refused my offered arm to help him on the icy walkways saying: "A little ice and snow never bothered a Swede."

In 1972, the American Society of Mechanical Engineers (2) awarded Dr. Weibull their gold medal citing Professor Weibull as "a pioneer in the study of fracture, fatigue, and reliability who has contributed to the literature for over thirty years. His statistical treatment of strength and life has found widespread application in engineering design." The award was presented by Dr. Richard Folsom, President of ASME, (in the middle of the picture below with Neil Armstrong, the astronaut, facing Professor Weibull). By coincidence the author received the 1988 ASME gold medal for statistical contributions including advancements in Weibull analysis.

Professor Sam C. Saunders took the first picture above in 1971 when Dr. Weibull was visiting Seattle. In correspondence with the author he comments: "During that visit he bragged to me that his second wife was younger than mine! (My wife was 39.) When I met his second wife in Stockholm during the cocktail party at the 1984 Symposium dedicated to his memory, she said her youngest daughter much resembled Waloddi and she called her over to introduce her to me. The daughter was then about 20 and the mother about 50, and both were very handsome women." I thank Professor Saunders for permission to reproduce the picture and for sharing his memories of this great man. Dr. Weibull would have been 84 years old in 1971.

The second picture above of a much younger Professor Weibull was obtained from Professor Nowak at the University of Warsaw, Poland. The other two pictures as well as some prose herein were taken from Dr. Glenn Bowie's excellent Website with his permission. http://glennbowie/tripod.com In correspondence between Dr. Bowie and Weibull's wife, Mrs. Ibby Weibull, the following is quoted, "We are happy to contribute to your idea of giving better information into Internet about Waloddi Weibull. His full name is **Ernst Hjalmar Waloddi Weibull.** The name **Waloddi** was shortened to **Doddi** in the family and by close friends." Weibull's son Jorden Weibull told me that in Sweden his name is pronounced, "Vayloddi Vaybull." Nils Weibull, Waloddi Weibull's grandnephew says that the family referred to Waloddi as "Uncle Doddi."

The US Air Force Materials Laboratory should be commended for encouraging Waloddi Weibull for many years with research contracts. The author is also indebted to WPAFB for contracting the original USAF Weibull Analysis Handbook (3) and Weibull video training tape, as he was the principal author of both. The latest version of that Handbook is the fourth edition of The New Weibull Handbook (4).

Professor Weibull's proudest moment came in 1978 when he received the Great Gold medal from the Royal Swedish Academy of Engineering Sciences which was personally presented to him by King Carl XVI Gustav of Sweden. Below is the photo with King Carl XVI Gustav of Sweden, Waloddi Weibull, and in the middle Gunar Hambræus, then President of the Royal Swedish Academy of Engineering. When Waloddi stood in front of the King he said: "Seventy-one years ago I stood in front of Your Majesty's grandfather's grandfather (King Oscar II) and got my officer's commission." The King then said: "That is fantastic!"

He was devoted to his family and was proud of his nine children and numerous grand and great-grandchildren.

Dr. Weibull was a member of many technical societies and worked to the last day of his remarkable life. He died on October 12, 1979 in Annecy, France at the age of 92.

The Weibull Distribution was first published in 1939, over 60 years ago and has proven to be invaluable for life data analysis in aerospace, automotive, electric power, nuclear power, medical, dental, electronics, every industry. Yet the author is frustrated that very few universities in the USA teach Weibull analysis to engineering students. To encourage the use of Weibull analysis the author provides free copies of The New Weibull Handbook and The New Weibull Handbook Student Version to university libraries in English speaking countries that request the book. The corresponding SuperSMITH software is available from Wes Fulton in demo version free from his web site. (www.weibullnews.com)

The author would appreciate comments and questions about Waloddi Weibull and Weibull analysis.

References: $ indicates the reference may be downloaded from Paul Barringer's Website (see References). Italics indicate a book rather than a paper.

1. S. Eggwertz & N.C. Lind, Editors: ***"Probabilistic Methods in the Mechanics of Solids and Structures,"*** Symposium Stockholm Sweden, June 1984, To the Memory of Waloddi Weibull, Springer-Verlag.
2. $Announcement of Waloddi Weibull receiving **1972 ASME Gold Medal** for distinguished engineering achievement.

3. Abernethy, Robert B.; J.E. Breneman; C.H. Medlin; G.L. Reinman; ***"Weibull Analysis Handbook,"*** U.S. Air Force AFWAL-TR-83-2079, November 1983, AD #A143100 Approximately 14,000 copies were distributed by the US Air Force, Pratt & Whitney, and the author.

4. Abernethy, Robert B. ***"The New Weibull Handbook,"*** Fourth Edition, November 2000, published by the author. ISBN 0-9653062-1-6. Over 14,000 copies are in circulation.

5. $ Waloddi Weibull,**"En Statistisk Teori För Utmattmingshallfastheten",** Report TM 344 written for AB Bofors (in Swedish) dated 2.9.1948 *If you could translate this to English, please send a copy to the author.*

6. $ Waloddi Weibull, **"A Statistical Representation Of Fatigue Failures In Solids,"** Transactions Of The Royal Institute Of Technology, Stockholm, Sweden, Number 27—1949,

7. $ Waloddi Weibull, **"A Statistical Distribution Function of Wide Applicability",** ASME Journal Of Applied Mechanics Paper, 1951. *The hallmark paper of Weibull analysis.*

8. $ Waloddi Weibull, **"Statistical Evaluation Of Data From Fatigue And Creep-Rupture Tests. Fundamental Concepts And General Methods",** Wright Air Development Center Report 59-400, Report # AD229450, September 1959.

9. $ Waloddi Weibull, **"Size Effects On Fatigue Crack Initiation and Propagation In Aluminum Sheet Specimens Subjected To Stresses Of Nearly Constant Amplitude",** Report 86 to Flygtekniska Försöksanstalten The Aeronautical Research Institute Of Sweden, 1960.

10. $ Waloddi Weibull, ***"Fatigue Testing And Analysis Of Results,"*** 1961, published for the Advisory Group For Aeronautical Research and Development North Atlantic Treaty Organization, Pergamon Press

11. $ Waloddi Weibull, **"Estimation Of Distribution Parameters By A Combination Of The Best Linear Order Statistic Method And Maximum Likelihood".** USAF Report AFML-TR-67-105 / AF 61(052)-522. 1967.

12. $ Waloddi Weibull, **"Composition And Decomposition Of Bounded Variates With Special Reference To The Gamma And The Weibull Distributions:** USAF report AF61(052)-522; indexed as report AD-816012, March 1967

13. $ Waloddi Weibull, **"The Order Statistics $y_i = \log(z_i^m)$, Their Properties And Use For Parameter Estimation (z = standardized Weibull variate),"** USAF AFML-TR-67-161; indexed as report AD-818104, June 1967

14. $ Waloddi Weibull, **"Moment Estimators For Weibull Parameters And Their Asymptotic Efficiencies"** 1969. USAF report AFML-TR-69-135; indexed as AD-690162

15. $ Waloddi Weibull, **"A Statistical Theory of the Strength of Materials,"** 1939, *This is the very first paper on the Weibull Distribution. This original paper written in English was kindly provided by Professor Isamu Yoshimoto, retired from the Tokyo Institute of Technology. He traveled to Stockholm in 1976 and obtained this original paper which will be of great interest to metallurgists. It provides Weibull's thoughts and derivation of the basic Weibull distribution with lots of case studies.*

References

[Books titles are in italics. The $ sign indicates the reference may be downloaded from Paul Barringer's Website ... see "Barringer" below]

Abernethy, Robert B., ***"The New Weibull Handbook,"*** 4th edition, (this text), see $Winterbottom for Royal Statistical Society review of the third edition listed below.*(Actually the 5th edition if the "**USAF Weibull Analysis Handbook"** is considered the original.)*

Abernethy, Robert B., J.E. Breneman, C.H. Medlin, G.L. Reinman, ***"Weibull Analysis Handbook,"*** U.S. Air Force AFWAL-TR-83-2079, November 1983, AD #A143100. Also with same co-authors, USAF video, "**An Introduction to Weibull Analysis,"** now available from the author with US Air Force permission.

$Abernethy, Robert B. and Jack Sammons, **"Three Applications of Monte Carlo Simulation to the Development of the F100 Turbofan Engine,"** AIAA/SAE 12th Propulsion Conference, 1976.

$Abernethy, Robert B., C.H. Medlin, B.G. Ringhiser, **"Analysis of Turbopump Failures Using Advanced Weibull Techniques,"** RAMS Proceedings available from IEEE, #83RM367, January 1983.

$Abernethy, Robert B. and Wes Fulton, **"New Methods for Weibull & Lognormal Analysis,"** ASME Winter Annual 1992, 92-WA/DE-14.

Abernethy, Robert B. et al,***"ICRPG Handbook for Estimating the Uncertainty in Measurement Made With Liquid Propellant Rocket Engine Systems,"*** 1969, JANNAF Report CPIA No. 180, AD 851 127. Earliest standard on measurement uncertainty analysis. The later SAE-ASME-ISO documents are better.

$Abernethy, Robert B. et al, ***"Handbook Uncertainty in Gas Turbine Measurements,"*** US Air Force AEDC-TR-73-5, 1973, Reprinted as The Instrument Society of America's "***ISA Measurement Uncertainty Handbook***."

$Abernethy, Robert B. and Barbara Ringhiser, **"The History and Statistical Development of the New ASME-SAE-AIAA-ISO Measurement Uncertainty Methodology,"** 1975 ASME-SAE-AIAA Propulsion Conference.

Abernethy, Robert B., R. Benedict and R.Dowell, "**ASME Measurement Uncertainty,"** ASME paper 83-WA/FM-3.

Abernethy, Robert B. and the ASME MFFCC S/C 1 Committee, "***Measurement Uncertainty for Fluid Flow in Closed Conduits,"*** ANSI/ASME MFC-2M-1983. *The standard for fluid flow measurement developed by the author and chairman of this sub-committee. This document lead to the ISO draft described below.*

Abernethy, Robert B. and the SAE E-33 Committee, ***"Uncertainty of In-Flight Thrust Determination,"*** SAE ARP 1678, August 1985-1, reaffirmed December 1992, *This is the standard for the aerospace industry including NATO Agard. The author was founding chairman of the committee.*

Abernethy, Robert B. and the ASME PTC 19.1 Committee, ***"Measurement Uncertainty,"*** ANSI/ASME PTC 19.1-1985-2, *This standard applies to all ASME activities. The author was the succeeding Chairman of this committee. The new version,* ***"Test Uncertainty,"*** *PTC19.1-1998 introduces more optimism in the uncertainty estimates. The author prefers the original.*

$Abernethy, Robert B. and the ISO TC30 SC9 Committee, ***"Fluid Flow Measurement Uncertainty,"*** 1988. *The author was the principal author of this draft revision of ISO/DIS 5168 that was approved by the committee and the world vote (17-19) 1988 but never published as the France controls the secretariat of TC30. The best draft was the 1982 version available on Paul Barringer's website (see Barringer below).*
Abernethy...see also Fulton and Marquart.

Alley, William and Stetson, Harold, A. **"Monte Carlo Simulation of the Engine Development Process," 1983** AIAA-SAE Propulsion Conference. *First use of Crow-AMSAA for modeling first failure observations. Simulation constructed by Barbara Ringhiser and Robert Abernethy.*

"AMSAA Reliability Growth Handbook, TR-652" Broehmm, William; Ellner, Paul; Woodworth, John; AMSAA Report HDBK-A-1, September 1999, *The latest upgrade of MIL-HDBK-189, the most complete book on Crow-AMSAA modeling. This is an excellent presentation. It may be downloaded from* http://web.amsaa.army.mil/AR/rel_growth_guide.htm

Ascher and Feingold, ***"Repairable Systems Reliability,"*** Dekker, 1984. *Heavy stuff, good cure for insomnia.*

ASTM "Manual on Presentation of Data and Control Chart Analysis," *Sixth Edition, contains excellent approximations for the C4 factor.*

"ASTM C1239-95 Standard Practice for Reporting Uniaxial Strength Data and Estimating Weibull Distribution Parameters for Advanced Ceramics" November 1995. *Illustrates the use of reduced bias adjustment for MLE estimates. Authors: Stephen F. Duffy, George Quinn, Curt Johnson, Bill Tucker and John Cuccio.*

Barringer, H. Paul, **"Pipe Wall Thickness Decisions Using Weibull Analysis,"** Equipment Inspection... Conference Organized Lake Area Industries, available on Paul Barringer's website (see below).
"Process Reliability and Six-Sigma," National Manufacturing Week Conference Track For Manufacturing Process and Quality, March 2000.
"Consider using a new reliability too: Weibull analysis for production data," coauthor W. T. Roberts, Dow Chemical, Hydrocarbon Processing, October 2001.
"Barringer1.com" website – reinstatement in process. Contains sample Weibull Library and Problem of the Month case studies.

Barmoav, Moshe Felix, **"Reduce Bias Factor Distribution: A new Distribution Reduces the Likelihood Estimates Bias of Small Sample Size"** Unpublished paper 2009

Bayes, Thomas, **"An Essay on solving a Problem in the Doctrine of Chances,"** as presented by Mr. Richard Price on December 23, 1763.

$Benard, A., and E.C. Bosi-Levenbach, **"The Plotting of Observations on Probability Paper,"** Statististica Neerlandica, Vol. 7, 1953, p. 163-173. *We have a translation from the Dutch to English available on Paul Barringer's website, thanks to Ronald Schop. Benard indicates the median approximation is actually an optimal plotting position.*

Berkson, Joseph, **"Are There Two Regressions?,"** Journal of the American Statistical Association, Vol. 45, 1950. See also, Mary Natrella, **"Characterizing Linear Relationships between Two Variables" in *NBS Special Handbook 91 - 1966 and NBS Special Publication 300, Volume 1.***

Bitker, Lyle, **"Use of Weibayes for Safe Lower Cost Overhaul Interval,"** ASQ Reliability Review, 2001 September, Volume 21, Number 3. *A good case study of using Weibayes to extend overhaul intervals.*

Billatos, S.B. and L.A. Kendall, **"A Replacement Model for Multi-Tool Transfer Lines,"** Journal of Engineering for Industry, August 1990. Also see **"Approximate Renewal Functions for Optimization,"** Proceedings, 40th Mechanical Failure Prevention Group, NBS, April 1985. *Weibull applications to machine tool wear, also provides approximate renewal function for block replacement.*

Blom, G., **"Statistical Estimates and Transformed Beta-variables,"** Stockholm, 1958

Box, Joan Fisher, ***"R.A. Fisher The Life of a Scientist,"*** John Wiley and Sons, *An excellent biography on perhaps the world's greatest statistician by his daughter, including material on his peers, Pearson and Gosset. See also* ***"The Lady Tasting Tea."*** *By David Salsburg.*

Bucci, R.M. et al, **"Failure Prediction of Underground Distribution Feeder Cables," IEEE 92 WM 197-4 PWRD.**

Canfield, David, SDC Computer Services under contract to Prestolite and Leece-Neville Heavy Duty Systems, 1999, **"Weibull Whitepaper on Using SuperSMITH Weibull to Analyze Warranty Data,"** Available by Phone 716-492-1700 from the author, ***Procedure for analysis of Warranty data using EXCEL and SSW.***

Cox, D.R, D. Oakes, ***"Analysis of Survival Data,"*** Chapman & Hall, 1984.

Crow, Larry H., **"Evaluating the Reliability of Repairable Systems,"** and **"Reliability Growth Estimation With Missing Data - II,"** 1990 RAMS Proceedings. See also MIL-HDBK-189, ***"Reliability Growth Management,"*** written by Larry Crow. **also** IEC Standard 61164, ***"Reliability Growth-Statistical Tests and Estimation Methods,"*** 1995. (Dr. Crow is the chairman of IEC Committee 56)

Cutler, Sidney J., Fred Ederer, **"Maximum Utilization of the Life Table Method in Analyzing Survival,"** Journal of Chronic Diseases, Volume 8, 1958, *Describes the Kaplan-Meier Model with the Actuarial Correction.*

Darveaux, Robert and Andrew Mawer, **"Thermal and Power Cycling Limits of Plastic Ball Grid Array Assemblies,"** 1995 SMI Proceedings, *Prize winning paper from Motorola on solder fatigue in electronic assemblies. See also Mawer.*

Duane, J.T. (1964), **"Learning Curve Approach to Reliability Monitoring,"** IEEE Transactions on Aerospace 2, No. 2, 563-566.

$Dumonceaux, Robert and Charles E. Antle, **"Discrimination Between the Lognormal and the Weibull Distributions,"** Technometrics Volume 15 No. 4, 1973.

Duncan, Acheson J., ***"Quality Control and Industrial Statistics,"*** Irwin, Fourth Edition, *Excellent text on SQC, contains good material on the C4 factor.*

Dzekevich, Joseph, **"Finding Defects Using Burn-in Analysis,"** Raytheon Analysis Laboratory, August 2002, http://www.reliabilityanalysislab.com/0208_burnin.asp *Presents Weibull plot interpretation that lead to optimal burn-in intervals.*

Edgeman, Rick L. **"Correlation Coefficient Assessment of NPPs,"** ASQC Reliability Review, Vol. 10, Number 2, June 1990.

Fisher, Sir Ronald and Frank Yates, ***"Statistical Tables for Biological, Agricultural and Medical Research,"*** Oliver and Boyd, Sixth Edition. *The hallmark statistical tables competing with Biometricka.* Fisher, Sir Ronald. **"Student,"** Annals of Eugenics, v. 9, p. 1-9 (1939), with permission of Cambridge University Press 1939, *Fisher's obituary on his best friend.*

$Fulton, Wes and Robert B. Abernethy, **"A Simple Method for Comparing Designs...Are Two Data Sets Significantly Different?,"** SAE 960544, February 1996, Included in Chapter 7. *Describes best practice for this popular problem.* , also **"Likelihood Adjustment: A Simple Method for Better Forecasting from Small Samples,"** presented at the 2000 RAMS Conference. *Describes corrections that allow the likelihood function, the likelihood test, and Weibull / lognormal parameters to be used with small samples accurately.*

Gan, F.F. and K.J. Koehler, **"A Goodness-of-Fit Test Based on P-P Probability Plots,"** Journal of Quality Technology, April 1992.

Glasser, Gerald J., **"Planned Replacement: Some Theory and its Application,"** Journal of Quality Technology, Vol. 1, No. 2, April 1969. *Hallmark paper on replacement intervals.*

Gonick and Woollcott, ***"The Cartoon Guide to Statistics,"*** *A good introduction to statistical methods with some humor.* See also: ***"The High-Stakes World of Statistics"*** by the Standard Deviants, a video published by Cerebellum Corporation.

Gorman, Jeff, Stavropoulos, R.W. Staehle, C.S. Welty, **"Selection of Statistical Distributions for Prediction of Steam Generator Tube Degradation,"** EPRI 1991 Conference on Environmental Degradation of Materials. *Interesting paper on distribution analysis in the nuclear field.*

Gumbel, E.J., ***"Statistics of Extremes,"*** Columbia University Press, New York, 1958.Hahn and Shapiro, ***"Statistical Models in Engineering,"*** John Wiley, 1967. *Excellent text!*

Harter, H. Leon and Moore, Albert H., **"Point and Interval Estimators, Based on m Order Statistics, for the Scale Parameter of a Weibull Population with Known Shape Parameter,"** Technometrics, Vol. 7, No. 3, August 1965. *Early description of Weibayes method.*

IEC Standard 61164, ***"Reliability Growth-Statistical Tests and Estimation Methods,"*** 1995. (Dr. Larry Crow is the chairman of IEC Committee 56: Dependability) The American version is ANSI/IEC/ASQ D61164-1997. *Excellent well-written document on component testing to measure reliability growth.*

IEC 61710, ***"Power law model – Goodness-of-fit tests and estimation methods,"*** This document applies the Crow-AMSAA model to tracking units in service, both repairable and non-repairable. A*nother well-written document describing this important application of Crow-AMSAA.*

IEC 61649 Edition 2, (2008-8), ***Weibull Analysis,*** This new world standard is absolutely consistent with this Handbook and the SuperSMITH software.

Jensen, Finn, ***"Electronic Component Reliability,"*** John Wiley Reprinted 1998. *Leading text on Electronic Reliability.*

Jackson, L.L. and J.D. Shepherd, **"V-Belt Reliability -- A statistical Study of Large Sample Size Fatigue Tests,"** Society of Automotive Engineers, Technical Paper Series No. 800446, 1980. *Early application of Weibayes to reduce testing sample sizes.* Also Jackson, L.L. and Elroy E. Schock, **"Manufactured Products like Fingerprints and Snowflakes are Never Identical,"** The Gates Rubber Company, May 22, 1968.

Johnson, Leonard G., **"Median Ranks of Sample Values in their Populations with an Application to Certain Fatigue Studies,"** Industrial Mathematics, Vol. 2, 1951. Also Johnson, Leonard G., "***The Statistical Treatment of Fatigue Experiments***," Elsevier Publishing Company, 1964. *Describes median rank plotting and the Beta-Binomial Confidence bounds.*

Juran, J.M., **"*Quality Control Handbook*,"** McGraw-Hill, 1951.*Includes table of C4.*

Kaplan, E.L. and P. Meier, **"Nonparametric Estimation from Incomplete Observations,"** American Statistics Association Journal, 1958, *The Kaplan-Meier product limit estimator.*

Kapur, K.C., and L.R. Lamberson, **"*Reliability in Engineering Design*,"** John Wiley, 1977, *Well respected older text.*

Lam, R and Spelt, J.K., **"Comparison of Weibull Small Samples using Monte Carlo Simulations."** Quality and Reliability Engineering International, 2007, 23.503-513

Lamberson, L.R. and G.S. Wasserman, **"Modified Growth Model Improved Spares Management,"** ASQC Reliability Review, September 1992, *Duane Model Case Study*. See also Wasserman.

Lawless, Jerald, **"*Statistical Models and Methods for Lifetime Data*,"** John Wiley, 1982. *Advanced text with excellent treatment of likelihood methods by the leading author on this subject.*
Also, **"Statistical Methods in Reliability,"** Technometrics, Vol. 25, 1983.

Lieblein, J. and M. Zelen, **"Statistical Investigation of the Fatigue Life of Deep-Groove Ball Bearings,"** November 1956, Journal of Research of the National Bureau of Standards, Vol 57, No.5, *Hallmark paper on Bearing Analysis and early application of Weibayes.*

$Liu, Chi-Chao (now Allen Liu), **"A Comparison Between the Weibull and Lognormal Models used to Analyze Reliability Data,"** Doctoral thesis submitted August 1997 at the University of Nottingham. *The most extensive and important study of these two distributions using several different methods and both Weibull and lognormal data sets with and without random suspensions. Liu made three major conclusions: (1) Median rank regression, X on Y, is more accurate then MLE except for very large sample sizes. (2) the two best measures of goodness of fit are the "p" value or r square and the likelihood ratio test. (3) With 20 failures or less the two parameter Weibull is the distribution of choice. See also [Wenham]. See picture of Dr. Chi-Chao Liu with the author below.*

Dr. Chi-Chao Liu with the author lecturing in Taiwan.

Maller, R.A. and S. Zhou, **"Testing for Sufficient Follow-up and Outliers in Survival Data,"** ASA Journal Vol. 89 1994. *Interesting study of the batch problem of perpetual survivors using sex crime data.*

Marquart, Todd A., **"Comparison Of Methods For Life Data Analysis Of Interval Data Using Monte-Carlo Simulations,"** This new study is the first comparison of the inspection option, Probit and interval MLE and othe methods for grouped data.

Mawer, Andrew, et al, **"Board-Level Characterization of 1.0 and 1.27 mm Pitch PBGA for Automotive Under-Hood Applications,"** 1999 ECTC Conference, San Diego, *Weibull analysis of accelerated tests of solid state electronic boards at Motorola.*

McCool, John I., **"The Analysis of Two-Factor Rolling Contact Fatigue Test Results,"** 50th Annual Meeting of STLE, Chicago, 1995. Also **"Evaluating Weibull Endurance Data by the Method of Maximum Likelihood,"** American Society of Lubrication Engineers, ASLE/ASME Conference Houston, Texas, 1969. *Early attempt to unbias the Weibull MLE Beta Estimates. And* **"Inference on the Weibull Location Parameter,"** *On the three parameter Weibull t-zero.*

McGlone, M.E., **"Reliability-Growth Assessment, Prediction, and Control for Electronic Engine Control (GAPCEEC),"** USAF AFWAL-TR-84-2024, April 1984.*The most complete comparison of reliability growth models ever made. Most of the analysis was done in the author's Pratt & Whitney RMS department. Conclusion: Duane-AMSAA is the best model with predictive intervals of +/- 10% for 12 months ahead for service data and +/- 20% for R&D data estimates of in-service reliability. ...Not good, but better than any other method.*

Meeker, William Q., Escobar, Luis A., ***"Statistical Methods for Reliability Data,"*** John Wiley & Sons, 1998. *Use MLE exclusively for all methods and includes USAF Weibull Analysis Handbook case study data sets. However, without median rank regression and without the MLE reduced bias adjustment this text has limited application for small samples.*

Medlin, C.H. and Elsaesser, F.L., **"Weibull/Weibayes Analysis of Hydraulic Pump Malfunction Data,"** SAE 831542, October 1983.

Menis, D.L. et al, **"Uniaxial Tensile Fatigue of Two Dental Acrylics and Simplex-P Bone Cement,"** Northwestern Dental Research News, Fall 1991. **Weibull analysis of dental bone cements. See Section 9.15.**

MIL-HDBK-189, ***"Reliability Growth Management,"*** Department of Defense, Washington, DC (1981). Available from the National Technical Information Service, Springfield, VA. "AMSAA" for latest upgrade.

$Mischke, C.R., **"A Distribution-Independent Plotting Rule for Ordered Failures,"** ASME Design Engineering Technical Conference, June 1979. *Excellent discussion of plotting positions.* Also, **"Some Property Data and Corresponding Weibull Parameters for Stochastic Mechanical Design,"** ASME Journal of Mechanical Design, March 1992. *Provides estimates of the_Weibull parameters for many metals.*

Mulready, Richard C., ***"Advanced Engine Development at Pratt & Whitney, The Inside Story of Eight Special Projects 146-1971,"*** *The history of hydrogen as a fuel for jet engines and rockets involving many of the secret projects that Mulready and the author worked on together.Excellent!*

Nelson, Wayne, ***"Applied Life Data Analysis,"*** John Wiley, 1982. *Hallmark text by the leading author.*
"Weibull Analysis of Reliability Data with Few or No Failures," Journal of Quality Technology, Vol. 17, No. 3, July 1985. *Good discussion of Weibayes theory including the derivation of confidence bounds for Weibayes with failures.*
"Accelerated Testing," John Wiley, 1990, *Best text on this subject.*
"Weibull Prediction of a Future Number of Failures," Quality and Reliability Engineering International, 2000; 16; 23-26. *Failure Prediction with assumed shape parameter.*
" Recurrent Events Data Analysis for Product Repairs, Disease Recurrences, and Other Applications," ASA SIAM 2003. ISBN 0-89871-522-9 *Excellent new text on* graphical repair analysis. *See Appendix M for this Nelson's paper on this subject.* **"Graphical Repair Analysis**

Nevell, David, "**Using Aggregated Cumulative Hazard Plots to Visualize Failure Data,"** Quality and Reliability Engineering International, 2000: 16: 209-219. *The ACH method (Appendix F) is the best tool for detecting batch problems. This paper emphasizes the use of the method to look at a graphic display of the data.*

Nicholls, David; Lein, Paul, "Weibayes Testing: What is the Impact if Assumed Beta is Incorrect?" 2009 RAMS Conference *Excellent new paper from RIAC*

O'Connor, Patrick D.T., "***Practical Reliability Engineering***" John Wiley, 1991. *Very readable.*

Oden, John L., **"A Weibull Shelf-Life Model for Pharmaceuticals,"** ASQC 32nd Annual Technical Conference, 1978.

Pearson, E.S. & H.O. Hartley, "***Biometrika Tables for Statisticians,***" Volume 1, Cambridge University Press. Also Pearson, E.S., "***Student,***" Clarendon Press Oxford, 1990. *This excellent biography of W.S. Gosset was completed by my professor, George Barnard, after Pearson passed on. See also Box and Salsburg.*

Quanterion Solutions website, **"Quanterion.com"**, has an extensive Weibull Library for engineering materials, etc. … Highly recommended.

Ran, A., M. Koch, and H. Mellander, Volvo Corporation, **"Fitting Injury Versus Exposure Data into a Risk Function,"** International Research Committee On BioKinetics of Impacts, 1984, Research Institute for Road Vehicles, Delft, Holland...and also Haffner, Mark, U.S. Dept. of Transportation, **"Synthesis of Pelvic Fracture Criteria for Lateral Impact Loading,"** 10th Experimental Safety Vehicles Conference, Oxford, England, 1985. *Weibull analysis of bone fractures from automobile accidents.*

"RAPTOR" software, Rapid Availability Prototyping for Testing Operational Readiness, Kenneth Murphy, HQ AFOTEC/SAL, Kirkland Air Force Base, NM 87117-5538. *Friendly simulation software available with a user's manual. [BoozAllen acquired RAPTOR from ARINC after a series of divestitures and mergers. As of 2018, BoozAllen, has discontinued sales of RAPTOR due to Windows compatibility issues. There are other programs named RAPTOR, but they are NOT this software.]*

Rheinfurth, Mario H., **"Weibull Distribution Based on Maximum Likelihood with Interval Inspection Data,"** NASA Report TM-86515, July 1985. *MLE methods for interval data.*

Royal Statistical Society review of **"The New Weibull Handbook,"** see Winterbottom below.

Salsburg, David, ***"The Lady Tasting Tea."*** W.H. Freeman and Company, 2001. *Excellent book on statisticians from Karl Pearson on, well written and readable.*

Sherwin, D.J. and F.P. Lees, **"An Investigation of the application of Failure Data Analysis to Decision-making in Maintenance of Process Plants,"** Parts I & II, Proceeding Institute of Mechanical Engineers, Vol. 194, 1980, Great Britain. *Suggests fitting the topmost point for interval inspection data, WeibullSMITHJ Inspection Option #1.*

Sica, Charles, and Luko, Steve, **"Correction Factors for Use with the Method of 'Sudden Death' Weibull Analysis,"** SAE paper 98031. *Provides a useful factor for situations where more than one failure occurs in Sudden Death testing. This technology is available in the SuperSMITH software.*

Smith, David B., **"Analysis of NERC-GADS Database for Steam Turbine Generator Reliability Trends,"** 1999 ASME Joint Power Generation Conference. *Describes Crow-AMSAA analysis to improve maintainability. This paper will received the Edison Electric Institute Award for best paper at the 2000 ASME Joint Power Conference.*

Spector, R.B., **"Reliability Prediction Techniques for Second Generation Marine and Industrial Gas Turbines,"** ASME-GE-P&W Gas Turbine Conference, Technion Institute, Haifa, Israel, 1980. *First published application of Duane-Crow-AMSAA for first time failure analysis.*

Staehle, Roger W., **"Lifetime Prediction of Materials in Environments,"** A chapter in the "**Corrosion Handbook,** 2000. Also, **"Framework for Predicting Stress Corrosion Cracking,"** for inclusion in **"Environmentally Assisted Cracking: Predictive Methods for Risk Assessment and Evaluation of Materials, Equipment, and Structures,"** ASTM STP 1401, 2000. Dr. Staehle may be reached at 651-482-9493, staeh001@tc.umn.edu.

Staton, Marie and Bruce Sharp, **"Reliability Centered Maintenance at Turkey Point,"** Nuclear Plant Journal, May-June 1991. *Good discussion of RCM and replacement intervals.*

Sterling, W.J., "Estimating Extra-long Balloon Tendon Lifetime," AIAA 2007-2616, *An excellent paper on using Weibull analysis to estimate creep life of advance plastics..*

D. Stienstra and T.L. Anderson, **"Statistical Inferences on Cleavage Fracture Toughness Data"** Journal of Engineering Materials and Technology, June 1988. *Weibayes application to material properties.*

Tallian, Tibor, **"Weibull Distribution of Rolling Contact Fatigue Life and Deviations Therefrom,"** ASLE/ASME Conference on Lubrication, Chicago, 1961.

Tarum, Carl D., **"Modeling Bathtub Curve Failure Data with Weibull Equations,"** SAE Weibull Users Conference, Detroit, March 1996. Also **"Determination of the Critical Correlation Coefficient to Establish a Good Fit for Weibull and Lognormal Failure Distributions"**, SAE Paper 1999-01-057,

Thoman, Darrel and Lee J. Bain, **"Two Sample Tests in the Weibull Distribution,"** Technometrics Vol. 11, 1969. Also Darrell Thoman, Lee J. Bain and Charles E. Antle, **"Inferences on the Parameters of the Weibull Distribution,"** same source. *Contains factors to reduce the bias in MLE Weibull parameter beta very similar to the MLE-RBA corrections in Chapter 6, (slightly larger than C4^60).*

Thompson, Geoffrey A., **"Determining the Slow Crack Growth Parameter and Weibull Two-Parameter Estimates of Bilaminate Disks by Constant Displacement Rate Flexural Testing,"** Presented at The Academy of Prosthodonics on May 6, 2002.*An outstanding paper on an extensive study on crack growth in ceramic materials used in dentistry*

Tobias, Paul A. and David Trindade, ***"Applied Reliability,"*** Van Nostrand Reinhold, 1986. *Good discussion of physical background of Weibull, Lognormal and Kaplan-Meier.*

Unkle, Richard, Ray Venkataraman, **"Relationship between Weibull and AMSAA models in reliability analysis: a case study,"** International Journal of Quality and Reliability Management, Vol 19, No 8/9, 2002. *An excellent paper on using both models together to provide the earliest warning of problems on locomotives in service at a reasonable cost by linking the parameters of the two models.*

Walton, Mark D. et al, **"Accelerated Cable Life Testing of EPR-Insulated Medium Voltage Distribution Cables,"** IEEE Transactions on Power Delivery, Vol. 9, 1994.*First paper on accelerated testing of high voltage cables.*

Wang, C. Julius, Chrysler Corporation, **"Sample Size Determination of Bogey Tests Without Failures,"** Quality and Reliability Engineering International, Vol. 7, (1991). *Applies Weibayes zero-failure testing, Chapter 6, to automotive Bogey testing. Also* **"Two Useful Equations In Reliability Demonstration,"** Presented At The (2004) Clarion Weibull User Conference. Excellent Demonstration Of Using Weibayes For Substantiation And Reliability Testing.

Wang, Wendai, **" Improving the Quality of Field Data Reliability Analysis,"** 2001 RAMS Proceedings, page 285. *Takes the Reduced Bias Adjustment (Chapter 5 herein) even further by reducing the bias in eta. Also suggests a method of further unbiasing median rank regression solutions.*

Wasserman, Gary S., "***Reliability Verification, Testing, and Analysis in Engineering Design,***"Marcel Dekker Inc. 2004. *This new text has numerous references to research including some to this Handbook. It contains many examples using Minitab and Excel as well as SuperSMITH Weibull. Recommended.*

Wenham, M., **"Comparison of Weibull Parameter Calculation Methods,"** GKN International Report No. 5647, 1997. An internal confidential document. *An excellent study. See also [Liu, Chi-Chao]*

$Waloddi Weibull, Please see Appendix N for Weibull's biography and list of references.

Weiss, George H., **"On the theory of replacement of machinery with a random failure time,"** Naval Research Logistics Quarterly, Volume 3, December 1956., *Provides the derivation of the MTBF function with planned replacement.*

$Winterbottom, Alan, Review of **"The New Weibull Handbook,"** The Royal Statistical Society Journal A (1997), 160, Part 2.

Yoshimoto, Isamu, **" On the original text of Weibull distribution,"** Journal of Japan Research Institute for Screw Threads and Fasteners, September 2003. *Professor Yoshimoto discusses Weibull's 1939 paper and The New Weibull Handbook, 4th edition.*

Index

Paul Barringer (1936 – 2016)

Dr. Bob & Paul Barringer at Wes Fulton's wedding August 2002.

Paul Barringer was a close associate of Wes Fulton and the author. He contributed original research results to this Handbook, including production process control, Crow-AMSAA goodness of fit and the binomial test tables all in Chapter 8. Paul was a leading reliability, manufacturing, and engineering consultant. His practice involved worldwide travel, consulting, and training most frequently with chemical processing and petroleum refining.. He is the author of the basic training courses Reliability Engineering Principles , Process Reliability, and Life Cycle Cost.

Barringer had more than forty years of engineering and manufacturing experience in design, production, quality, maintenance, and reliability of technical products. Barringer is a Licensed Professional Engineer (Texas-51508). He is named as inventor in six USA Patents. Barringer's education includes a MS and BS in Mechanical Engineering from North Carolina State University.

Barringer has been:

- Director of Manufacturing, Director of Engineering and Plant Manager in an ISO 9001 facility for highly specialized mechanical devices used as safety valves for containing pressures up to 20,000 psi during drilling of high pressure, high temperature oil and gas wells in sub-sea and land applications.
- Operations Manager in a multi-plant manufacturing operation for high-speed equipment used in textile machinery.
- Manufacturing Manager and Plant Engineer in nickel and cobalt based high alloy mills, foundries, machine shops, tube mills, and fabricating operations.
- Development Engineer for ultra high-speed centrifuges used in separating uranium isotopes and biological materials on a production basis.

The author highly recommends his Website for reliability, maintainability, life cycle cost, case studies, problems of the month, and a Weibull library. Paul lived in Houston, Texas. His legacy lives on in the material at his website (see References).